Editor-in-Chief

Dr Raj Pal Singh, Principal, Gandhi Memorial National College, Ambala Cantt. has research, teaching and administrative experience of more than 33 years. He has published/edited 18 Books and 38 Research Papers in National/International Journals and proceedings of Seminars/Conferences. He has acted as Resource Person in more than 24 orientation Programmes/Refresher Courses etc. He is member, General Council, NAAC. He has organized more than 20 National Conference/Seminars. This year on 26th Jan, 2018, he has been honored by H.E. Governor of Haryana for excellent services in the field of education.

The Editors

Dr Kuldeep Yadav is working as Head, Department of Botany, Gandhi Memorial National College, Ambala Cantt, Haryana. He has done M.Sc., M.Phil. and Ph.D. from Kurukshetra University, Kurukshetra. He has contributed extensively in the field of Plant Biotechnology, Molecular Biology and Mycology. He has published sixty research papers in reputed peer reviewed journals and has also authored six book chapters. He has also edited one book entitled, "Mycorrhizal Fungi".

Dr Shikha Jaggi is presently working as Head, Department of Zoology at GMN College, Ambala Cantt. She has done his Ph.D. degree in subject of Zoology from Kurukshetra University, Kurukshetra in the field of Human Cytogenetics. She has published eight research papers in reputed journals.

Dr Meenu Rathi is presently working as Assistant Professor at Gandhi Memorial National College Ambala Cantt. She has done Ph.D. from Choudhary Charan Singh University, Meerut in subject of Botany. She is having 8 years of teaching experience and has published seven research papers and four book chapters.

Emerging Trends and Challenges in Biosciences

Emerging Trends and Challenges in Biosciences

Editor-in-Chief

Dr Raj Pal Singh

Editors

Dr Kuldeep Yadav

Dr Shikha Jaggi

Dr Meenu Rathi

2019

Daya Publishing House®

A Division of

Astral International Pvt. Ltd.

New Delhi – 110 002

ISBN: 978-93-5124-988-7 (HB)

Published by : **Daya Publishing House®**
A Division of
Astral International Pvt. Ltd.
– ISO 9001:2015 Certified Company –
4736/23, Ansari Road, Darya Ganj
New Delhi-110 002
Ph. 011-43549197, 23278134
E-mail: info@astralint.com
Website: www.astralint.com

Digitally Printed at : Replika Press Pvt. Ltd.

This copy is for sale in Indian Sub-Continent.

Dr. Ramesh Kumar Yadava
Chairman
Haryana Ksian Ayog
Sector 20, Panchkula- 134116

Foreword

The science is considered as an endless endeavour since concern not only sustaining but also improving the quality of life for the living beings around us. Biology being the mother of all sciences is increasingly poised to transform the ways in which future challenges for sustaining the environment, agriculture and human health at the mother Earth are to be addressed. The combination of multidisciplinary research in plants, animals, microorganisms and their interaction has taken the life to the greatest extent for the wide horizons of life. Advances in biological sciences go beyond molecular base science therefore interest continues to grow still higher and more.

This book "Emerging Trends and Challenges in Biosciences" provides an excellent, up-to-date discussion of the state of knowledge and fills an important information gap. This book will be useful to a wide array of scientists including biologists, academicians and other industrialists to encourage integration of life sciences for better understanding of the complexity and diversity of life processes.

I am pleased to note that the faculty of Gandhi Memorial National College, Ambala Cantt, Haryana, the premier Institute in North India has brought up the present book and the editors namely Raj Pal Singh, Kuldeep Yadav, Shikha Jaggi and Meenu Rathi have taken a lot of pain in providing a comprehensive source of current information and have elaborated the foundation for applied research and development. Their efforts are appreciable on account of contents of the book that encompass advanced information on various topics of biological sciences at single place in an uncomplicated way and have Aworth in reading for understanding of basics of life.

Dr. Ramesh Kumar Yadava
Chairman
Haryana Kisan Ayog
Sector 20, Panckula-134116

Preface

India becomes an International contributor in the area of biological sciences, powered by its current financial development and aspiration in the field of Botany, Zoology, Biotechnology, Biochemistry, Microbiology and Bioinformatics researches. To complete its aspirations, India's furthermost challenge is in educating, recruiting and supporting its next generation. Rapid advances along all the frontiers of biological sciences necessitate frequent review of the state of our knowledge not only to keep our self up to date but also to ponder over the various strategies ought to be adopted for further resolution of the science in all its aspects. "Emerging trends and challenges in biosciences" is an attempt to bring under one cover and up to date account of the challenges before us and progress achieved in different fields of biological sciences viz. Biodiversity, Biotechnology, Molecular Biology, Food Science, Genetics, Biochemistry, Waste Management, Reproductive Physiology, Immunology, Developmental Biology, Global Climate Change & Sustainable Agriculture. The editors of the book are please to host wide range of research article from various disciplines of biological sciences on current advances and emerging trends in biosciences and its future prediction. This book emphasized the significance of modern biological tools and innovations for the development of Eco-friendly and sustainable bioprocesses and products. The all chapters bring advanced information with current references. It is hoped that the book will ensure the relevant and latest comprehensive information useful to postgraduate students, researchers, scientists, and teachers and help them to pursue their research and teaching programmes with better technical skills and competence in the field of life sciences.

Raj Pal Singh

Kuldeep Yadav

Shikha Jaggi

Meenu Rathi

Contents

Chapter 1

Microbial Interactions in Mycorrhizosphere of Spices and their Significance for Sustainable Development

Esha Jangra[1*], Anju Tanwar[2], Neetu[3] and Ashok Aggarwal[1]

[1]*Department of Botany, Kurukshetra University, Kurukshetra, Haryana 136119*

[2]*Department of Botany, Govt. P. G. College, Ambala Cantt., Haryana*

[3]*Department of Botany, Govt. P. G. College for Women, Sector-14, Panchkula, Haryana*

**Corresponding Author E-mail: ashokbotanykuk@gmail.com; eshujangra13@gmail.com*

Abstract

Spices which have long been the basis of traditional medicine in many countries have also been the subject of study, particularly by the pharmaceutical, chemical and food industries, because of their potential use for improving health of human beings. To increase the production of various spices, different types of fertilizers and chemical fungicides are applied to increase the production of spices but these fertilizers and fungicides pollute the ecosystem and non-target organisms. Another way to increase the production of spices is the use of microorganisms as biofertilizers and arbuscular mycorrhizal fungi is one group of fungi which are being used as biofertilizer. AM fungi symbiosis help the spices in improving (i) rooting and plant establishment (ii) uptake of nutrients especially phosphorus (iii) nutrient cycling (iv) enhanced plant tolerance to biotic and abiotic stress and (v) improving the quality of soil. It is now clear that AMF are highly beneficial to spice cultivation and AMF inoculation must be an important factor for yield and higher biomass of spices. But a comprehensive effort to bring the literature together and provide information to how to make AM application more effective in spice cultivation is the need of the hour. Hence, an effort was made in this review article to collect information of different spices benefitted by its use and aspects of integration of AMF in spice cultivation system.

Keywords: *Arbuscular mycorrhizal fungi, spices, nutrient uptake and biomass*

1. Introduction

Spices and herbs have been used both for medicinal and culinary purposes for centuries. Man has utilized spices and herbs as food to keep them healthy and harness their secondary metabolites to protect himself from various diseases. Over the last few decades, research into their role as contributors of dietary polyphenols known to possess a number of properties associated with reducing the risk developing chronic diseases has increased. However, keeping in mind how these spices are consumed in small quantities and in combination with other foods, it is unclear what their true benefit is from a health perspective.

The word 'spice' has been derived from the word 'species' which was used to groups of exotic food stuffs. Aromatically scented spice products have been used since ancient times to flavour foods. Rare spices were used in cooking as a sign of wealth in Rome and the privileged developed an exaggerated taste for spicy foods. A spice can be broadly defined as a tropical herbal plant that is valued for providing typical colour and aromatic flavouring along with stimulating odour for use in cooking and in condiments. Spices are also used as preservatives, appetizers, digestives and aphrodisiacs by many communities. These spices are derived from specific parts of a plant such as flowers, leaves, rhizomes, seeds, roots, buds, fruits, and exudates of plant part and even the bark of a tree.

The spice trade probably is the most ancient trade started by man. The annual global spice trade is estimated to be in the order of 2000 million dollars involving a quantity of 500000 tons. India has about half the share of world trade in spices because of its excellence reputation. The taste of Indian food largely depends on the spices used that impart the food various flavour. Besides giving the various flavour to food, all spices also impart medicinal properties which keep the human body in a healthy condition and also cure some of the diseases according to the Indian system of medicine called Ayurveda. The medicinal value of any spice depends upon the potency of the plant part that acts on digestive system, circulatory system and respiratory system etc.

Spices are not only used individually but also in the form of mixture i.e. 'curry powders' to suit different dishes and taste. Interestingly, the fat level varies from 0.6% in garlic to 42.6% in mustard, protein content varies from 4.5% in rosemary leaves to 31.5% in mustard and ash content from 2.3 in Marjoram to 16.7 in Basil leaves reflecting high material content. Apart from this, spices do contain significant number of vitamins and some important minerals which cannot be ignored. A few spices like red pepper, black pepper, coriander, cumin, and fennel are also rich source of fiber (Srinivasan, 2005). Extensive studies have indicated that even at higher dietary level of red pepper, black pepper and turmeric have no adverse effects on growth, organ weight and blood composition. The spices like fenugreek, red pepper, turmeric, garlic, onion and ginger were found to be effective as hypocholesterolemia. Garlic and onion are two other spices that have been widely used for their antidiabetic potential. Turmeric is another spice claimed to possess beneficial hypoglycemic effect and to improve glucose tolerance. Spices such as ginger, mint, ajwain, cumin, fennel, coriander and garlic are used as ingredients of home remedies for digestive disorder like flatulence and intestinal disorder. Spices such as turmeric, red pepper, ginger and mint have also been reported to stimulate pancreatic digestive enzymes like lipase, amylase, trypsin which play a crucial role in food digestion (Platel and Srinivasan,1996, 2001).

The antioxidant activities of spices have been reported by many workers (Sharma, 1976; , Reddy *et al.*, 1994). The antioxidant properties of capsaicin and curcumin inhibited lipid peroxidation by quenching oxygen free radicals and by enhancing the activity of endogenous antioxidant enzymes. Turmeric is considered as an anti-inflammatory drug. Spices that have antioxidant property can function as antimutagens. The curcumin the active principle in turmeric, Eugenol found in cloves are known to produce antimutagenic effect by protecting the cell from damage to its DNA.

Although cancer is considered to be an incurable disease, but research during the last few decades have shown that cancer is preventable by dietary intervention (Aggarwal *et al.*, 2003). Several studies indicate that curcumin can suppress both tumour initiation and tumour promotion. Garlic is yet another spice widely used for its chemo-preventive potential. *Asafeotida* that is obtained from the roots of *Ferula asafoetida* is used as antimicrobial agent in traditional medicine. The extracts of garlic, onion, nutmeg, saffron, cumin and thyme have antimicrobial properties.

Spices which have long been the basis of traditional medicine in many countries have also been the subject of study, particularly by the pharmaceutical, chemical and food industries, because of their potential use for improving health of human beings. To increase the production of various spices, different types of fertilizers and chemical fungicides are applied to increase the production of spices but these fertilizers and fungicides pollute the ecosystem and non-target organisms. Another way to increase the production of spices is the use of microorganisms as biofertilizers and arbuscular mycorrhizal fungi is one group of fungi which are being used as biofertilizer.

Biofertilizers are those substances that contain living organisms and they colonize the roots of the plants and increase the supply of nutrients and stimulate the growth of the plant. There are numerous soil microorganisms that colonize the roots of the plants. Soil microorganisms such as arbuscular mycorrhizal fungi (AMF or AM fungi) represent a key link between roots of higher plants and soil mineral nutrients. Thus, they are collecting growing interest as natural biofertilizers. AMF are obligate symbionts belonging to the phylum Glomeromycota (Schuβler *et al.*, 2001) which form mutualistic symbiosis with majority (about 80%) of land plant species. The German plant pathologist, A. B. Frank, first noticed the existence of such a characteristic association in the roots of Cupulifereae in 1885 and coined the term 'Mycorrhiza'.

In these associations there is a bidirectional flow of nutrients. Carbon flows out from the plant host to the fungus and mineral nutrients flow from the fungus to the plant. There are a number of plant families that are non-mycorrhizal including Brassicaceae, Caryophyllaceae, Chenopodiaceae, Cyperaceae, Juncaceae and Proteaceae.

The AMF mycelium that comes from the root system can acquire nutrients especially phosphorus from soil that are inaccessible to roots (Jakobsen, 2003). The AM fungal hyphae are much thinner than roots and are therefore, able to penetrate small pores of soil (Allen, 2009). These AM fungal hyphae extensively colonize the cortex of root and form branched structures inside the cells called arbuscules, which are considered the site of nutrient exchange (Balestrini *et al*, 2015). In AM mycorrhizae the fungal hyphae also develop structures called vesicles within the root cortical cells. Vesicles are thin walled, spherical to oval in shape, borne on the tip of the hyphae either in intercellular spaces or in the cortical cells of the root (Aggarwal *et al*, 2005, a, b). These vesicles are considered as food storage organs of the fungus. However, the arbuscules which is brush like dichotomously branched haustoria developed within the cortical cells.

Through widely distributed geographically, the AM Fungi are not found in flooded sites. The establishment of the mycorrhizal network offers a large number of advantages for the uptake of mineral nutrients:

- ☆ AM fungal hyphae extend beyond the area of nutrient depletion surrounding the roots. Plants are heavily infected with AM fungi in phosphorus deficient soils and mycorrhizal are poorly developed when phosphorus supply is high. It is thus, a self-regulating system, increasing phosphorus uptake when this element is in short supply.
- ☆ AM fungal hyphae greatly increase the surface area for the absorption of nutrients relative to non-mycorrhizal roots. AM fungi colonized roots increase the effectiveness of absorbing capability of surface host root as much as ten times in comparison to non-mycorrhizal one (Pellegrino and Bedini, 2014).
- ☆ AM extra radical hyphae of AM fungi being able to extend into soil pores that are too small for roots to enter. Glomalin, a new factor of presumably great importance in soil aggregation was discovered by Wright and Upadhaya (1996). Glomalin is a glycoprotein, produced by AMF and its concentration helps in soil aggregates and soil correlates with the percentage of water stable aggregates.
- ☆ Some AM fungi can access forms of N and P that are unavailable to non-mycorrhizal plants particularly organic forms of these nutrients.

It has been seen that mycorrhizal roots absorb much higher amounts of phosphorus that did non-mycorrhizal plants in soil (Bucher, 2006). This suggests that mycorrhizal fungus hyphae have higher affinity for phosphate ion and lower threshold concentration for absorption than do plant roots. Mycorrhizal hyphae can store larger amounts of absorbed Phosphorus than plant roots, thus facilitating the continued movement of P into the hyphae. Inorganic phosphorus absorbed by AM hyphae is stored in three forms i.e. soluble orthophosphate granule (Harley, 1978) and phosphate granules (Chilvers and Harley, 1980) and soluble polyphosphate. The bulk of the phosphate is stored as polyphosphate. Polyphosphates are synthesized by an inducible polyphosphate kinase. Due to hydrolysis of polyphosphate, a high concentration of phosphorus uptake is maintained in the fungus.

It has been seen that mycorrhizal treated plants use more of nutrients efficiently than non-mycorrhizal plants. There are three approaches to identify the sources of soil Phosphorus for mycorrhizal and non-mycorrhizal plants. Using radioactive phosphorus (P^{32}), Bolan, (1991) have tried to differentiate between the fractions of soil phosphorus taken by mycorrhizal and non-mycorrhizal plants and found that mycorrhizal plants took up more phosphorus than did non-mycorrhizal plants. Secondly the addition of phosphorus fertilizers of different solubilities for phosphorus uptake of mycorrhizal and non-mycorrhizal plants have been compared and it was found that the increase in plant growth due to mycorrhizal colonization was found to be large with poorly soluble phosphorus than with the soluble phosphorus. This lead to the conclusion that AM fungi have the ability to solublize the poorly soluble phosphorus due to various types of enzymatic activity. It was also found that mycorrhizal plants grow better than non-mycorrhizal plants in poorly soluble rock phosphate (Sabannavar and Lakshman, 2009). The simple explanation for better uptake of phosphorus by mycorrhizal plants is that the mycorrhizas explore the soil more thoroughly and found more of sources of phosphorus. Another explanation is that AM hyphae may produce certain chemicals that modify the availability of iron phosphate by producing organic compounds with chelating properties (Bolen *et al.*, 1987).

In addition to phosphate ion, nitrogen, sulphur and potassium can also be transferred to plants through AM hyphae (Sieh *et al.*, 2013). On the AM fungal hyphae an ammonium transporter has been found in arbuscules which suggest that fungus may absorb nutrients (Perez-*Tienda et al.*, 2011). The accumulation of potassium in AM fungi has been reported in spores, hyphae and vesicles (Olsson *et al.*, 2008). This uptake of potassium by AM fungi can be correlated to an improved plant tolerance to abiotic stress like salinity and drought (Garcia *et al.*, 2016). Apart from crop better growth, AM symbiosis due to its positive role in better plant nutrition could also have a positive impact on crop quality (Pellegrino and Bedini, 2014). Lehmann *et al.* (2014) also reported that the cooperation in AM interaction is related to the type of fungus and host involved in the symbiosis and also depends on various environmental factors, availability of nutrients and functional diversity.

2. PGPR (Plant Growth Promoting Rhizobacteria)

Apart from AM fungi present in the soil, there are rhizospheric bacteria that are favourable and beneficial for plant growth and yield are known as 'Plant Growth Promoting Rhizobacteria' (PGPR) (Kloepper *et al.*, 1980). There are several reports depicting the ability of different bacterial species which operate by a wide variety of mechanisms to solubilize inorganic phosphate compounds, nitrogen fixation and phytochrome production (Barea *et al.*, 2005). Traditionally, pseudomonas have been considered to be important organisms but other bacteria included are *Bacillus*, *Rhizobium*, *Aereobacter* and *Erwinia*. While many studies have reported *Pseudomonas* as dominant communities in the rhizosphere but in some cases *Bacillus* species may dominate (MacRae et al., 2001). Phosphorus as native in soil and when applied in organic fertilizers, becomes unavailable to the plant because of its low mobility and solubility and its ability to become fixed in soil. Here the Phosphate Soluble Bacteria (PSB) help in improving phosphate uptake of plants in various ways. The soil bacteria *Pseudomonas* species play a major role by which insoluble phosphorus compounds are solubilized and mobilized by the production of organic acids along with change of pH of soil. Thus, the PGPR namely *Azospirillum, Azotobacter, Pseudomonas, Bacillus* and *Rhizobium* promote the plant growth by producing plant growth promoting substances or by eliciting root metabolic activities by supplying biologically fixed nitrogen and indirectly by suppressing the growth of the phytopathogenic microorganisms (Kloepper, 1993). According to Rashid et al (2004), PSB are biofertilizers because they have the ability to solubilize organic and inorganic phosphorus compounds by various enzymatic activities. In terms of agricultural importance, the best example of interactions are the Fabaceae- *Rhizobium* spp./ *Bradyrhizobium* spp. root nodule symbiosis (Kadian *et al.*, 2013, a,b). Simultaneous colonization by dominant AM fungi stimulates the Fabaceae, *Rhizobium* symbiosis, thereby enhancing the number and size of nodules and increasing nitrogen fixation. This plant fungi-bacterial association has been exploited for many crops like soybean, bean and peanut (Deaker *et al.*, 2004).

According to Raaijmakers *et al.* (2009), microbes like the genus *Trichoderma* spp. and large variety of PGPR as mentioned above are able to suppress soil-borne plant pathogens and stimulate plant growth by different mechanisms like mycoparasitism and competence with soil plant pathogens, production of phytohormones, enhancing the bioavailability of essential nutrients and mineralization of organic matter. According to Czarnes *et al.* (2000), mucilage consisting of polysaccharides is secreted by root cells as the secondary and tertiary root grows through the soil. In addition to this, mucilage is also secreted by the rhizosphere microbes. When the mucilage cones in contact with soil, it forms a gel and this gel binds soil particles and microbes together with the root to form rhizosheath. So, it binds the soil and root together and lubrication of roots to move through the soil for absorption of nutrients. Further, mucilage also promotes soil stability and increasing resistance to erosion.

3. Beneficial Fungi

Fungi are more operational in growing through the soil and rhizosphere therefore they have the advantage over bacterial inoculants. A variety of mechanisms involved in plant growth promotion by fungi include antibiotic production, competition for nutrients and space and induced resistance. Apart from direct inhibition of soil borne plant pathogens, *Trichoderma* spp. have been reported to enhance crop growth and health. *Trichoderma harzianum* is a good solubilizer of phosphorus.

4. Vesicular Arbuscular Mycorrhizae (VAM)

Vesicular arbuscular mycorrhizal associations (VAM) which are also called arbuscular mycorrhizas (AM) or Glomeromycota mycorrhizas (GM) are the most common and cosmopolitan root fungus association (Brundrett, 2004). The term Mycorrhiza (meaning fungus-root) was coined by German Plant Pathologist A. B. Frank (1885) who was sure that these symbiotic or mutualistic plant fungus associations were required by the both partner for the nutrition. Recently Brundrett (2004) defined Mycorrhiza is a symbiotic association essential for both the partners, between a soil fungus and a root of a living plant that is primarily meant for nutrient absorption. These fungi are ubiquitous soil organic material (Joner and Jakobsen, 1995). The mycorrhizal associations provide substantial benefits to both partners. Brundrett (2004) proposed that those mycorrhizas be called 'balanced' to imply that exchange processes are in dynamic equilibrium and both organisms receive essential nutrients through reciprocal exchange. The majority of mycorrhizae are known to be obligate symbionts, so the benefits mycorrhizal fungi received from balanced association with plants are not in doubt. All AM and most ectomycorhizal fungi are obligate symbiotic incapable of independent life without plants but Ericoid mycorrhizal fungi are not obligate plant symbionts, and Orchid fungi are fully independent of their hosts (Brundrett, 2004).

4.1 Types of Mycorrhizas

In theory, mycorrhizas are categorized on the basis of structural or physiological characteristics, but in practice only anatomical characters can be used to designate types of these associations because links between mycorrhizal colonization and plant physiological parameters are not known fully.

4.1.1 Vesicular Arbuscular Mycorrhizas (VAM)

Generally, arbuscules are used to define VAM associations and their abundance is usually correlated with the degree of colonization of the roots by VAM fungi (Toth *et al.*, 1990). However, arbuscules are ephemeral structures that are difficult to see with standard microscopic procedures in the roots. Vesicles may be present or absent in this category. There is disagreement whether AM or VAM association is the most appropriate term for these mycorrhizas. Now the term AM has gradually become more common because some fungi do not form vesicles in roots. There are some problems with the use of arbuscules alone to define VAM fungi because (i) these associations are generally seen in old roots without intact arbuscules (ii) their role in nutrient transfer has not fully understood (Smith and Read, 2010). The terminology for describing VAM depends on the phytogenetic, structural and functional approach used to classify their components. The various arguments given by various workers suggest that the term VAM is as accurate term but the name AM is now more widely used (Morton and Redecker, 2001).

According to Gallaud (1905), there are two types of VAM, namely '*Arum*' and '*Paris*' type. In plants where Paris VAM associations are seen, these hyphae grow as coils within cells, while those with Arum VAM have colonies that expand by linear hyphal growth along longitudinal axis between cortex cells. The 'Arum' and Paris categories of VAM can be designated as 'linear' and 'coiled' VAM respectively. There may be physiological differences between linear and coiled VAM as it has been said that substantantial host fungus exchange of nutrients may occur within plant cells that contain only hyphal coils (Smith and Smith, 1997). Auxillary cells are cluster of thin walled cells having different shape, size and surface ornamentation and are restricted to the subfamily Gigasporaceae (Morton, 1990). Functionally, they act as temporary storage structures of carbon compounds. All VAM spores may be azygospores or chlamydospores formed singly or aggregated in sporocarp. The VAM forming genera are *Glomus* sp., *Acaulospora* sp., *Gigaspora* sp., *Entrophospora* sp., *Sclerocystis* sp. and *Scutellospora* sp.

4.1.2 Ectomycorrhizas

According to Frank (1885), the presence of a Hartig net, consisting of hyphae between root cells is designated as Ectomycorrhizal (ECM) association. This Hartig net is the primary zone of nutrient transfer (Burgess *et al.*, 1994). ECM are associations where Basidiomycota and few members of Ascomycota form short swollen lateral roots covered by mantle hyphae. There are two basic types of ECM: (i) associations with angiosperms such as *Betula, Populus* and *Eucalyptus* with a Hartig net confined to epidermal cells (ii) those of Gymnosperms (family Pinaceae) where the Hartig net occupies multiple layers of cells in cortex (Alexander and Hogberg, 1986). Some of the important fungal partners are *Amanita, Muscaria, Boletus edulis, Lactaricus deliciosus, Pisolithus tinctorius, Rhizopogon roseolus* etc.

4.1.3 Arbutoid, Monotropoid and Ectendomycorrhizal Association

Arbutoid mycorrhizae are variation of the ECM and are seen in autotrophic and partially achlorophyllous trees (Smith and Read, 2008). This symbiosis association is characterized by the presence of external hyphae which ramify between the cortical cells and forms Hartig net. Plants exhibiting Arbutoid association are found in genera like *Arbutus, Arctous, and Arctostaphylos* while fungal partners include species of *Amanita, Lactarius* and *Boletus* (Largent *et al.*, 1980).

White Monotropoid symbiosis is also a variation of ECM, formed by the members of Monotropoideae (*Monotropa uniflora*) having achlorophyllous plants. This symbiosis is characterized by the presence of fungal mantle sheath and Hartig net confined to the epidermis.

Ectendomycorrhiza was first called as 'Ectendotrophic mycorrhizas' by Melin (1923) but now called as Ectendomycorrhizas (Egger and Fortin, 1988). It is similar to ECM except that the mantle sheath may be reduced or absent in some genera but fungal hyphae penetrate into the root cortical cells. There are reports where hyphae penetrating host cells belong to a different fungus than the one forming the Hartig net (Yu *et al.*, 2001). The fungal partners involved are members of Ascomycota involving *Wilcoxina* spp. and *Sphaerosporella* and to a lesser extent *Phialophora* and *Chloridium* spp. (Trevor *et al.*, 2001). The hosts involved are primarily *Pinus* and *Larix* spp.

4.1.4 Other Mycorrhizas

Morphological categories have not been found within Ericoid mycorrhizas but may exist. Ericoid mycorrhizae are found in Ericales and fungi to Ascomycota. In this association, the hair like roots are enmeshed in a extensive weft of hyphae, which later enters the root cells and inner cortex but no mantle sheath is formed (Smith and Read, 2008). The plats showing Ericoid mycorrhizal association includes *Erica, Rhododendron* and *Vaccinium*.

In Orchidaceous mycorrhiza association, the fungus grows within the host cell and later forms hyphal coil called as pelotons (Gallaud, 1905). The fungal partner belongs to Basidiomycota with *Rhizoctonia* (perfect stage: *Thanetophorus* spp.) to be the dominant fungal partner. The other members included are *Armillaria, Fomes* and *Marasmius* spp.

5. Significance of AM Fungi in Global Sustainable Development

During the past four to five decades, we have seen the doubling of human population and a concurrent doubling of food production. The application of nitrogenous fertilizer has increased almost nine-fold that of phosphorus (P) more than four times. But the increasing use of fertilizers have created environmental problems such as reduced biodiversity, suppressed ecosystem, deterioration of soil quality, surface water and ground water (Schultz *et al.*, 1995; Socolow, 1999). To overcome these problems, microorganisms that allow more efficient nutrients use can provide sustainable solution for present and future agriculture practices. AMF fall into this category, associating with the majority of vascular plant roots, providing them with nutrients and protection from environmental stress. Here, a brief account of role of AMF in sustainable development is discussed.

5.1 AMF in Soil Fertility

Many workers have been shown that AMF can overcome nutrient limitation to plant growth by better uptake of nutrients (Clark and Zeto, 2000). Phosphorus is an essential plant nutrient required in large amounts and plays a vital role in all biological functions in energy transfer through the formation of energy transfer through the formation of various important macromolecules like nucleotides and phospholipids (Marschner, 1995). Similarly, nitrogen is also required for the formation of various important macromolecules like amino acids, proteins and nucleic acid synthesis. The AM hyphae have the ability to take nitrogen and transport it from soil to plants. AM fungi improve growth, nodulation and nitrogen fixation in legume-*Rhizobium* symbiosis (Kadian *et al.*, 2013b). Recently, a plant ammonium transporter, which is activated by AM association has been identified and indicated that the way by which N_2 is transferred in plant may be similar to phosphorus transfer (Guether *et al.*, 2009). Apart from this, AMF extrametrical hyphae also take up and transport some macronutrients like Zn, Cu, Co under conditions of low soil nutrient availability (Faber *et al.*, 1990). The AMF hyphae also transport potassium (K), calcium and sulphate and AM colonization affects the concentration and accumulation of K in shoots.

5.2 Uptake of Water

According to Safir and Nelson (1985), AM association improves the hydraulic conductivity of roots and improves water uptake by plant roots. Mycorrhizal plants show better survival than non-mycorrhizal plants because mycelia network extends deeper in the soil in search of water and nutrients. The permeability of cell membrane to water may also be changed by mycorrhizal colonization (Osonubi *et al.*, 1991). The probable explanation can be due to better distribution of absorbing hyphal network, greater surface of roots, increased functional longevity of roots, altered rhizosphere microbial population, greater hydraulic conductivities and lower transpiration rates per unit leaf area.

5.3 Soil Stabilization and Soil Aggregation

Arbuscular mycorrhizal fungi help in the soil conservation by binding the soil particles and improve soil aggregation. AM fungi produce 'Glomalin' which upon accumulation acts as a backbone for soil aggregation and soil stabilization. It also produces certain exudation in the soil and thus promotes aggregate stability and boost up the microbial growth (Johnson *et al.*, 2002).

5.4 Reclamation of Wasteland

AMF have a great potential in the recovery of waste land and disturbed lands. It has been suggested that many plants require dominant mycorrhizal colonization in order to survive in disturbed land. AM fungi have been shown to improve revegetation of coal spoils, arid or semi-arid areas, road sites and other disturbed areas (Jha *et al.*, 1994). Rani *et al.*, (1999) from our laboratory had worked on the effect of *G. mosseae*, *G. fasiculatum* along with *Rhizobium* and *Trichoderma* on establishment and better biomass of *Acacia nilotica* and *Prosopis cineraria* and reported that co-inoculation of AM and *Rhizobium* resulted in maximum growth and better establishment.

5.5 Role of AMF in Agriculture

The extensive use of fertilizers causes pollution problems and health hazards. So, the use of AMF in agriculture has been encouraged. The commercial exploitation of AM fungi is not easy because large scale production of AM fungi is not easy. Majority of the crops are highly dependent on AMF to meet early phosphorus requirement (Aggarwal *et al.*, 2012).

Crop rotation is used for managing nutrients supply and disease management. This phenomenon 'rotation effect' cannot be explained entirely by nutritional effects and other factors like AMF may play an important role in the success of crop rotation. It is well known that AMF colonization is decreased by non-mycorrhizal crop and highly mycorrhizal host crop increase AMF inoculum potential of the soil and colonization of next crop (Karasawa *et al.*, 2002). An increase in AMF colonization and better growth in maize occurred following sunflower (a mycorrhizal crop) when compared to maize following mustard (a non-mycorrhizal crop). Here non-mycorrhizal plants in the rotation reduce the rate of AMF colonization in the following crops. These results suggest that AMF population can be built up and the negative effect of non-mycorrhizal can be reversed after cropping with a mycorrhizal crop (Gavito and Miller, 1998). The benefits of AMF are greatest in systems where Phosphorus in the soil is low. AMF also play a vital role in reforestation and there are many reports of increased establishment of forest seedlings (Dickie *et al.*, 2013).

5.6 Alleviation of Environmental Stress

The roles played by AM fungi in alleviating drought, water, salinity and metal pollution are discussed briefly.

Water stress is a major agricultural constraint in the semi-arid tropics. AMF improve the better uptake of nutrients like phosphorus and nitrogen in water stress conditions (Tobar *et al.*, 1994). AMF alters the non-hydraulic root to shoot signaling of soil drying by eliminating the leaf response (Augé, 2004). The extra radical hyphae of AMF increase the absorptive surface area of roots which in turn reduces the resistance of water uptake (Hampp *et al.*, 2000). AMF have been observed to evolve strategies which can alleviate heavy metal threats in polluted system and thus from the food chain (Kramer, 2005) which involve precipitation of polyphosphate granules in the soil, immobilization of metal compounds and finally chelation of heavy metal (Gaur and Adholeya, 2004). Endomycorrhizal colonization of roots can reduce translocation of heavy metals to the cells of the fungal hyphae in root. This is how, mycorrhizae can help higher plants to adopt and survive in heavy metal contaminated soil. *Glomus* spp. seem to be a promising mycorrhizal fungus for bioremediation of heavy metal from contaminated soil (Liao *et al.*, 2003).

According to Giri *et al.* (2003), Kadian *et al.* (2013) salinization of soil is a major problem and is increasing steadily in many parts of the world in arid and semiarid areas. Overall, salinity leads to many harmful effects on plant at various stages of life cycle. To overcome this problem, many strategies were proposed such as removing excessive salt accumulation in groundwater and desalinization water for irrigation (Flowers, 2005) but microbes particularly AMF plays a vital role in alleviating salinity. Mycorrhizal colonization in plants grown in saline soil causes improvement in morphology of the plant, better phosphorus uptake, better phosphate activity, antioxidant enzyme activities and decreased electrolyte leakage as different physiological and morphological parameters are vegetative affected by salinity stress (Aggarwal *et al.*, 2017). Mycorrhizal associations are widely used in saline soil and have an ability to utilize water and mineral salts more efficiently than the non-mycorrhizal plants.

Fungicides are used to control various plant pathogens but farmers seldom realize the harmful effect of fungicides on the non-target microorganism especially AM fungi (Aggarwal *et al.*, 2007). Fungicides may affect AMF directly through the soil or indirectly through systemic responses in the plants (Aggarwal *et al.*, 2009). According to Kjoller and Rosendahl (2000), the extra mycelium could be expected to be more sensitive to the direct effect of fungicides than intraradical hyphae as the root surface may protect the internal hyphae.

5.7 AM Fungi as Biocontrol Agent

According to Manoharachary (2004), the beneficial role of AM fungi in plant biomass production is associated with their capacity to reduce plant disease development. Mycorrhizal plants usually suffer less damage from infection than non-mycorrhizal plants (Dehne, 1982). Several mechanisms have been proposed by different workers to clarify the protection extended by AM fungi to host plant against pathogens (Kaushish *et al.*, 2011). The known interaction includes a number of mechanisms such as lignification of plant cell wall, improved nutrition may also protect the plant against plant pathogens, exclusion of pathogen, activation of antimicrobial phenyl propanoid metabolism in roots. Apart from this, production of phytoalexins (Glyceollin and Daidzein), chitinase β 1-3 glucanase and PR proteins are important components of AM induced changes leading to resistance of host against pathogens (Arya and Chatterjee, 2007).

5.8 Biohardening Tool

According to Adholeya *et al.* (2005) the inoculation of AM fungi to nursery plants has been proved beneficial but it has been extended to micropropagated plants.

Inoculation of micropropagated plantlets with dominant culture of AM fungi appeared to be useful for their growth and survival (Yadav *et al.*, 2011). This phenomenon avoids 'transient transplant shock', stunted growth and mortality on transfer in the field. AM fungi can modify root architecture of micropropagated plants which are better adopted for better uptake of nutrient and water (Jangra *et al.*, 2017).

5.9 Increasing Vase Life of Ornamental Flowering Plants

Vase life is an important parameter for choosing flowers. Colonization with AM fungi has been shown to increase the vase life of many cut flowers (Karishma *et al.*, 2011) but the mechanism involved is not properly understood. Some of the reasons of having better and prolonged vase life of cut flowers in mycorrhizal treated plants can be due to better vascular development by AM fungi (Chang 1994); or due to decreased ethylene production (Besmer and Koide, 1999). The increase of peroxidase activity is the most important factor for increasing vase life. These antioxidative enzymes play an important role in defense mechanism of cells against superoxide free radicals produced under stress conditions (Bowler *et al.*, 1992).

5.10 Review of Literature

A survey of literature reveals that bioinoculations of AM fungi into the roots of plants play a beneficial role. Facilitated better nutrient uptake particular with respect to immobile nutrient like phosphorus is believed to be the main benefit of mycorrhizal symbiosis for plants. The concept of improving the secondary plant metabolites through AM is gaining momentum. The initial work on the accumulation of secondary compounds in the foliar parts of AM treated culinary and aromatic herbs such as Basil and Fennel has been reported by many workers (Copetta *et al.*, 2006; Gupta *et al.*, 2002; Kapoor *et al.*, 2002; Toussaint *et al.*, 2007). As per information about these AM treated plants mainly the quantity of the essential oil was affected and sometimes some changes in the quality and composition of essential oil were also found (Kapoor *et al.*, 2002 and Gupta *et al.*, 2002). The alternations of secondary metabolites are not restricted to terpenes but some other chemical compounds are also affected. For example, an increased content of rosmarinic acid, an antioxidant phenolic compound was observed in AM colonized Basil plants (Toussaint *et al.*, 2008). According to Toussaint (2007), the reasons for change secondary metabolites due to AM colonization is not properly understood but speculations attribute that this effect to be due to improved phosphorus status or a changed hormonal balance of the plants.

Recently, beneficial effect on secondary metabolites and biomass in *Echinacea purpurea* inoculated with *Glomus intraradices* were detected in aerial and underground parts. In the shoots, caffeic and its derivates were unaffected by AM colonization while in roots highly significant increase of these compounds were found which could be related to be increase phosphorus content (Araim *et al.*, 2009). In another study, AM fungi can induce the synthesis of caffeic acid and rosmarinic acid in sweet basil (Toussaint *et al.*, 2007). The positive effect on the production of secondary metabolites in medicinal plants including spices through mycorrhization would mean a higher benefit and at the same time would contribute to a more sustainable product for the conservation of species (Parkash *et al.*, 2011).

Mixed inoculation with bacteria and AM fungi creates synergistic results that may result in a significant increase in growth and an enhancement in the uptake of essential mineral nutrients (Linderman, 1992). Mohan and Rajendran (2014) found that the shoot length, root length and biomass were increased in combined inoculation of *Azospirullum* spp. and AM fungi and *Pseudomonas* spp. in *Feronia elephantum*. Rajendran (2012) reported an increase of shoot length and biomass in *Erythrina indica* as a result of different bioinoculants i.e. consortium application. Kumar *et al.* (2009) conducted an experiment to analyze the effect of two dominant AM fungi (*G. mosseae* and *A. laevis*) along with *Trichoderma viride* on different growth parameters of *Salvia officinalis* and found that dual treatment (*A. laevis* and *Trichoderma*) was most effective in increasing growth parameters including AM spore and root colonization. Recently Dutta and Neog, (2015) reported that application of *G. fasiculatum* along with *Trichoderma* spp. has increased biomass of rhizome of *Curcuma longa*. Fini *et al.* (2011) reported that the consortium AM fungi i.e. *Gigaspora decipiens, Acaulospora koskei, Entrosphora colmbiania* and *Scutellospora heterogama* resulted in better growth of rhizome of *Zingiber officinale.* Tanwar *et al.* (2013) from our laboratory have also reported that mixed treatment of AM fungi along with *Trichoderma* spp. resulted in better shoot growth, phosphorus and nitrogen content in tomato.

A review of the literature revealed that to cope with environmental hazards and struggle against water stress, plants develop various defense approaches, mutualistic association of AM fungi with roots of plant is an example (Wang and Qiu, 2006). Under water stress conditions, AM fungi ameliorate water status of plant by better water uptake through extraradical hyphae, regulation of stomatal conductance as well as osmotic regulation of different osmolytes. Higher proline concentration has been observed in mycorrhizal plants by Smith and Read (2010) under water stressed conditions in *Ocimum sanctum*. Increased phosphorus uptake and curcumin concentration in mycorrhizal plants under water stress helps in tolerance of *Curcuma longa* to water stress by increasing the availability of phosphorus (Daei *et al.*, 2009). Increased in essential oil content and yield up to 31% on mycorrhizal inoculation in *Coriandrum sativum* has also been confirmed under water stress conditions. Similar results have been reported on *Mentha* by Fan and Liu (2011), *Foeniculum vulgare* by Apel and Hirt (2004) and *Zingiber* by Mittler (2002), *Oreganum vulgare* by Asensio *et al.* (2012) and Barea *et al.* (2011) and *Cumminum cyminum* by Subramaniam and Charest (1999) respectively.

According to Manchanda and Garg (2008) salinity is a result of long term natural build up of salts in the soil and is widely responsible for increasing the concentration of salts in the soil that impairs plant growth. Several studies have demonstrated that AM symbiosis not only contributes to plant growth but also includes adaptability of plant under varying salt conditions (Garg and Aggarwal, 2012 and Garg *et al.*, 2017). Among various *Glomus* spp. *Glomus mosseae* is reported to survive in salt contaminated soil with some other mycorrhizal species including *Acaulospora* and *Gigaspora* (Khade and Adholeya, 2017). AM mediates alleviation of stress effects and improves tolerance of plants by enhancing uptake of nutrients, accumulation of osmolytes, photosynthetic efficiency and some structural adaptations (Kapoor *et al.*, 2013). According to Suhail and Mahdi (2013), *Alium cepa* induced uptake in phosphorus when inoculated with the *Glomus mosseae* and *Acaulospora laevis* under varying condition of salinity stress by reducing the negative effect of Na^+ and Cl^- ion by maintaining membrane negativity. In addition, studies have also revealed that *Glomus mosseae, Glomus intraradices* counteracts the adverse effect of salinity stress in *Ocimum basilicum* due to uptake of essential micronutrients. Furthermore, Sowmya, (2014) credited the enhancement of biomass and fruit yield in *Capsicum annum* to the *G. macrocarpum* mediated mobilization of nutrients. Several studies have been reported that salt tolerant AM fungi exclude Na+ during its transfer to the plants thus maintaining a balance of Ca^+/Na^+ are Na^+/K^+ (Garg and Bhandari, 2016). Recently Porcel *et al.* (2016) reported that AM fungi upregulated the expression of genes encoding transporters involved in Na^+/K^+ homeostasis in rice under saline conditions. Additionally, different workers have validated that by increasing shoot and root biomass of spices, AM fungi also protects these spices from toxic effects of Na^+ thereby causing salt dilution effect (Evelin *et al.*, 2013; Garg and Pandey, 2016; Suhail and Mahdi, 2013).

The above literature revealed that bioinoculations studies has been manly carried on various aspects like the influence of bioinoculations on growth, mycorrhizal biodiversity in the rhizosphere, alleviating water and salt stress in various species. Referring to the availability and benefits of the spices, it is essential to take up the study on various spices grown in India to produce healthy and to conserve target plant. Species by eco-friendly methods like AM fungi to produce healthy and superior quality stock.

References

Adholeya, A., Tiwari, P. and Singh, R. 2005. Large-scale inoculum production of arbuscular mycorrhizal fungi on root organs and inoculation strategies. In: *In vitro* culture of mycorrhizas. Springer Berlin Heidelberg. pp. 315-338.

Aggarwal, A., Kadian, N., Karishma., Neetu., Tanwar, A. and Gupta, K.K. 2012. Arbuscular mycorrhizal symbiosis and alleviation of salinity stress. *J. Appl. Nat. Sci.*, 4: 144-155.

Aggarwal, A., Prakash, V., Sharma, D., Sharma S., Kaushik, S. and Mehrotra, R. S. 2009. Mycoflora flora of sunflower rhizosphere in relation to soil fumigation. *Helia*, 32: 77-84.

Aggarwal, A., Prakash, V., Sharma, D. and Sharma, S. 2005a. Vesicular Arbuscular Mycorrhizal (VAM) aids in growth of *Prosopis juliflora* and *Tecomella undulata* in arid, semi-arid climate. *J. Mycol Pl. Pathol.*, 35(1): 184-187.

Aggarwal, A., Prakash, V., Sharma, D. and Sharma, S. 2005b. Mycorrhizal occurrence, status and diversity of some ethno botanical plants of Himachal Pradesh. In: The Fungi- Diversity and Conservation in India (Eds.) J.S. Dargan G.S. Atri, and J.S. Dhingra, Bishen Singh Mahendara Pal Singh. Dehradun, (U.A) India, pp. 241-251.

Aggarwal, A., Sharma, N., Kumar, A., Esha and Yadav, K. 2017. Salinity stress in plants and its amelioration by arbuscular mycorrhizal fungi In: Advances of Life Sciences (Eds.) Som Dutt, Ashu Tyagi, Harish Pal Bhati and Harjeet Singh, S.R. Scientific Publication, India, pp. 77-96.

Aggarwal, A., Sharma, S., Sharma, D. and Prakash, V. 2007. Effect of fungitoxins on non-target microorganisms. In: Advances in Mycology and Plant Pathology (Eds.) I.B. Prasher & M. P. Sharma, Bishen Singh Mahendra Pal Singh Publishing. Dehradun, (U.A). pp. 227-243.

Aggarwal, B. B., Kumar, A., Bharti, A. C. (2003). Anticancer potential of curcumin: Preclinical and clinical studies. Anticancer Res., 23: 363–398.

Alexander, I.J. and Högberg, P. 1986. Ectomycorrhizas of tropical angiospermous trees. *New Phytol.*, 102: 541-549.

Allen, J.W. and Shachar-Hill, Y. 2009. Sulfur transfer through an arbuscular mycorrhiza. *Plant Physiol.*, 149: 549-560.

Allen. and Michael, F. 2011. Linking water and nutrients through the vadose zone: a fungal interface between the soil and plant systems. *J. Arid land*, 3: 155-163.

Apel, K. and Hirt, H. 2004. Reactive oxygen species: metabolism, oxidative stress, and signal transduction. *Annu. Rev. Plant Biol.*, 55: 373-399.

Araim, G., Saleem, A., Arnason, J.T. and Charest, C. 2009. Root colonization by an arbuscular mycorrhizal (AM) fungus increases growth and secondary metabolism of purple coneflower, *Echinacea purpurea* (L.) Moench. *J Agric Food Chem*, 57: 2255-2258.

Arya. A., Chatterjee. T. 2007. Management of Fungal Disease by Application of Arbuscular Mycorrhizal Fungi. In: Biocontrol of Plant Diseases, (Ed.) P.C. Trivedi, Aavishkar Publishers and Distributors, Jaipur, pp. 193-206.

Asensio, D., Rapparini, F. and Pen˜uelas, J. 2012. AM fungi root colonization increases the production of essential isoprenoids vs. nonessential isoprenoids especially under drought stress conditions or after jasmonic acid application. *Phytochem.*, 77:149–161.

Augé, R.M. 2004. Arbuscular mycorrhizae and soil/plant water relations. *CAN J SOIL SCI*, 84(4), 373-381.

Balestrini, R., Lumini, E., Borriello, R. and Bianciotto, V. 2015. Plant-soil biota interactions. Soil Microbiology, Ecology and Biochemistry, (Ed.) Eldor A. Paul. Academic Press. Pp. 311-338.

Barea J. M., Palenzuela J., Cornejo P., Sánchez-Castro I., Navarro-Fernández C., Lopéz- García A.2011. Ecological and functional roles of mycorrhizas in semi-arid ecosystems of Southeast Spain. J. Arid Environ. 75 1292–1301.

Barea, J. M., Pozo, M. J., Azcon, R. and Azcon-Aguilar, C. 2005. Microbial co-operation in the rhizosphere. *J. Exp. Bot.*, 56: 1761-1778.

Besmer, Y.L. and Koide, R.T. 1999. Effect of mycorrhizal colonization and phosphorus on ethylene production by snapdragon (*Antirrhinum majus* L.) flowers. *Mycorrhiza*, 9: 161-166.

Bolan, N.S. 1991. A critical review on the role of mycorrhizal fungi in the uptake of phosphorus by plants. *Plant Soil*, 134: 187-207.

Bowler, C., Montagu, M.V. and Inzé, D. 1992. Superoxide dismutase and stress tolerance. *Ann. Rev. Plant Biol.*, 43: 83-116.

Brundrett, M.C. 2004. Diversity and classification of mycorrhizal associations. *Biol. Rev.*, 79: 473-495.

Bucher, M. 2007. Functional biology of plant phosphate uptake at root and mycorrhiza interfaces. *New Phytol.* 173: 11–26.

Burgess, T., Dell, B. and Malajczuk, N. 1994. Variation in mycorrhizal development and growth stimulation by 20 *Pisolithus* isolates inoculated on to *Eucalyptus grandis* W. Hill ex Maiden. *New Phytol.*, 127: 731-739.

Chang, D.C. 1994. What is the potential for management of vesicular-arbuscular mycorrhizae in horticulture? In: Management of mycorrhizas in agriculture, horticulture and forestry, (Eds.) Robson A.D., Abbot L.K. and Malajczukn, Kluwer Academic Publishers, Netherlands. pp. 187-190.

Chilvers, G.A. and Harley, J.L. 1980. Visualization of phosphate accumulation in beech mycorrhizas. *New Phytol.*84: 319-326.

Clark, R.B. and Zeto, S.K. 2000. Mineral acquisition by arbuscular mycorrhizal plants. *J Plant Nutr*, 23: 867-902.

Copetta, A., Lingua, G. and Berta, G. 2006. Effects of three AM fungi on growth, distribution of glandular hairs, and essential oil production in *Ocimum basilicum* L. var. Genovese. *Mycorrhiza*, 16(7), 485-494.

Czarnes, S., Hallett, P. D., Bengough, A. G. and Young, I. M. 2000. Root-and microbial-derived mucilages affect soil structure and water transport. *"Eur J Soil Sci."*, 51(3), 435-443.

Daei, G., Ardekani, M. R., Rejali, F., Teimuri, S. and Miransari, M. 2009. Alleviation of salinity stress on wheat yield, yield components, and nutrient uptake using arbuscular mycorrhizal fungi under field conditions. *J Plant Physiol.*, 166: 617-625.

Deaker, R., Roughley, R.J. and Kennedy, I.R. 2004. Legume seed inoculation technology-a review. *Soil Biol Biochem*, 36: 1275-1288.

Dehne, H.W. 1982. Interaction between vesicular-arbuscular mycorrhizal fungi and plant pathogens. *Phytopathol*, 72, 1115-1119.

Dickie, I.A., Martinez-Garcia, L.B., Koele, N., Grelet, G.A., Tylianakis, J.M., Peltzer, D.A. and Richardson, S.J. 2013. Mycorrhizas and mycorrhizal fungal communities throughout ecosystem development. *Plant Soil*, 367: 11–39.

Dutta, S.C. and Neog, B. 2015. Effects of AM Fungi and Plant Growth-promoting Rhizobacteria on Enzymatic Activities of Soil under Turmeric (*Curcuma longa* L.) Cultivation. *Indian Society of Soil Science*, 63: 442-448.

Egger, K.N., Fortin and J.A. 1988. Ectendomycorrhizae: diversity and classification. In Canadian workshop on mycorrhizae in forestry. (Ed.) M. Lalonde and Y. Piché. CRBF, Université Laval, Ste-Foy, Que. pp. 113-114).

Evelin, H., Giri, B. and Kapoor, R. 2013. Ultrastructural evidence for AMF mediated salt stress mitigation in *Trigonella foenum-graecum*. *Mycorrhiza*, 23: 71-86.

Faber., B.A., Zasoski., R.J., Burau., R.G. and Uriu., K. 1990. Zinc uptake by corn affected by vesicular-arbuscular mycorrhizae. *Plant Soil*, 129: 121-131.

Fan, Q.J. and Liu, J.H. 2011. Colonization with arbuscular mycorrhizal fungus affects growth, drought tolerance and expression of stress-responsive genes in *Poncirus trifoliata*. *Acta Physiol Plant*, 33: 15-33.

Fini, A., Frangi, P., Amoroso, G., Piatti, R., Faoro, M., Bellasio, C. and Ferrini, F. 2011. Effect of controlled inoculation with specific mycorrhizal fungi from the urban environment on growth and physiology of containerized shade tree species growing under different water regimes. *Mycorrhiza*, 21: 703-719.

Flowers, T.J. and Flowers, S.A. 2005. Why does salinity pose such a difficult problem for plant breeders? *Agr. Water. Manage*, 78: 15-24.

Frank, A.B. 1885. The mode of nutrition of certain trees through soil fungi based on root symbiosis. *Ber. Dtsch. Bot. Ges*, 3: 128-145.

Gallaud, I. 1905. Etudes sur les mycorhizes endotrophes. *Rev. Gen. Bot.*17: 15-48.

Garcia, K., Doidy, J., Zimmermann, S.D., Wipf, D. and Courty, P.E. 2016. Take a trip through the plant and fungal transportome of mycorrhiza. *Trends Plant Sci*, 21: 937-950.

Garcia, K. and Zimmermann, S.D. 2014. The role of mycorrhizal associations in plant potassium nutrition. *Front Plant Sci*, 5: 337.

Garg, N. and Aggarwal, N. 2012 Effect of interactions between cadmium and lead on growth, nitrogen, phytochelatin and glutathione production in mycorrhizal Cajanus cajan (L.) Millsp. J. Plant Growth Regul., 30: 286-300.

Garg, N. and Pandey, R. 2016. High effectiveness of exotic arbuscular mycorrhizal fungi is reflected in improved rhizobial symbiosis and trehalose turnover in *Cajanus cajan* genotypes grown under salinity stress. *Fungal Ecol*, 21, 57-67.

Garg, N. and Bhandari, P. 2016. Silicon nutrition and mycorrhizal inoculations improve growth, nutrient status, K+/Na+ ratio and yield of *Cicer arietinum* L. genotypes under salinity stress. *Plant Growth Regul*, 78: 371-387.

Garg, N., Bhandari, P., Kashyap, L. and Singh, S. 2017 Arbuscular mycorrhizal symbiosis: A boon for sustainable legume production under salinity and heavy metal stress. In: Mycorrhizal Fungi (Eds) A. Aggarwal and K. Yadav, Astral International Pvt. Ltd., New Delhi, pp. 247-273.

Garg, N. and Manchanda, G. 2008. Effect of arbuscular mycorrhizal inoculation on salt-induced nodule senescence in *Cajanus cajan* (pigeonpea). *J. Plant Growth Regul.*, 27: 115.

Gaur, A. and Adholeya, A. 2004. Prospects of arbuscular mycorrhizal fungi in phytoremediation of heavy metal contaminated soils. *Current Sci.*, 86: 528-534.

Gavito, M.E. and Miller, M.H. 1998. Changes in mycorrhiza development in maize induced by crop management practices. *Plant Soil*, 198: 185-192.

Giri B., Kapoor, K. and Mukerji, K.G. 2003. Influence of arbuscular mycorrhizal fungi and salinity on growth, biomass and mineral nutrition of *Acacia auriculiformis. Biol. Fert. Soils*, 38: 170-175.

Guether, M., Neuhäuser, B., Balestrini, R., Dynowski, M., Ludewig, U. and Bonfante, P. 2009. A mycorrhizal-specific ammonium transporter from *Lotus japonicus* acquires nitrogen released by arbuscular mycorrhizal fungi. *Plant Physiol.*, 150: 73-83.

Gupta, M.L., Prasad, A., Ram, M. and Kumar, S. 2002. Effect of the vesicular–arbuscular mycorrhizal (VAM) fungus *Glomus fasciculatum* on the essential oil yield related characters and nutrient acquisition in the crops of different cultivars of menthol mint (*Mentha arvensis*) under field conditions. *Bioresour Technol.*, 81: 77-79.

Hampp, R., Mertz, A., Schaible, R., Schwaigerer, M. and Nehls, U. 2000. Distinction of *Araucaria angustifolia* seeds from different locations in Brazil by a specific DNA sequence, *Trees*, 14: 429-434.

Harley, J.L. 1978. Ectomycorrhizas as nutrient absorbing organs. *Proc. R. Soc. Lond.* B, 203: 1-21.

Jakobsen, I., Smith, S.E. and Smith, F.A. 2003. Function and diversity of arbuscular mycorrhizae in carbon and mineral nutrition. *Mycorrhizal ecology*. Springer, Berlin, Heidelberg. Pp. 75-92.

Jangra, E., Yadav, K. and Aggarwal, A.2018. Potential Role of Arbuscular Mycorrhizal Fungi andPseudomonas fluorescens on Growth, Physiological Parameters, and Yield of Capsicum frutescens L., International Journal of Vegetable Science, 24:2, 137-145.

Jha, D.K., Sharma, G.D. and Mishra, R.R. 1993. Mineral nutrition in the tripartite interaction between *Frankia, Glomus* and *Alnus* at different soil phosphorus regimes. *New Phytol.,* 123: 307–311.

Johnson, D., Leake, J. R. and Read, D. J. 2002. Transfer of recent photosynthate into mycorrhizal mycelium of an upland grassland: short-term respiratory losses and accumulation of 14C. *Soil Biol. Biochem,* 34: 1521-1524.

Joner, E.J. and Jakobsen, I. 1995. Growth and extracellular phosphatase activity of arbuscular mycorrhizal hyphae as influenced by soil organic matter. *Soil Biol. Biochem,* 27: 1153-1159.

Kadian, N., Badda, N., Yadav K. and Aggarwal, A. 2013. Influence of Arbuscular mycorrhizal fungi and salinity on growth, biomass, mineral nutrition and yield of *Cicer arietinum. Russ. Agric. Sci,* 39: 321-329.

Kadian, N., Yadav, K. and Aggarwal, A. 2013. Significance of bioinoculants in promoting growth, nutrient uptake and yield of *Cyamopsis tetragonoloba. Eur. J. Soil Biol.*, 58: 66-72.

Kapoor, R., Evelin, H., Mathur, P. and Giri, B. 2013. Arbuscular mycorrhiza: approaches for abiotic stress tolerance in crop plants for sustainable agriculture. In: Plant acclimation to environmental stress. Springer, New York, pp. 359-401.

Kapoor, R., Giri, B. and Mukerji, K. G. 2002. *Glomus macrocarpum*: a potential bioinoculant to improve essential oil quality and concentration in Dill (*Anethum graveolens* L.) and Carum (*Trachyspermum ammi* (Linn.) Sprague). *World J. Microbiol. Biotechnol.*, 18: 459-463.

Karasawa, T., Kasahara, Y. and Takebe, M. 2002. Differences in growth responses of maize to preceding cropping caused by fluctuation in the population of indigenous arbuscular mycorrhizal fungi. *Soil Biol. Biochem*, 34: 851-857.

Karishma., Tanwar A., Neetu, and Aggarwal A. 2011. Diversity of endomycorrhizal fungi associated with some ornamental flowering plants of family *Asteraceae* of Kurukshetra. *Coherence*, 1: 69-77.

Kaushish, S., Kumar, A., Aggarwal A. and Parkash V. 2011. Influence of inoculation with the Endomycorrhizal Fungi and Trichoderma viride on morphological and physiological growth parameters of *Rauwolfia serpentina* Benth. Ex. Kurtz. *Indian J. Microbiol.* 52: 295-299.

Khade, S. W. and Adholeya, A. 2009. Arbuscular mycorrhizal association in plants growing on metal-contaminated and non-contaminated soils adjoining Kanpur tanneries, Uttar Pradesh, India. *Water Air Soil Pollut*, 202: 45-56.

Kjøller, R. and Rosendahl, S. 2000. Effects of fungicides on arbuscular mycorrhizal fungi: differential responses in alkaline phosphatase activity of external and internal hyphae. *Biol. Fertil. Soils*, 31: 361-365.

Kloepper, J.W., Leong, J., Teintze, M. and Schroth, M.N. 1980. Enhanced plant growth by siderophores produced by plant growth-promoting rhizobacteria. *Nature.*, 286: 885.

Kumar, A., Mangla, C., Tanwar, A. and Aggarwal, A.2009-. Interaction of Arbuscular mycorrhizal fungi (AMF) with Withania somnifera (L.) Dunal. Acta Botanica Indica, 37-38:71-80.

Kumar, A., Mangla, C., Tanwar, A. and Aggarwal, A.2009-. Interaction of Arbuscular mycorrhizal fungi (AMF) with Withania somnifera (L.) Dunal. Acta Botanica Indica, 37-38:71-80.

Kloepper, J.W., Tuzun, S., Liu, L. and Wei, G. 1993. Plant growth-promoting rhizobacteria as inducers of systemic disease resistance. Pest management: biologically based technologies. American Chemical Society Books, Washington, DC, pp. 156-16.

Largent, D. L., Sugihara, N. and Brinitzer, A. 1980. *Amanita gemmata*, a non-host-specific mycorrhizal fungus of *Arctostaphylos manzanita*. *Mycologia*, 72: 435-439.

Lehmann, A., Veresoglou, S.D., Leifheit, E.F. and Rillig, M.C. 2014. Arbuscular mycorrhizal influence on zinc nutrition in crop plants–A meta-analysis. *Soil Biol. Biochem*, 69: 123-131.

Liao J.P., Lin, X.G., Cao, Z.H., Shi, Y.Q. and Wong, M.H. 2003. Interactions between arbuscular mycorrhizae and heavy metals under sand culture experiment. *Chemosphere.*, 50: 847-853.

Linderman, R.G. 1992. Vesicular-Arbuscular Mycorrhizae and Soil Microbial Interactions In: Mycorrhizae in sustainable agriculture, Bethlenfalvay, G.J. and Linderman, R.G. (Eds.) ASA Special Publication Number 54, pp. 45-70.

Macrae, A., Lucon, C.M., Rimmer, D. L. and O'donnell, A. G. 2001. Sampling DNA from the rhizosphere of *Brassica napus* to investigate rhizobacterial community structure. *Plant Soil,* 233: 223-230.

Maliha, R., S. Khalil., N. Ayub., S. Alam. and F. Latif. 2004. Organic Acids Production and Phosphate Solubilizaion by Microorganisms (PSM) under in vitro Conditions. *Pak. J. Biol. Sci.,* 7: 187-196.

Manchanda, G. and Garg, N. 2011 Alleviation of salt induced ionic, osmotic and oxidative stress in Cajanus cajan nodules by AM inoculation. Plant Biosyst., 145: 88-97.

Manoharachary, C. 2004. Biodiversity, taxonomy, ecology, conservation and biotechnology of arbuscular mycorrhizal fungi. *Indian Phytopathol,* 57(1): 1-6.

Marschner, H. and Dell, B. 1994. Nutrient uptake in mycorrhizal symbiosis. *Plant Soil* 159: 89–102.

Melin, E. 1923 Experimentelle Untersuchungen uber die Konstitution und Okologie der Mykorrhizen von Pinus sylvestris L. und Picea abies L. Karst.Mykol.Unters Ber.,2: 73-221.

Miller, R. M. and Jastrow, J. D. 1992. The role of mycorrhizal fungi in soil conservation. In: Behtlenfalvay, G.J., Linderman, R.G. (Eds.), Mycorrhizae in Sustainable Agriculture. ASA Special Publication 54, *Am. Soc. of Agronomy,* Madison, WI., pp. 29–44.

Mittler, R. 2002. Oxidative stress, antioxidants and stress tolerance. *Trends Plant Sci,* 7: 405-410.

Mohan, E. and Rajendran, K. 2014. Effect of Plant growth-promoting Microorganisms on Quality Seedling Production of *Feronia elephantum* (Corr.) in Semi-Arid Region of Southern India. *Int. J. Curr. Microbiol. Appl. Sci.,* 3: 103-116.

Morton, J.B. 1990. Evolutionary relationships among arbuscular mycorrhizal fungi in the Endogonaceae. *Mycologia,* 82: 192-207.

Morton, J.B. and Redecker, D. 2001. Two new families of Glomales, Archaeosporaceae and Paraglomaceae, with two new genera *Archaeospora* and *Paraglomus,* based on concordant molecular and morphological characters. *Mycologia,* 93: 181-195.

Olsson, P.A., Hammer, E.C., Wallander, H. and Pallon, J. 2008. Phosphorus availability influences elemental uptake in the mycorrhizal fungus *Glomus intraradices,* as revealed by particle-induced X-ray emission analysis. *Appl. Environ. Microbiol,* 74(13), 4144-4148.

Ordóñez-Valencia, C., Alarcón, A., Ferrera-Cerrato, R. and Hernández-Cuevas, L.V. 2009. *In vitro* antifungal effects of potassium bicarbonate on *Trichoderma* sp. and *Sclerotinia sclerotiorum. Mycoscience.,* 50: 380-387.

Osonubi, O., Mulongoy, K., Awotoye, O.O., Atayese, M.O. and Okali, D.U.U. 1991. Effects of ectomycorrhizal and vesicular-arbuscular mycorrhizal fungi on drought tolerance of four leguminous woody seedlings. *Plant Soil,* 136: 131-143.

Parkash, V. and Aggarwal, A. 2011. Interaction of VAM fungi with *Trichoderma viride* and *Rhizobium* species on establishment and growth of *Eucalyptus saligna* Sm. *Int. Sci. Res. J.,* 3: 200-209.

Pellegrino, E. and Bedini, S. 2014. Enhancing ecosystem services in sustainable agriculture: biofertilization and biofortification of chickpea (*Cicer arietinum* L.) by arbuscular mycorrhizal fungi. *Soil Biol. Biochem.,* 68: 429-439.

Pérez-Tienda, J., Testillano, P. S., Balestrini, R., Fiorilli, V., Azcón-Aguilar, C. and Ferrol, N. 2011. GintAMT2, a new member of the ammonium transporter family in the arbuscular mycorrhizal fungus *Glomus intraradices. Fungal Genet. Biol.,* 48: 1044-1055.

Platel K, and Srinivasan K. 1996. Influence of dietary spices or their active principles on digestive enzymes of small intestinal mucosa in rats. Int J Food Sci Nutr ., 47: 55–59.

Platel, K. and Srinivasan, K. 2001. Studies on the influence of dietary spices on food transit time in experimental rats. *Nutrn. Res.,* 21: 1309-1314.

Porcel, R., Aroca, R., Azcon, R. and Ruiz-Lozano, J. M. 2016. Regulation of cation transporter genes by the arbuscular mycorrhizal symbiosis in rice plants subjected to salinity suggests improved salt tolerance due to reduced Na+ root-to-shoot distribution. *Mycorrhiza,* 26: 673-684.

Raaijmakers, J. M., Paulitz, T.C., Steinberg, C., Alabouvette, C. and Moënne-Loccoz, Y. 2009. The rhizosphere: a playground and battlefield for soil borne pathogens and beneficial microorganisms. *Plant Soil,* 321: 341-361.

Rajendran, K. 2012. Effects of Bioinoculants on Seedling Growth, Biochemical Changes and Nutrient Uptake of *Erythrinaindica* L, in Semi-Arid Region of Southern India, *J. Biomet. Biostat.,* 3: 1-6.

Rani, P., Aggarwal, A. and Mehrotra, R.S. 1999. Growth responses in *Acacia nilotica* inoculated with VAM fungi (*Glomus mosseae*), *Rhizobium* sp. and *Trichoderma harzianum. Indian Phytopathol.,* 52: 151–153.

Rashid, M., Khalil, S., Ayub, N., Alam. S. and Latif, F. 2004. Organic acids production and phosphate solubilization by phosphate solubilizing microorganisms (PSM) under *in vitro* conditions. *Pak. J. Biol. Sci.,* 7: 187-196.

Reddy, A.C.P. and Lokesh, B.R. 1994 Alterations in Lipid Peroxides in Rat Liver by Dietary n-3 Fatty Acids: Modulation of Antioxidant Enzymes by Curcumin, Eugenol, and Vitamin E. Journal of Nutritional Biochemistry, 5, 181-188.

Reddy, A., Pulla, C. and Lokesh, B.R. 1994. Effect of dietary turmeric (*Curcuma longa*) on iron-induced lipid peroxidation in the rat liver. *Food Chemical. Toxicol.,* 32: 279-283.

Sabannavar, S. J., and Lakshman, H.C. 2009. Effect of rock phosphate solubilization using mycorrhizal fungi and phosphobacteria on two high yielding varieties of Sesamum indicum L. World J. Agic. Sci., 5: 470-479.

Safir, G.R. and Nelson, C.E. 1985. VA mycorrhizas: plant and fungal water relations. In: Proceedings of the 6th North American Conference on Mycorrhizae (Ed. by R. Molina), Forest Research Laboratory, Corvallis, Oregon, pp. 161-164.

Schüßler, A., Schwarzott, D. and Walker, C. 2001. A new fungal phylum, the Glomeromycota: phylogeny and evolution. *Mycol. Res.*, 105: 1413-1421.

Schultz, R.C., Colletti, J.P., Isenhart, T.M., Simkins, W.W., Mize, C.W. and Thompson, M.L. (1995). "Design and Placement of a Multi-species Riparian Buffer Strip System", Agroforestry system, 29:1-16.

Sharma, O.P. 1976. Antioxidant activity of curcumin and related compounds. *Biochem. Pharmacol.*, 25.15: 1811-1812.

Sieh, D., Watanabe, M., Devers, E. A., Brueckner, F., Hoefgen, R. and Krajinski, F. 2013. The arbuscular mycorrhizal symbiosis influences sulfur starvation responses of *Medicago truncatula. New Phytol.*, 197(2): 606-616.

Smith, S. E. and Read, D. J. 2010. Mycorrhizal symbiosis. Academic Press, London, p.769.

Socolow, R. H. 1999. Nitrogen management and the future of food: lessons from the management of energy and carbon. Proceeding of National Academy of Sciences, 96(11), 6001-6008.

Sowmya, A. 2014. Interactions between glomus macrocarpum and the best compatible biocontrol agent on growth and yield of chilli (*Capsicum annuum* L., Doctoral dissertation, University of Agricultural Sciences, Dharwad.

Srinivasan, K. 2005. Plant foods in the management of diabetes mellitus: spices as beneficial antidiabetic food adjuncts. *Int. J. Food Sci. Nutr.* 56: 399-414.

Subramanian, K. S. and Charest. C. 1999. Acquisition of N by external hyphae of an arbuscular mycorrhizal fungus and its impact on physiological responses in maize under drought-stressed and well-watered conditions. *Mycorrhiza,* 9: 69–75.

Suhail, F.M. and Mahdi, I.A. 2013. Test the efficiency of mycorrhizal fungi (*Glomus fasciculatum*) and magnetic water to reduce the effect of salinity on plant onion (*Allium cepa* L.). *Bulletin UASMV Serie Agriculture,* 70: 325-333.

Tanwar, A., Aggarwal, A. and Panwar, V. 2013. Arbuscular mycorrhizal fungi and *Trichoderma viride* mediated *Fusarium* wilt control in tomato. *Biocontrol Sci. Technol.*, 23: 485-498.

Tobar, R., Azcón, R. and Barea, J. M. 1994. Improved nitrogen uptake and transport from 15N-labelled nitrate by external hyphae of arbuscular mycorrhiza under water-stressed conditions. *New Phytol,* 126: 119-122.

Toth, R., Toth, D., Starke, D. and Smith, D. R. 1990. Vesicular–arbuscular mycorrhizal colonization in *Zea mays* affected by breeding for resistance to fungal pathogens. *Can. J. Soil Sci.,* 68: 1039-1044.

Toussaint, J. P., Kraml, M., Nell, M., Smith, S. E., Smith, F. A., Steinkellner, S. and Novak, J. 2008. Effect of *Glomus mosseae* on concentrations of rosmarinic and caffeic acids and essential oil compounds in basil inoculated with *Fusarium oxysporum* f. sp. *basilici*. *Plant Pathol.*, 57: 1109-1116.

Toussaint, J. P., Smith, F. A., Smith and S. E. 2007. Arbuscular mycorrhizal fungi can induce the production of phytochemicals in sweet basil irrespective of phosphorus nutrition. *Mycorrhiza.*, 17: 291-297.

Trevor, E. Y., Egger, K. N. and Peterson, L. R. 2001. Ectendomycorrhizal associations– characteristics and functions. *Mycorrhiza.*, 11: 167-177.

Wang, B. and Qiu, Y. L. 2006. Phylogenetic distribution and evolution of mycorrhizas in land plants. *Mycorrhiza.*, 16: 299-363.

Wright, S. F. and Upadhaya, A. 1996. Extraction of an abundant and unusual protein from soil and comparison with hyphal protein from arbuscular mycorrhizal fungi. *Soil Sci.* 161: 575–586.

Wright, S. F. and Upadhyaya, A. 1998. A survey of soils for aggregate stability and glomalin, a glycoprotein produced by hyphae of arbuscular mycorrhizal fungi. *Plant and soil*, 198: 97-107.

Yu,T.E.J.C., Egger, K.N. and Peterson, R.L. 2001. Ectendo-mycorrhizal associations-characteristics and functions . Mycorrhiza, 11: 167-177.

Chapter 2

Enhanced Production of Cellulase and Xylanase Enzyme from *Bacillus Subtilis* by 'One-Factor-at-a-Time-Approach'

Chakarvati Sango and Jitender Sharma*

Department of Biotechnology, Kurukshetra University, Kurukshetra-136119, Haryana, India.

**Corresponding Author E-mail: jksharmakuk@rediffmail.com*

Abstract

Present study involves the production of cellulase and xylanase enzymes from bacterial species through submerged fermentation. Various statistical as well as one-factor-at-a-time approaches have been used for production of enzymes. Bacterial species Bacillus subtilis used in present research carries co-production of both enzymes. Various carbon substrates such as wheat bran, rice bran, coconut powder, rice husk, saw dust and waste papers were used to enhance enzyme production. Optimization of various parameters such as inoculum size, inoculum age, pH, production time, carbon substrate, and nitrogen source with temperature and time profile enhanced the enzyme production as much as 14.5 IU/ml of cellulase activity and 294 IU/ml of xylanase.

Keywords: *Crude Cellulase, Bacillus subtilis, Submerged fermentation, Xylanase, Rice bran*

1. Introduction

Lignocellulosic waste is probably the most abundent renewable resources on earth consisting mainly 23% lignin, 40% cellulose and 33% hemicellulose by dry weight. Cellulose is an unbranched glucose polymer in which anhydro-D-glucose units linked by 1, 4-β-D-glucoside bonds, which can be hydrolyzed by cellulolytic enzymes comprising of three enzyme subunits; Endoglucanase, Exoglucanase and β-glucosidases (Robson *et al.*, 1989; Holker *et al.*, 2005). Hemicellulose is a polymer of xylan subunits attached by β-1, 4-D-xylopyranose residues, present between lignin and cellulose residues and its hydrolysis is carried out by xylanases and xylosidases (Lama *et al.*, 2004; Collins *et al.*, 2005). However, enzymatic hydrolysis of lignocelluloses involves high concentrations of both cellulase and xylanase although it has been reported that fewer concentrations of both enzymes carries out optimum degradation of cellulose biomass (Goncalves *et al.*, 2015). Cellulolytic enzymes have been produced by many micro-organisms consisting bacteria, fungi and actinomycetes. Fungal species such as *Trichoderma, Humicola, Penicillium* and *Aspergillus* are known to produce cellulases as well as xylanases in abundent amount with degradation of various carbon sources such as wheat bran, rice bran, rice husk and waste paper (Jun *et al.*, 2011; Wen *et al.*, 2005; Olsson *et al.*, 2003; Ahamed and Vermette, 2008; 2010). Many bacterial species such as Bacillus spp. have already been studied and extensively used as a source of industrial enzymes by biotechnology companies. Isolation of bacteria and fungi from soil samples and screening of these micro-organisms carried out in laboratory in order to isolate industrially important microbial strain (Battan *et al.*, 2007). *Bacillus subtilis* is gram positive bacterial species which is delibrately used for enzyme production. Spores of *Bacillus subtilis* species germinated in the gastrointestinal tract of various animals such as poultry and livestock leading to increase in nutritional values due to its ability to tolerate extreme environments as well as extracellular enzyme production such as cellulase, xylanase, phytase, protease, lipase, and keratinase (Hendricks *et al.*, 1995; Sen *et al.*, 2012). Various ligno-cellulosic materials such as wood, waste paper, fruit pomace, bagasse, wheat straw, corn cob, wheat bran, aspen wood, coconut powder, waste newsprint, and waste paper sludge have been used as probable substrates for production of cellulase and xylanase by *Bacillus subtilis* sp. (Wen *et al.*, 2005; Reczey *et al.*, 1996; Liming and Xueliang, 2004). Apart from *Bacillus sp.*, cellulase and xylanase enzymes have also been produced by other bacterial species such as *Clostridium, Ruminococcus* and *Cellulomonas.* The potential applications of celluloxylanolytic enzymes involves recycling of waste paper, fruit juice clarification, bread making, textile industry, bioethanol preparation, animal feed industry

2. Material and methods

2.1 Isolation of Bacterial Strain

Soil samples were collected from paper mill campus, Yamuna Nagar. Screening studies showed the presence of clear zones in cellulose and wheat bran enriched petri plates. Characterization studies showed the specific strain to be the alkalophilic *Bacillus subtilis* sp.

2.2 Enzyme Production

Crude cellulase enzyme was produced in Erlenmeyer flask (250 ml) containing 50 ml of basal medium having (g/L): yeast extract, 5.0; $MgSO_4.7H_2O$, 0.2; KH_2PO_4, 1.0 supplemented with rice bran, 5.0 and peptone, 5.0; pH 9.0. The flasks were inoculated with 2% of the 24 h grown inoculum and incubated at 37°C under shaking conditions (180 rpm) for incubation period of 48 h. The enzyme was harvested by centrifuging at 10,000×g for 15 min.

2.3 Enzyme Assay

Both cellulase and xylanase activities were determined by measuring the amount of reducing sugars using Bailey's method (1992) and Miller's method (1959) respectively. For CMCase activity, the reaction mixture of 500μl of 2% w/v Carboxy methyl cellulose, as a substrate prepared in 0.2 M glycine NaOH buffer (pH 9.0), was incubated at 50°C in water bath for 30 min which includes 10μl of crude enzyme filtrate incubated in same water bath with proper dilution. Control for both, heat killed substrate and enzyme were included simultaneously which were treated for 20 min at 100°C. After the proper incubation, 1.5ml of DNS reagent was added to the reaction mixture, and was kept in boiling water for 15 min for proper color development indicating the reaction between substrate and enzyme. After cooling down, the reaction mixture was shaken and optical density was measured at 540 nm at UV-VIS spectrophotometer. The values were compared with the values of standard curve for D-glucose.

For xylanase activity, 10 μl of suitable enzyme dilution was added to the 490 μl of 1% birchwood xylan which was prepared in 0.2M glycine NaOH buffer (pH 9.0) at 50°C. Controls were also included as described above. The reaction mixture was incubated at 50°C for 15 min. After that, 1.5 ml of DNS reagent was added to the reaction mixture and boiled for 15 min. Optical density for xylanase was determined under UV-VIS spectrophotometer and reducing sugar was determined by comparing the values using standered curve prepared for D-xylose.

2.4 Optimum pH and Temperature for Cellulo-xylanolytic Enzyme Production

The optimum pH for Cellulase and Xylanase enzyme activity was determined by performing enzyme assays from the pH range of 4-11. The effect of temperature on both enzyme activities was checked at different temperatures ranging from 30-65°C at optimized pH. Substrates were dissolved in citrate buffer (pH 4-6), phosphate buffer (pH 6-8), Tris-HCl buffer (pH 8-9) and glycine-NaOH buffer (pH 9-11) for pH optimization.

2.5 Parametric Optimization of Cellulo-xylanolytic Enzymes Production under Submerged Fermentation

Various factors were optimized for enzyme production during submerged fermentation from *Bacillus subtilis*:

2.5.1 Effect of Carbon Substrate

Different cheap and easily available carbon sources such as wheat bran, rice bran, rice husk, saw dust, coconut powder and waste paper were taken in 2% w/v for 1 ml of 24 h old inoculum with production time of 48 h. After incubation, media in the flask was centrifuged and crude supernatant was used to calculate enzyme activity.

2.5.2 Effect of Inoculum Size and Inoculum Age

Erlenmeyer flasks (250 ml) containing 50 ml of production medium were inoculated with 2% (v/v) inoculum of different age (2-14 h). The growth of *Bacillus subtilis* at different time intervals was estimated by taking the absorbance of the inoculated nutrient broth at 540 nm. The production flasks were incubated at 37°C under shaking conditions (120 rpm). The contents of the flasks were harvested after 48 h and assayed for cellulase and xylanase activities by dinitrosalicylic acid method to determine the optimal inoculum age for concurrent production of Cellulo-xylanolytic enzymes. To study the effect of inoculum size, the production media (50 ml) were inoculated with different amounts (0.25-2%) of 6 h old culture and incubated at 37°C at 120 rpm for 48 h. The contents of the flasks were then centrifuged at 10,000 g for 15 min and supernatant was assayed for enzymes activity.

2.5.3 Effect of Incubation Time for Enzyme Production

Erlenmeyer flasks (250 ml) containing 50 ml of production medium were inoculated with optimized inoculums(age and size) at 37°C at 120 rpm, production flasks were incubated for different time intervals from 18 to 72h. After incubation, the media of the flasks were centrifuged at 10,000 g for 15 min and supernatant was assayed for enzymes activity.

2.5.4 Effect of Various Nitrogen Sources

Different nitrogen sources such as peptone, yeast extract, beef extract, tryptone, magnesium sulphate, potassium nitrate, sodium chloride and their different combinations on cellulase and xylanase production was studied by supplementing various nitrogen sources (0.5% w/v) in the medium containing 2% (w/v) each of the optimized carbon source and devoid of any nitrogen source. Each 250 ml flask containing 50 ml medium (pH 9.0) was inoculated and incubated at 37°C for 48 h under shaking conditions (120 rpm). Cellulase and xylanase activities were determined by dinitrosalicylic acid method as described above. Further, the selected nitrogen source was investigated at different concentrations (0.1% to 0.5%) in the production medium (pH 9.0) to optimize the concentration of nitrogen source.

2.5.5 Optimization of Assay Temperature and Time for Cellulo-xylanolytic Ezyme Activity

Effect of temperature on the activity of Cellulase and Xylanase was determined by incubating the crude enzyme extract at different temperatures ranging from 30 to 70°C.Optimized temperature at which, crude extract showed its maximum enzyme activity was taken for time profile optimization. It involved the gradual change in enzyme activity with increase in time. The range for time profile was taken from 5 min to 120 min to check the change in enzyme activity at specific temperature.

3. Results

Initially, enzyme production carried out by adding 2% of 24 h old inoculum into 50 ml of modified horikoshi medium incubated for 48 hrs at 37°C under shaking conditions (120 rpm) taking wheat bran as carbon substrate. Enzyme activity for cellulase and xylanase was found out to be 1.58 IU/ml and 19.5 IU/ml respectively. The results of various optimized cultural conditions for obtaining maximum production of cellulo-xylanolytic enzymes by *Bacillus subtilis* are given as under:

3.1 Optimization assay pH and Temperature for Cellulo-xylanolytic Enzyme Activity

During optimization of pH and temperature, both enzyme activities remained almost same as we moved from pH 4 to 6. Thereafter a sudden increase in activity was observed from pH 7 to 9 after which it started to decrease. The maximum enzyme activity was recorded at pH 9 (Fig. 2.1). Also, 37°C was the optimum temperature for the growth of cellulo-xylanolytic enzyme producing bacteria *Bacillus subtilis*. There was continuous decrease in enzyme activity with further increase in temperature while no production of enzyme at 60°C. Optimum pH for enzyme production found out to be 9. There was decrease in enzyme activity at lower and higher values from optimum value because of effect of pH on the charge distribution over enzyme and substrate molecules (Shah *et al.*, 2005).

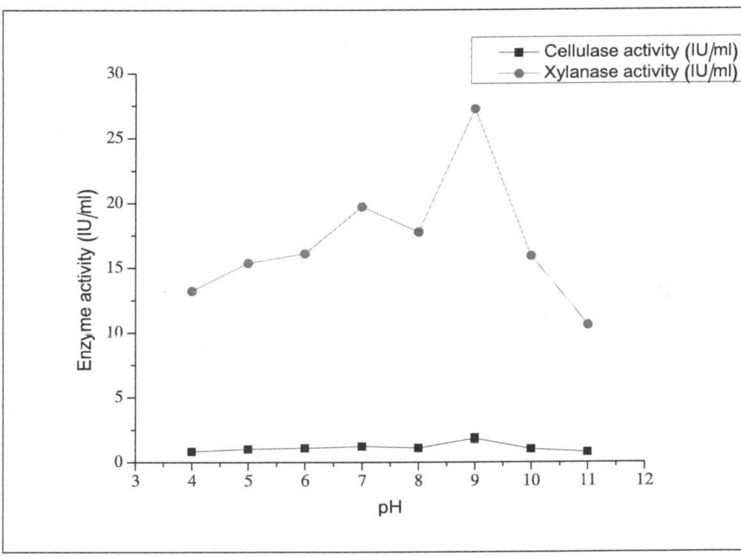

Fig. 2.1: Effect of different pH on cellulo-xylanolytic activity.

3.2 Effect of Carbon Substrates

Apart from commercially available carboxy methyl cellulose and xylan substrates, cheap and easily available carbon sources were used for enhancement in enzyme activity. Among all the used substrates, rice bran gave maximum enzyme activities for both cellulase and xylanase (Fig. 2.2). Wheat bran also showed improved enzyme production at alkaline pH while saw dust showed improved xylanase activity over cellulase activity. Coconut powder and rice husk produced medium cellulo-xylanase enzyme titer while waste paper showed lowest activities for both enzymes.

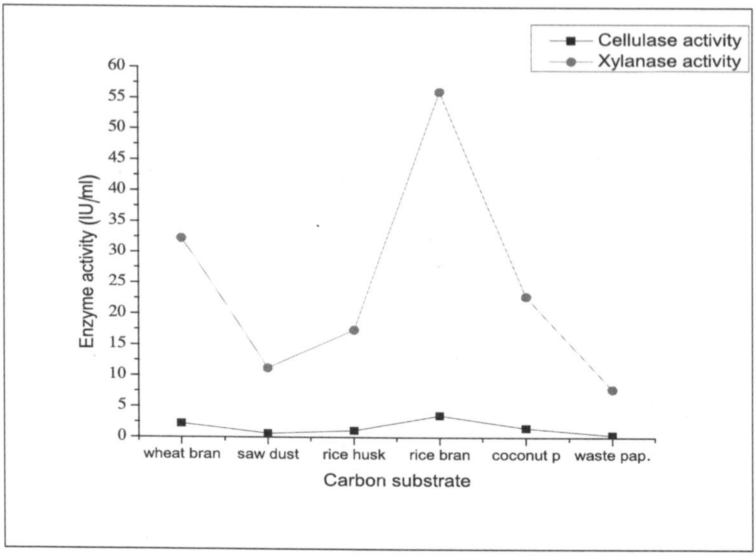

Fig. 2.2: Effect of different carbon substrates on cellulo-xylanolytic activity.

3.3 Effect of Inoculum Age and Size

Highest coproduction production of cellulase and xylanase enzymes from *Bacillus subtilis* was obtained when 6 h old inoculum was used (Fig. 2.3). The optimization studies showed the maximum growth rate of the bacteria was between 6 to 8 hrs. As inoculum for enzyme production used was from log phase, maximum enzyme production occurred. The range of inoculum size was 0.25 to 2% in which sudden rise of activity was noticed as moved from 0.25 to 0.5% while from 0.5 to 2% there was steadiness in enzyme activity. Both cellulo-xylanolytic enzyme activities were found to be maximum with optimum inoculum size of 0.5% (Fig. 2.4).

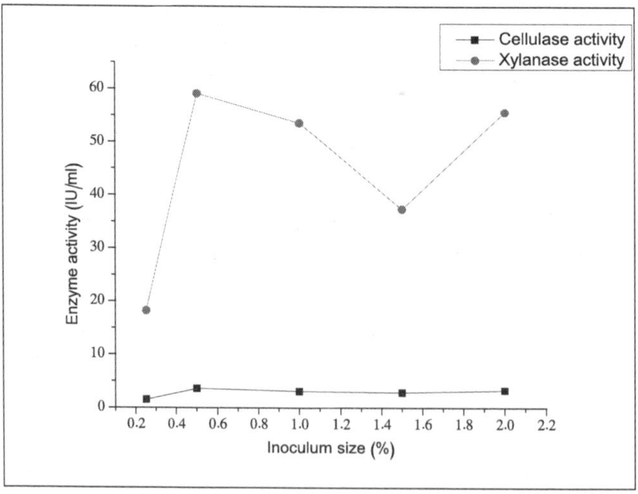

Fig. 2.3: Effect of different inoculum size on cellulo-xylanolytic activity.

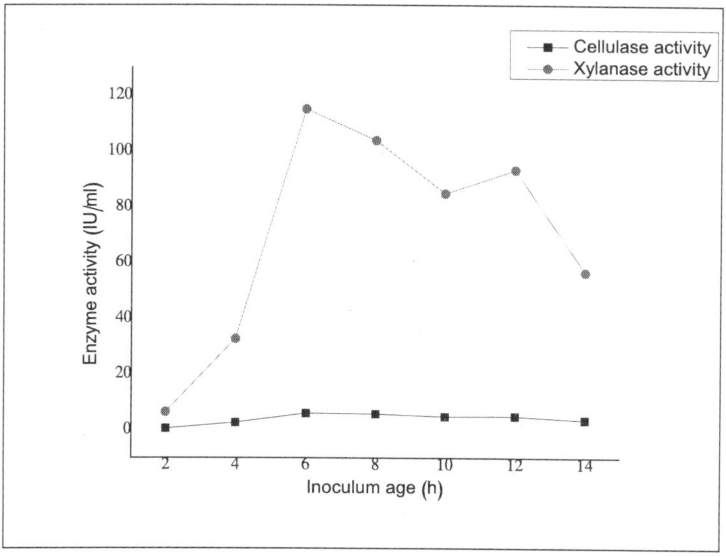

Fig. 2.4: Effect of different inoculum age on cellulo-xylanolytic activity.

3.4 Effect of Incubation Time on Enzyme Production

The range of incubation time for both Cellulase and Xylanase enzymes showed variations as we moved from 18 h to 70 h. Enzyme activity started to increase with increase in incubation time from 18 h upto 42 h while it remained steady at 48 h and started to decrease thereafter. Maximum recorded activity for both enzymes was after 42 h (Fig. 2.5). The loss of enzyme activity with increase of production time might be due to instability of the crude enzyme.

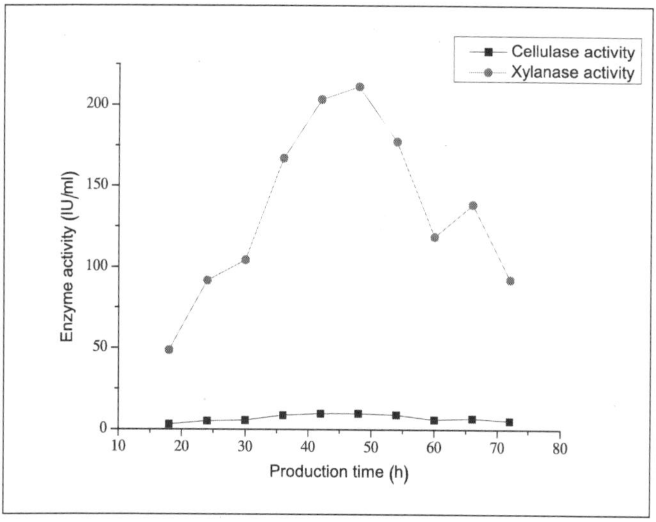

Fig. 2.5: Effect of incubation time on cellulo-xylanolytic activity.

3.5 Effect of Various Nitrogen Sources and Shaking Speed

Various nitrogen sources such as peptone, yeast extract, KNO_3 were optimized according to their concentrations for enzymes production. Among five concentrations of peptone, 1.5% was found to be the significant for Cellulo-xylanolytic enzymes activity as higher concentrations showed significant decline in enzyme activity. Yeast extract and KNO_3 also found to affect enzyme activity significantly. Organic nitrogen sources and their combinations with inorganic nitrogen sources have been found to enhance the xylanase production from *Bacillus* species (Sanghi *et al.*, 2009). Maximum Cellulase and Xylanase activity recorded were 9.9 IU/ml and 215 IU/ml with 1.5% peptone concentration (Fig. 2.6 & 2.7). Shaking speed of rotatory shaker also affected enzyme production considerably. The agitation speed of 150 rpm was selected for carrying out further experiments as enzyme activity remained nearly similar at 180 rpm while it tends to decrease with further increase in agitation speed.

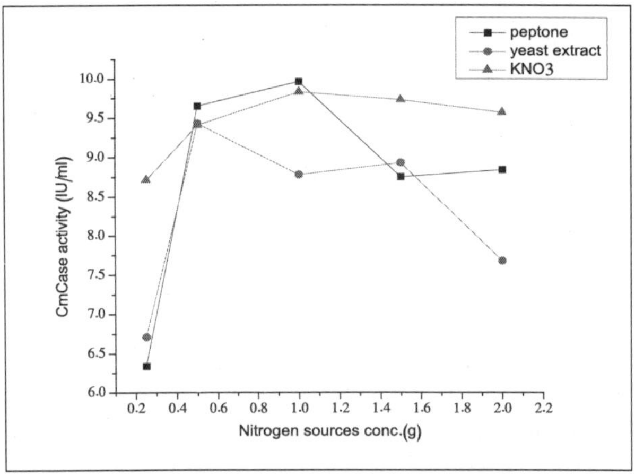

Fig. 2.6: Effect of different nitrogen sources on cellulase (CMCase) activity.

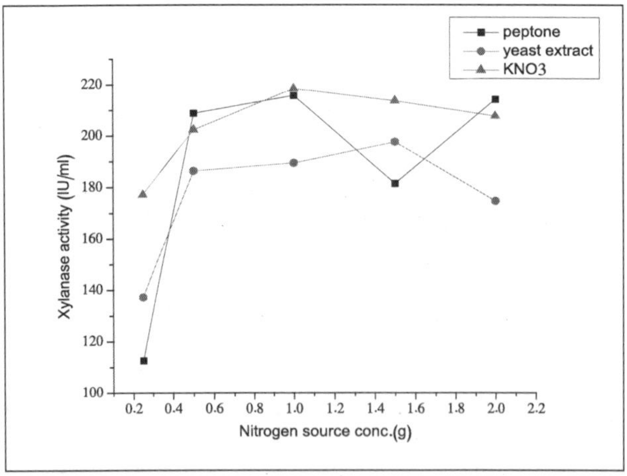

Fig. 2.7: Effect of different nitrogen sources on xylanase activity.

3.6 Optimization of Enzyme Assay Temperature

Crude enzyme extract (supernatant after centrifugation) was diluted with glycine-NaOH buffer (pH 9.0) and was placed in a water bath and assayed at different temperatures from 30°C to 70°C. Enzyme activity at 30°C was 8.5 IU/ml for Cellulase and 174.6 IU/ml for Xylanase which further increased with temperature increase. Maximum enzyme activity recorded for cellulase and xylanase were 12.85 IU/ml and 267.82 IU/ml at 50°C respectively, which further declined with increase in temperature (Fig. 2.8). The reason behind such decrease in enzyme activity is denaturation of protein due to increase in temperature of enzyme assay.

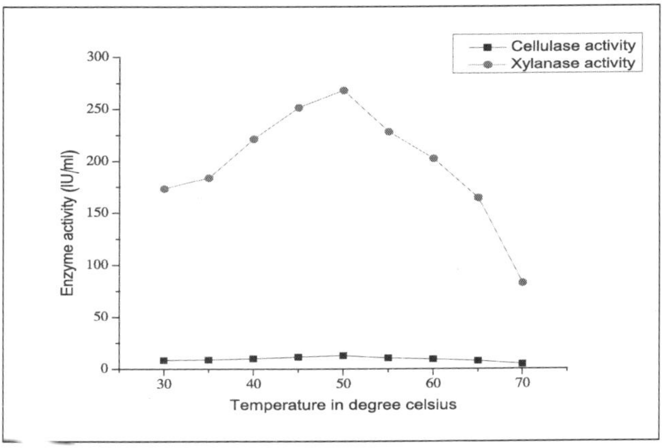

Fig. 2.8: Optimization of enzyme assay temperature for cellulo-xylanolytic activity.

3.7 Optimization of enzyme assay time

Crude enzyme extract was diluted with glycine-NaOH buffer (pH 9.0) and incubated in water bath at optimum temperature and assayed for enzyme activity after proper interval of time such as 0, 15, 30, 45, 60, 90 and 120 min. Maximum enzyme activity was recorded between 15 min to 30 min.

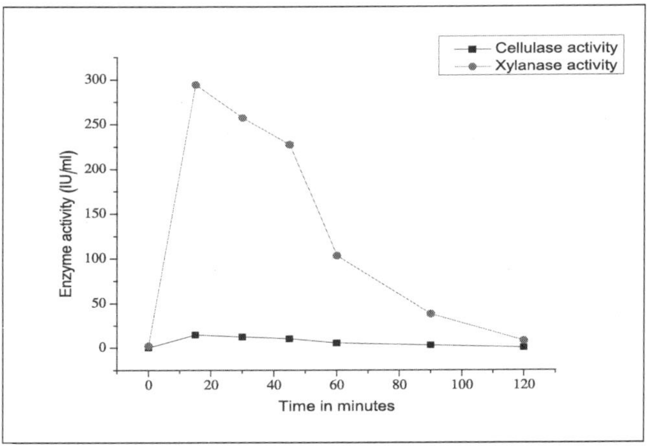

Fig. 2.9: Optimization of reaction time for enzyme assay for cellulo-xylanolytic activity.

Further, there was slight loss in enzyme activity upto 120 min. Enzyme titre recorded were maximum at 15 min with 14.53 IU/ml and 294.37 IU/ml for Cellulase and Xylanase respectively (Fig. 2.9).

4. Discussion

Above work on *Bacillus subtilis* for cellulase and xylanase enzymes showed clear hydrolysis zones on carboxy methyl cellulose as well as on wheat bran Petri plates. After isolation, both cellulo-xylanolytic enzyme activities were measured by keeping slandered conditions such as, 2% inoculum size, 24 h inoculum age, 48 hrs production time, 2% wheat bran as carbon substrate in modified horikoshi media. In recent years, *Streptomyces sp.* showed maximum xylanase production at pH 9 (Techapun *et al.*, 2001) while *Bacillus SSP*-34 showed maximum xylanase with small cellulase activity at pH 9 and 10 (Balakrishnan *et al.*, 1992; 1997) and also maximum cellulase activity with traces of xylanase was recorded at pH 7 (Paul *et al.*, 1993). The optimum inoculum size and age reported were 0.5% and 6 h, respectively. This might be due to decrease in enzyme titre with increase in inoculum size (Battan *et al.*, 2007).The effect of inoculum age was studied from 2 h to 12 h. Maximum enzyme activity of cellulase and xylanase recorded were 5.972±0.93 IU/ml and 115.149±4.42 IU/ml respectively for 6 h inoculum which showed association between enzyme production process and growth of micro-organisms (Kumar *et al.*, 2012; Nagar *et al.*, 2010). Various carbon substrates showed different range of enzyme activities for both cellulase and xylanase in which rice bran showed maximum cellulo-xylanolytic enzyme activity as it contained significant amount of nutrients and remained loose for moist conditions rather than other cheap organic sources (Dhillon and Khanna 2000; Santos *et al.*, 2002; Techapun *et al.*, 2001). However, previous studies showed commercial substrates and wheat bran as the cheap carbon source (Battan *et al.* 2007; Van *et al.* 2011). Lu *et al.*, (2003) reported that incubation temperature between 30-45°C was the optimum temperature range for Cellulo-xylanolytic enzyme production. Vyas *et al.*, (2005) also reported incubation temperature of 40°C for enhanced cellulase and xylanase production. Different nitrogen sources in media components were optimized according to their specific concentrations. Peptones, yeast extract, KNO_3 were optimized according to their concentrations which showed positive effects on enzyme activity enhancement (Haltrich *et al.*, 1997; Hoq *et al.*, 1994).

5. Conclusion

In this study, we have optimised modified horikoshi media and growth conditions for production of cellulase and xylanase enzymes from *Bacillus subtilis* under submerged fermentation by growing bacteria at 37°C and pH 9. Production rate of cellulase was found higher than that of xylanase. Production of cellulo-xylanolytic enzymes has been directly proportional to biomass concentration indicating that specific medium composition and optimum cultural conditions are the key parameters for higher biomass production. Growth of microorganisms requires primary metabolites i.e. Cellulo-xylanolytic enzymes, that convert complex substrates into reducing sugars that is absorbed into the cells. Rice bran found out to be the optimum carbon source stimulating initial growth of micro-organisms in submerged fermentations. Optimization of various parameters such as inoculum size, inoculum age, pH, production time, carbon substrate, and nitrogen source with temperature and time profile enhanced the enzyme production as much as 14.5 IU/ml of cellulase activity and 294 IU/ml of xylanase.

Acknowledgement

The author greatfully acknowledges Department of Biotechnology, Kuruksketra University, Kurukshetra for providing facilities while carrying out experimental work.

References

Ahamed, A. and Vermette, P. 2008. Culture-based strategies to enhance cellulase enzyme production from *Trichoderma reesei* RUT-C30 in bioreactor culture conditions. *Biochem Eng J.*, 40: 399–407.

Ahamed, A. and Vermette, P. 2010. Effect of mechanical agitation on the production of cellulases by *Trichoderma reesei* RUT-C30 in a draft-tube airlift bioreactor. *Biochem Eng J.*, 49: 379–387.

Balakrishnan, H., Dutta-Choudhury, M., Sreenivasan, M.C. and Rele, M.V. 1992. Cellulase-free xylanase production from an alkalophilic *Bacillus* sp. *World J. Microbiol. Biotechnol.*, 8: 627–631.

Bailey, M.J., Biely, P. and Poutanen. K. 1992. Interlaboratory testing of methods for assay of xylanase activity. *J. Biotechnol*, 23: 257.

Battan, B., Sharma, J., Dhiman, S.S. and Kuhad, R.C. 2007. Enhanced production of cellulase-free thermostable xylanase by Bacillus pumilus ASH and its potential application in paper industry. *Enzym Microb Technol*, 41: 733–739.

Collins, T., Gerday, C., Feller, G. 2005. Xylanases: xylanase families and extremophilic xylanases. *Microbiology Reviews*, 29: 3–23.

Dhillon, A. and Khanna, S. 2000. Production of a thermostable alkalitolerent xylanase from Bacillus circulans AB 16 grown on wheat straw. *World J. Microbiol Biotechnol*, 16: 325–327.

Gonc¸alves, G.A., Takasugi, Y., Jia, L., Mori, Y., Noda, S., Tanaka, T., Ichinose, H. and Kamiya, N. 2015. Synergistic effect and application of xylanases as accessory enzymes to enhance the hydrolysis of pretreated bagasse. *Enzyme Microb. Technol*, 72: 16–24.

Haltrich, D., Nidetzky, B., Kulke, K.D., Steiner, W. and Zupanie, S. 1997. Production of fungal xylanases. *Bioresource Technol*, 58: 137 -161.

Hendricks, C.W., Doyle, J.D. and Hugley, B. 1995. A new solid medium for enumerating cellulose-utilizing bacteria in soil. *Appl. Environ. Microbiol*, 61: 2016–19.

Holker, U., Hofer, M. and Lenz, J. 2004. Biotechnological advantages of laboratory scale solid state fermentation with fungi. *Appl. Microbiol. Biotechnol*, 64: 175-86.

Hoq, M., Hempel, C., and Deckwer, W. D. 1994. Cellulasefree xylanase by Thermomyces lanuginosus RT9: effect of agitation, aeration, and medium components on production. *J. Biotechnol.,* 37: 49-58.

Jun, H., Kieselbach, T. and Jo¨nsson, L.J. 2011. Enzyme production by filamentous fungi: analysis of the secretome of Trichoderma reesei grown on unconventional carbon source. *Microbial Cell Factor,* 10:1.

Kumar, A., Gupta, R., Shrivastava, B., Khasa, Y.P. and Kuhad, R.C. 2012. Xylanase production from an alkaliphilic actinomycete isolates Streptomyces sp. RCK2010, its characterization and application in saccharification of second generation biomass. *J. Mol. Catal. B: Enzym,* 74: 170–177.

Lama, L., Calandrelli, V., Gambacorta, A. and Nicolaus, B. 2004. Purification and characterization of thermostable xylanase and β-xylosidase by thermophilic bacterium *Bacillus thermantarcticus. Res. Microbiol,* 155: 283-89.

Liming, X. and Xueliang, S. 2004. High-yield cellulase production by Trichoderma reesei ZU-02 on corn cob residue. *Bioresour Technol,* 91(3):259–262.

Lu, W., Li, D. and Wu, Y. 2003. Influence of Water Activity and Temperature on Xylanase Biosynthesis in Pilot-Scale Solid-State Fermentation by *Aspergillus sulphureus. Enzyme and Microbial Technology,* 32: 305-311.

Miller, G.L. 1959. Use of dinitrosalicylic acid reagent for determination of reducing sugar. *Ana. Chem.,* 31: 426-428.

Nagar, S., Gupta, V.K., Kumar, L. and Kuhad, R.C. 2010. Production and optimization of cellulase-free, alkali-stable xylanase by *Bacillus pumilus* SV-85S in submerged fermentation. *J. Ind. Microbiol. Biotechnol,* 37: 71–83.

Olsson, L., Christensen, T.M., Hansen, K.P. and Palmqvist, E.A. 2003. Influence of the carbon source on production of cellulases, hemicellulases and pectinases by Rut C-30. *Enzyme Microb Technol,* 33: 612–619.

Paul, J. and Varma, A.K. 1993. Characterisation of cellulose and hemicellulose degrading *Bacillus* sp. from termite infested soil. *Curr. Sci.,* 64: 262–266.

Reczey, K., Szengyel, Z., Eklund, R. and Zacchi, G. 1996. Cellulase production by T. reesei. *Bioresour Technol,* 57: 25–30.

Robson, L.M.and Chambliss, G.H. 1989. Cellulases of bacterial origin.*Enzyme Microb Technol,* 11: 626-644.

Sanghi, A., Garg, N., Kuhar, K., Kuhad, R.C. and Gupta, V.K. 2009. Enhanced production of cellulase-free xylanase by alkalophilic *Bacillus subtilis* ASH and its application in Biobleaching of kraft pulp. *Bioresources,* 4: 1109-1129.

Santos, E.S.D., Guirardello, R. and Franco, T.T. 2002. Preparative chromatography of xylanase using expanded bed adsorption. *J Chromatogr,* 944: 217–224.

Sen, S., Ingale, S.L., Kim, Y.W., Kim, J.S., Kim, K.H. and Lohakare, J.D. 2012. Effect of supplementation of Bacillus subtilis LS 1-2 to broiler diets on growth performance, nutrient retention, caecal microbiology and small intestinal morphology. *Res Vet Sci,* 93: 264–268.

Shah, A.R. and Madamwar, D. 2005. Xylanase production by a newly isolated *Aspergillus foetidus* strain and its characterization. *Proc. Biochem.*, 40: 1763-1771.

Subramaniyan, S., Prema, P. and Ramakrishna, S.V.1997. Isolation and screening for alkaline thermostable xylanases. *J. Basic Microbiol.* 37: 431–437.

Techapun, C., Sinsuwongwat, S., Poosaran, N., Watanabe, M. and Sasaki, K. 2001. Production of a cellulase-free xylanase from agricultural waste materials by a thermotolerant *Streptomyces* sp. *Biotechnol Lett*, 23: 1685–1689.

Van, H.V., Tuan, A.P. and Keun, K. 2011. Improvement of Fungal Cellulase Production by Mutation and Optimization of Solid State Fermentation, *Mycobiology*, 39: 20-25.

Vyas, A., Vyas, D. and Vyas, K.M. 2005. Production and Optimization of Cellulases on Pretreated Groundnut Shell by *Aspergillus terreus* AV49. *Journal of Scientific and Industrial Research*, 64: 281-286.

Wen, Z., Liao, W. and Chen, S. 2005. Production of cellulase by Trichoderma reesei from dairy manure. *Bioresour Technol*, 96: 491–499.

Chapter 3

In vitro Study and Micropropagation of a Thalloid Liverwort: *Reboulia Hemisphaerica* (L.) Raddi.

Vishal Awasthi, Neerja Pande and Virendra Nath

[1]*Botany Department, University College, Kurukshetra University, Kurukshetra –136119*

[2]*Department of Botany, Kumaun University, Nainital – 263002*

[3]*Bryology Laboratory, CSIR – National Botanical Research Institute, Lucknow – 226001*

**Corresponding Author E-mail: vawasthi1979@rediffmail.com*

Abstract

Reboulia hemisphaerica (L.) Raddi, the only species of genus Reboulia belonging to the family Aytoniaceae of the order Marchantiales is one of the potential liverworts that have medicinal importance. Well differentiated thalli of R. hemisphaerica have been raised and multiplied in vitro in order to facilitate the availability of the living material in pure population for other experimental studies and maintenance of its germplasm as a mean of ex-situ conservation. The ex plants (spores), inoculated into half strength Knop's macronutrients and Nitsch trace elements along with 10 ppm freshly prepared ferric citrate produced dichotomously branched thalli, while the same when supplemented with 1% sucrose, produced a cluster of many well developed thalli within 12 weeks of culture. Multiplication of thalli was readily achieved by transferring of the apical portion of cultured thalli or innovations produced from cultured plants on fresh basal medium with above constituents. An enhanced growth has been observed in continuous light of 4000-5500 lux at a temperature of 21 ± 2ºC. MS medium responded improper growth led to the formation of twisted and deformed thalli, while Gamborg B-5 medium produced callus like pulpy mass of undifferentiated tissue. Thalli developed in culture were acclimatized and transferred to soil for their further propagation where sexual phase induced at low temperature (16–18ºC) simultaneously with alternate light (1600-2000 lux) and dark condition of 10 hours and 14 hours respectively.

Keywords: *Acclimatization, germplasm, Reboulia hemisphaerica, propagation, spores*

1. Introduction

In recent years, importance of Bryophytes in medicine has come into light on account of the presence of some novel compounds like terpenoids, bis-bibenzyls, sesquiterpenoids etc. (Asakawa, 2007). *Reboulia hemisphaerica* (L.) Raddi, the only species of genus *Reboulia* belonging to the family Aytoniaceae of the order Marchantiales is one of the potential liverworts that have medicinal importance. It is used as antipyretic, antidotal, diuretic, for cure of cuts, fracture, snake bite, burns, scalds, haemostasis, external wounds and bruises. Marchantin –A a bis bibenzyl and Riccardin-C isolated from *R. hemisphaerica* display cytotoxic activity against KB cells (Asakawa and Matsuda, 1982; Dixit and Banerji, 2007). However, liverworts are generally tiny species often grow together with mosses. This makes it difficult to collect a large quantity of one single species sufficient for the isolation of compounds required for structure elucidation and biological assay. Implementation of tissue culture technique has been proved helpful in isolation of novel chemical compounds from bryophytes (Sauerwein and Becker, 1990; Sabovljevic *et al.*, 2009). Well differentiated thalli of *R. hemisphaerica* have been raised and multiplied *in vitro* in order to facilitate the availability of the living material in pure population for other experimental studies and maintenance of its germplasm as a mean of *ex-situ* conservation.

2. Materials and Methods

The thalli with mature sporophytes were collected from Dhanaulti (Uttarakhand). Voucher specimen has been deposited in the Bryophyte Herbarium of National Botanical Research Institute, Lucknow (LWG-248745) India. Different range of Knop's macronutrients (Knop, 1865), Knop's macronutrients and Nitsch trace elements with ferric citrate (Kaul *et al*, 1962) have been used as basal medium. Sucrose was used as carbon source. Standard culture protocols have been established by using other inorganic media viz. MS (Murashige and Skoog, 1962), Gaborg B5 (Gamborg *et al*, 1968), Hoagland medium (Hoagland and Arnon, 1950) and growth regulators. All the media and culture vessels were sterilized by autoclaving at 15 lb/sq. in. for 15 minutes before inoculation in Laminar Air Flow cabinet. Standard aseptic conditions were maintained throughout the work. Surface sterilization of the explants took place with the application of 1% sodium hypochlorite solution to the undehished capsule for 4-6 minutes. After treating with sodium hypoclorite solution, the ex-plants were rinsed with sterilized double distilled water twice. Cultures were maintained under controlled and aseptic conditions. Optimum temperature range 15-25°C was provided with the help of Air conditioners. Cultures were provided continuous illumination of 4000-5500 lux as well as alternate light and dark period of 16 hours and 8 hours respectively with the help of fluorescent tubes.

In vitro raised plants were transferred to soil in two successive steps. In first step, soil was sterlized and saturated with sterilized mineral salts solution. Then the plants were detached from agar media and washed properly with sterilized distilled water. The washed plants were transferred to soil saturated with mineral salts solution filled in sterilized pots and kept in the controlled laboratory conditions. In second step, well established and hardened thalli in pots were transferred with their substrata to the soil beds in the Moss House of Botanic Garden, NBRI, Lucknow where humidity (60-100%) and temperature ± 25°C were maintained with the help of an advance fogger and sprinkler system.

3. Results

The spores of *R. hemisphaerica* germinated after 7-10 days of inoculation. The germination was exosporous. Spores germinated at the proximal face by rupturing spore coat at tri-radiate marking from which protrusion of a germ papilla took place. At this time in the germ papilla, numerous oil globules of different size and abundant chloroplasts were observed. A small rhizoidal cell cut off laterally from the germ cell (Fig. 3.1, a), which later elongated to form first rhizoid, but in many spores the first rhizoid may be absent. Germ cell soon undergoes elongation into a germ tube containing abundant chloroplasts and few oil globules like inclusions (Fig. 3.1, b). From the tip of the germ tube there were a few cells cut off by transverse divisions usually one to three. Further division of the terminal cell in the formation multicellular gerling (Fig. 3.1, c), from which several rhizoids emerged from the lower side. Dorsiventrality of the young thalli was established by the development of rhizoids on their ventral side (Fig. 3.1, d).

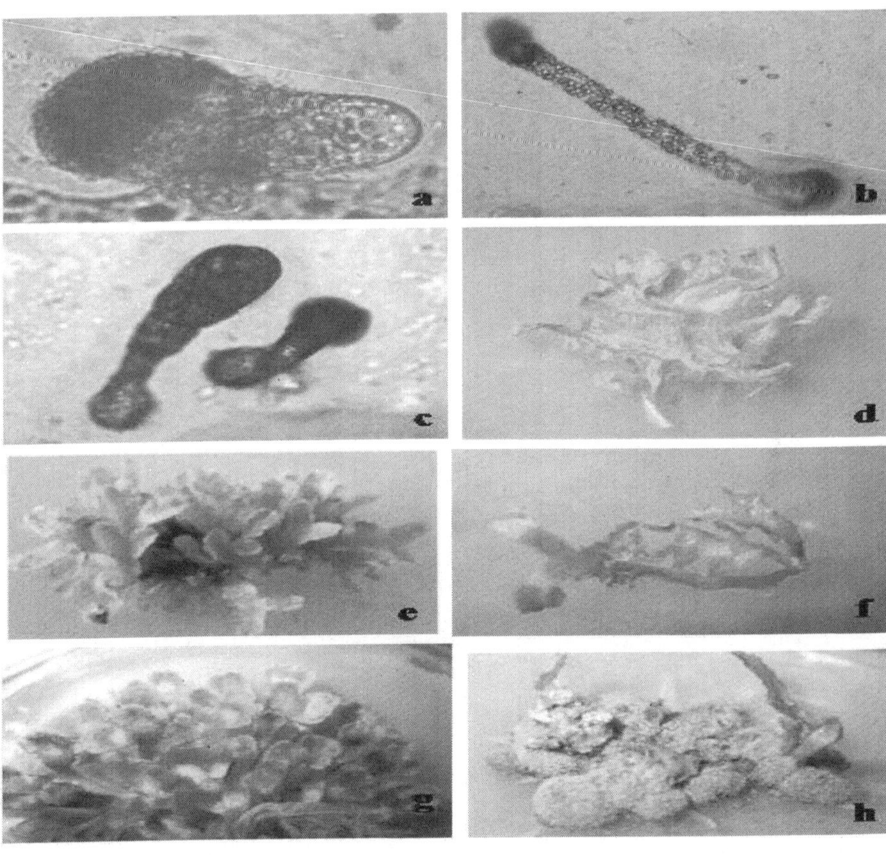

Fig. 3.1: ***In Vitro*** **growth of** ***Reboulia hemispherica.***

(a-b) Germinated spore showing short germ tube and laterally cut off rhizoidal cell; (c) Multicellular sporeling; (d) Thalli growing in half strength Knop's macronutrients medium (70 days old); (e) Growing thalli in half strength Knop's macronutrients medium + Nitsch trace elements (60 and 70 days old repectively); (f) Formation of adventitious shoots from a single intact thallus; (g) Vigorous growth of thalli in half strength Knop's macronutrients medium + Nitsch trace elements + 1% sucrose (100 days old); (h) Formation of abundant very small innovations or miniature of thalli from hardy callus like tissue in the same medium having 2% sucrose

Half strength Knop's macronutrients medium suited best for spore germination as in this medium, maximum number of spores was found to be germinated. In MS medium, Gamborg B-5 medium and Hoagland medium spores either did not germinate or if germinated, died at subsequent stage. In continuous light of 4000-5500 lux maximum germination occurred followed by alternated light and dark periods of 14 hours and 10 hours respectively (Table – 3.1). In continuous dark, no spore germination occurred.

Table 3.1: Percent spore germination in *R. hemisphaerica*, as observed in different culture media and light conditions after 10 days of inoculation.

Culture medium	Percentage germination of spores ± Standard Error	
	In continuous light (4000-5500 Lux)	In alternate light/dark condition of 16h/8h
½ K	82.30 ± 2.80	52.23 ± 2.86
½ K + N	45.04 ± 1.60	29.50 ± 2.77
½ K + N + 1S	8.62 ± 2.04	3.22 ± 1.39
MS	–	–
B-5	–	–
H	–	–

Abbreviations: ½ K: Half strength Knop's macronutrients; ½ K + N: Half strength Knop's macronutrients + Nitsch trace elements with 10 ppm ferric citrate; ½ K + N + 1S: Half strength Knop's macronutrients + Nitsch trace elements with 10 ppm ferric citrate + 1% sucrose

Under continuous illumination and at 21 ± 2ºC temperature, length and width of thalli of *R. hemispherica* exhibited marked difference. In controlled (in wild) conditions, the average length and width of the thalli were 20.57 ± 0.99 mm and 5.68 ± 0.15 mm, respectively (Table 2). In comparison to control, the average length and width of the thalli increased in the Knop's nutrient medium supplemented with Nitsch trace elements and 1% sucrose whereas, in half strength Knop's medium, a pronounced decline in length (16.39 ± 0.98 mm) and width (2.13 ± 0.12 mm) of thalli was observed. The best performance of the cultured thalli was observed in half strength Knop's macronutrients + Nitsch trace elements + 1% sucrose with an average length of 27.06 ± 1.04 mm and width of 5.97 ± 0.09 mm.

Table 3.2: Growth performance of *R. hemisphaerica* in different growth media as observed after 100 days of inoculation under continuous illumination (4000-5500 lux) at 21 ± 2ºC temperature.

Growth media	Average length of thalli ± SE (in mm)	Average breadth of thalli ± SE (in mm)
Growth response of cultured thalli raised from spore		
Control	20.57 ± 0.99	5.68 ± 0.15
½ K	16.39 ± 0.98	2.13 ± 0.12
½ K + N	20.02 ± 0.57	4.96 ± 0.08
½ K + N + 1S	27.06 ± 1.04	5.97 ± 0.09
½ K + N + 2S	1.67 ± 0.22	0.7 ± 0.07
Growth response of cultured thalli after sub culturing		
MS	8.34 ± 0.64	3.63 ± 0.28
MS + 1S	6.81 ± 0.42	3.91 ± 0.14
B-5	1.84 ± 0.13	0.98 ± 0.09
H	17.59 ± 0.74	5.27 ± 0.09
H + 1S	22.16 ± 0.49	5.67 ± 0.09

Abbreviations: ½ K: Half strength Knop's macronutrients; ½ K + N: Half strength Knop's macronutrients + Nitsch trace elements with 10 ppm ferric citrate; ½ K + N + 1S: Half strength Knop's macronutrients + Nitsch trace elements with 10 ppm ferric citrate + 1% sucrose; ½ K + N + 2S: Half strength Knop's macronutrients + Nitsch trace elements with 10 ppm ferric citrate + 2% sucrose; MS: Murashige and Skoog medium; MS + 1S: Murashige and Skoog medium supplemented with 1% Sucrose; B-5: Gamborg B-5 medium; H: Hoagland medium; H + 1S: Hoagland medium supplemented with 1% Sucrose.

Sporelings in half strength Knop's medium devoid of any trace elements could not well differentiated, instead light yellowish green, small, thin ribbon like thalli were produced in about 70 days (Fig. 3.2, d). In half strength Knop's macronutrients + Nitsch trace elements with 10 ppm ferric citrate, rapid growth and proper differentiation of sporelings into normal thalli occurred (Fig. 3.2, e). In half strength Knop's macronutrients + Nitsch trace elements with 10 ppm ferric citrate supplemented with 1% sucrose, abundant growth of rhizoids and vigorous growth of thalli occurred along with the development of many new innovations from the basal part of thalli (Fig. 3.2, g). When an intact thallus is transferred to half Knop's macronutrients + Nitsch trace elements + 10 ppm ferric citrate + 1% sucrose, many adventitious shoots were emerged out from entire surface of the thallus in 15-20 days (Fig. 3.2, f) that can be used as secondry inocula for rapid micropropagation. Sub culturing of apical part or intact thallus into same medium supplemented with 2% sucrose resulted into formation of a large number innovations over a hardy callus like tissue, from which only 1-3 thalli could be developed (Fig. 3.2, h).

When an intact thallus was transferred from half strength Knop's macronutrients to MS medium, decolourization occurred in entire thallus, while in some thalli the apical part remained green (Fig. 3.2, a), from which dark green coloured callus like undifferentiated mass of tissue developed in about 1 month. Subsequently, some short, twisted and deformed thalli regenerated from callus like mass but they remained as such even after 50 days (Fig. 3.2, b) and did not differentiated into mature, well developed thalli. Similar response was observed in the cultures in Gamborg B-5 medium (Fig. 3.2, c). In Hoagland medium, sub culturing of intact thalli or apical part of thalli resulted in formation of highly curled or twisted darker green thalli in compact cluster. Margins of these thalli were highly fringed and purplish red in colour (Fig. 3.2, d). In Hoagland medium supplemented with 1% sucrose, the average length and width of thalli were 22.16 ± 0.49 mm and 5.67 ± 0.09 mm which were almost similar to those of thalli grown in wild.

Fig. 3.2: ***In vitro* growth of *Reboulia hemispherica*.**

2 (a-b) Thalli, after sub culturing in MS medium (10 and 50 days respectively); (c) thallus, sub cultured in Gamborg B-5 medium; (d) Curled thalli with purplish fringed margin, sub cultured in Hoagland medium; (e) *In vitro* raised thalli, transferred on pots containing sterilized soil; (f) Regeneration of new thalli from apical part of thalli; (g, h) Population of well developed thalli, on soil (60 days old and 120 days respectively); (i) Formation of male receptacle under short day and low temperature condition

Exogenous supplied auxin (IAA) and cytokinin (Kinetin) retarded the growth of ex plant (apical part of cultured thalli) at all the concentrations tried (10^{-4}, 10^{-5} and 10^{-6} mole/litre). Addition of IAA or kinetin or both to the basal medium at 10^{-4} mole/litre proved lethal while the same at 10^{-5} mole/litre concentration arrested the growth and induced marked twisting of thalli. At 10^{-6} mole/litre concentration, the growth was much slow and thalli exhibited marked curling and stunted growth even after 90 days of inoculation. Thus, implementation of growth hormones could not be promising for *in vitro* propagation point of view.

Continuous light of 4000-5500 lux enhanced the growth of culture in rapidness as well as vigorousness. Best vegetative growth of the plant was observed at temperature 21ºC in both cases i.e., on agar medium and on soil saturated with minerals. The plants could sustain the rise in temperature up to 27ºC and survive even for 2 days. However, such condition induced the shortening of thalli and senescence. Similarly, onset of dry condition and upon exhaustion of nutrients resulted in shortening of thalli and emergence of innovations. When the 120 days old *in vitro* raised thalli were transferred on pots containing sterilized soil (Fig. 3.2, e) saturated with half strength Knop's macronutrients + Nitsch trace elements with 10 ppm ferric citrate, new thalli were regenerated from the apical green part while older basal region decayed (Fig. 3.2, f). In constant low temperature of 21ºC and high humidity (90-100%) and alternate light of 4000-5500 lux for 12 hours and dark period of the same duration, uniform population of well differentiated, broader and vigorously growing thalli were developed in pot within 65-70 days of transferring (Fig. 3.2, g). In these conditions thalli remained vigorous and broader, while at moderate temperature 25ºC and low humidity (40-60%) the thalli started to decay at the basal part and tended to be shorter and swollened (Table – 5). Onset of low temperature (18-21ºC) and higher humidity (90-100%) resulted in emergence of several innovations from the decaying basal region, the apical and from the mid-ventral side that in turn yielded another population over the first one. Such fluctuation in temperature and humidity gave rise the population in bulk (Fig. 3.2, h). When the thalli acclimatized on soil in pots were kept under low temperature (15-18ºC) and short day conditions (receiving 1600-2000 lux for 10 hours and dark period of 14 hours), disciform, cushion like male receptacles were differentiated at the apical part of thalli (Fig. 3.2, i).

4. Discussion

The spore germination pattern *viz.*, rupturing of spore coat at proximal side (at the tri-radiate mark), germ tube formation and cutting off of rhizoidal cell was found similar as in previous sporeling studies carried out by Mehra and Kachroo (1951) in the family Aytoniaceae. Lacking the waxy cuticle, bryophytes are often susceptible to osmotic shock. Voth (1943) used *M. polymorpha* to show that a concentrated nutrient solution could kill the tip and wing of thalli while reducing dry biomass and production of gemma cups. Preference of dilute culture medium has been reported in other liverworts also (Awasthi *et al.* 2013).

The poor growth response of cultures in MS and Gamborg B-5 media may be due to higher level of nitrogen in both ammonium ($NH4^+$) and nitrate (NO_3^-) form and comparatively less amount of calcium (Ca^{++}) in these media in comparison to half strength Knop's macronutrients. As a result, cell wall formation was not in congruence with rapid cell division and metabolic activity because of high nitrogen level. Kumra (1984) and Awasthi *et al.* (2013) also observed callus formation in *Asterella* and *Cryptomitrium* grown in MS medium. Calcium deficiency also manifested in production of blackish green callus like pulpy mass of twisted and deformed thalli in Gamborg B-5 medium . Healthy growth and normal differentiation of thalli in half strength Knop's macronutrient indicate the requirement of optimum low level of nitrogen and enough amount of calcium for the growth of this species. It also correlates the preference of the calcarious rocks as suitable habitat among majority of the Marchantiales including this species. Similarly relative optimum dose of nitrogen (in nitrate as well as in ammonium form) and calcium was also available in Hoagland medium thus more or less equally well differentiated thalli were produced in Hoagland medium, although thalli were darker green in colour and in compact cluster due to slightly higher concentration of nutrients and additional nitrogen in form of ammonia in comparison to half strength Knop's macronutrients. Addition of 2% sucrose into optimal basal medium induced callus formation as has been noted previously in other liverworts (Kaul *et al.*, 1962; Mehra and Pental, 1976). However, a large number of innovations were produced from such calli that can be used as secondary inocula for micropropagatin though sub culturing.

With reference to photoperiodic reaction, *R. hemisphaerica* may be categorized under short day plants. Temperature also induces a variety of response in bryophytes. Some bryophytes induce gametangial development in a low temperature like vernalization. Some taxa viz., *Lunularia cruciata* (L.) Dumort., *Aneura mirabilis* (Malmb.) Wickett & Goffinet, *Riccia crystallina* L. and *Leptobryum pyriforme* (Hedw.) Wils. required either low temperature alone or first low then higher temperature (Bopp, 1983). In the present experiment, alone low temperature in continuous light did not induce gametangial formation. The optimal low temperature (15-18ºC at day and 10-15ºC at night) for gametangial induction depended simultaneously with alternate light and dark period of 10 hours and 14 hour respectively. The consequence is that the species produces its gametangia during a short period of year (winter) in nature providing such conditions. The requirement of dark period for a certain period and low temperature below 20ºC for onset of sexual face have been reported in many other bryophytes like *Targionia hypophylla* and many liverworts (Chopra and Bhatla, 1983; Bapna *et al.*, 1984).

In *R. hemisphaerica,* shortening of thalli in elevated temperature up to 27ºC or in drier culture condition due to prolonged culture and development of innovations and subsequently new thalli occurred during onset of high humidity and low temperature. Such reaction was helpful in allowing the taxon to survive and withstand the dry conditions by reducing the surface area of thalli to limit the water loss from pores. Similarly, in culture of another thalloid liverwort *Cryptomitrium himalayense* when the dryness appeared and nutrients were exhausted, the thalli tended to be smaller in size and prominent tuber like swelling appeared at the thallus apices while remaining part of thallus started to decay due to senescence. Such a strategy to develop tuber like bodies also correlate the ability of this taxon to perenate and multiply themselves upon failure of sexual stage in adverse conditions.

Acknowledgements

One of the authors (VA) is grateful to the Director, National Botanical Research Institute (C.S.I.R., New Delhi), Lucknow for encouragement and providing facilities. Thanks are due to the Ministry of Environment and Forests, Govt. of India, New Delhi and University Grant Commission, New Delhi for providing financial assistance.

References

Asakawa, Y. 2007. Biologically active compounds from bryophytes. *Pure Appl. Chem.* 79(4): 557-580.

Asakawa, Y. and Matsuda, R. 1982. Riccardin C, a novel cyclic bibenzyl derivative from *Reboulia hemispherica*. *Phytochemistry* 21(8): 2143-2144.

Awasthi, V. Asthana, A. K. and Nath, V. 2013. *In vitro* study on the reproductive behavior of the endemic and threatened Indian liverwort: *Cryptomitrium himalayense* Kash. *Cryptogamie Bryologie* 34(3): 313-323.

Bapna, K. R., Singh, R. P. and Chaudhary, B. L. 1984. Induction of sex organs in *Targionia hypophylla* L. *Bryologist* 87(4): 340-342.

Bopp, M. 1983. Developmental physiology of Bryophytes. In: Schuster, R. M. (ed.) *New Manual of Bryology. Vol. 1.* The Hattori Botanical Laboratory, Nichinan, Miyazaki, Japan. pp. 276-324.

Chopra, R. N. and Bhatla, S. C. 1983. Regulation of gametangial formation in bryophytes. *Botanical Review* 49 (1): 29-63.

Dixit, B. S. and Banerji, R. 2007. Bryophytes- A boon for herbal medicine. In: Nath, V. and Asthana, A.K. (eds.) *Current Trends in Bryology*.. Bishen Singh Mahendra Pal Singh, Dehradun, India. pp. 347-351.

Gamborg, O. L., Miller, R. A. and Ojima, K. 1968. Nutrient requirements of suspension cultures of Soybean root cells. *Exp. Cell. Res.*, 50: 151-158.

Hoagland, D. R. and Arnon, D. I. 1950. The water culture method for growing plants without soil. *California Agr. Exp. Stat Circular*, 347: 1-32.

Kaul, K. N., Mitra, G. C. and Tripathi, B. K. 1962. Responses of *Marchantia* in aseptic culture to well known auxins and antiauxins. *Ann. Bot.*, 26: 447-466.

Knop, W. 1865. Quantitative Untersuchungen uber die Ernahrungsprozesse der Pflanzen. *Landwirtseh. Vers. Stn.*, 7 : 93-107.

Kumra, S. 1984. Callus induction in the liverwort *Asterella wallichiana* (L. *et* L.) Grolle. *J. Plant Physiol.*, 115: 263-269.

Mehra, P. N. and Kachroo, P. 1951. Sporeling germination studies in Marchantiales. I. Rebouliaceae. *Bryologist* 54: 1-16.

Mehra, P. N. and Pental, D. 1976. Induction of apospory callus and correlated morphogenetic studies in *Athalmia pusilla* Kash. *J. Hattori Bot. Lab.*, 40: 151-183.

Murashige, T. and Skoog, F. 1962. A revised medium for rapid growth and bioassay with tobacco tissue culture. *Physiologia Plantarum,* 15: 466-473.

Sabovljevic, A., Sabovljevic, M. and Jockovic, N. 2009. *In vitro* culture and secondary metabolite isolation in bryophytes. In: *Protocols for in vitro cultures and secondary metabolite analysis of aromatic and medicinal plants.* (ed. S. K. Jain SK and P. K. Saxena) Humana Press, Springer science. pp. 117-128.

Sauerwein, M. and Becker, H. 1990. Growth terpenoid production and antibacterial activity of an *in vitro* culture of a liverwort *Fossombronia pusilla. Planta Med.,* 56: 364-367.

Voth, P. D. 1943. Effect of nutrient solution concentration on the growth of *Marchantia polymorpha. Bot Gaz.,* 104: 591-601.

Chapter 4

Effect of Reduced Tillage on Crop Yield and Nitrogen Use Efficiency in Rice-Wheat Systems in North-western India

Pushpa Devi*, S. R. Gupta and Ashok Aggarwal

Department of Botany, Kurukshetra University, Kurukshetra-136119, India

**Corresponding Author E-mail: pushpa.devi21@gmail.com*

Abstract

The objective of the study was to evaluate the impact of conservation tillage viz. reduced tillage (RT) and conventional tillage (CT) on yield and yield components and nitrogen uptake in rice and wheat crop rotation (located at Bhanokheri (30º 17′ 00.2″N, 76º 45′18.7″ E, 244 m) in Ambala district on northern Haryana. The wheat crop was grown under conventional tillage (CT) and reduced-tillage (RT) systems. The rice crop was grown under conventional tillage and reduced-tillage wheat plot. The reduced-tillage in wheat had a favorable effect on crop biomass production, carbon accumulation and nitrogen use efficiency. As influenced by tillage practice, the accumulation of carbon in grain, straw and roots of wheat varied from 4691.08 to 3818.05 kg C ha^{-1}. The carbon accumulation in rice ranged from 4484.31 to 5237.10 kg C ha^{-1}. Total nitrogen uptake in wheat under different tillage systems was (kg N ha^{-1}): 66.95 CT; 71.03 RT). The efficiency of nitrogen use in rice crop was 44.52 to 53.98% and for the wheat crop it varied from 52.49 to 74.5%. The soil N pool was greatest in 0-30cm soil depth, the values being 1269.33 kg N ha^{-1}(RT) and 1205.99 kg N ha^{-1}(CT).

Keywords: *Conventional Tillage (CT), Reduced-tillage (RT), Crop Yield, Nitrogen Uptake, Soil Nitrogen Pool, nitrogen use efficiency.*

1. Introduction

In Asia, the rice-wheat system has been practiced for over 1000 years. The rice-wheat cropping system is the backbone of India's food security. During the past 30 years, agricultural production has been able to keep pace with population demand for food. This came about through significant area and yield growth. The rice–wheat cropping system being the oldest and most prevalent agricultural practices in India, is also practiced in many other regions of the world and wetland culture is the predominant soil management system adopted. Rice occupies 153 m ha land throughout the world. In India, out of the 43 m ha area under rice cultivation, puddled rice culture occupies 24 m ha, about 56% of the area (Anonymous, 2005). This involves ploughing the soil when wet, puddling it and keeping the area flooded for the duration of the rice crop. Wetland rice culture thus destroys soil structure and creates a poor physical condition for the following wheat crop. The efficiency of nitrogen is often low due to loss of nitrogen from flooded soil or leaching of nitrogen into ground water (De Datta and Buresh, 1989). This soil condition can reduce wheat yield presumably by limiting root growth and distribution (Oussible *et al.*, 1992). For regeneration and maintenance of soil structure in rice-wheat cropping system, the retention of plant residue is very important (Verma and Bhagat, 1992). Rice grown with conservation tillage can produce yields similar to that under conventional puddling with minimized expenses on field preparations (Sharma and De Datta, 1986).

Besides declining soil fertility, low wheat yields in rice-wheat cropping system are also obtained due to a short turnover period between rice harvest and delayed wheat sowing due to a number of factors, including delayed rice transplanting resulting in delayed rice harvest, high soil moisture content after the rice harvest, delay in removal of rice straw (a large part of it is being burned *in situ*, which besides the loss of precious organic C creates environmental and health problems), etc. Plant biomass is a source of C and N, which can replenish SOC and TN. Changes in soil conditions (e.g., temperature, moisture, O_2, pH, and nutrient availability), can alter the decomposition rate of plant biomass and the mineralization rate of soil organic matter.

Kumar and Goh (2000) examined that crop residues have positive effects on soil quality, soil nitrogen dynamics and recovery and crop yield. They concluded that crop residue is a significant factor for crop production through their effects on soil physical, chemical and biological properties as well as water and soil quality. The previous crop residues when retained/incorporated rather than burnt have been found to exert positive effects regarding yield, enhanced soil organic matter and total soil nitrogen. Mechanical tillage disturbs this process. Therefore, ZT is the most important tillage method to conserve resources and enhance wheat yield (Usman *et al.* 2013).

The objective of the study in this chapter was to evaluate the impact of two tillage systems viz. reduced tillage (RT) and conventional tillage (CT) on yield and yield components and nitrogen uptake in rice and wheat crop rotation.

2. Materials and Methods

2.1 Study Site

The study was carried out in a rice-wheat agricultural system on the farmer's field located at Bhanokheri (30º 17′ 00.2″N, 76º 45′18.7″ E, 244 m in Ambala district Haryana (Fig. 4.1). The agriculture field selected for the study has been under conventional tillage (CT) and reduced tillage (RT) practices for the last ten years.

The study was carried out during April 2011 to October, 2011 (I-yr) &April 2012 to October, 2012 (II-yr). For this study, the rice–wheat crop rotation was selected on the farmer's field.

Fig. 4.1: Google imagery of site of Bhanokheri, Ambala.

2.2 Climate and Soil

Climate of the study area is tropical monsoonal and semiarid. The year is divisible into a warm wet period (June to September; rainy season), a cool dry period (October to February; winter season), and a hot dry period (March to May; summer season). During the study period from January 2011 to December 2012, the average monthly rainfall varied from 0.2 mm to 274.2mm during different months. About 87.64 to 88.29% of the total annual rainfall (932mm) was received in the rainy season during January 2011 to December 2012 (Fig. 4.2).

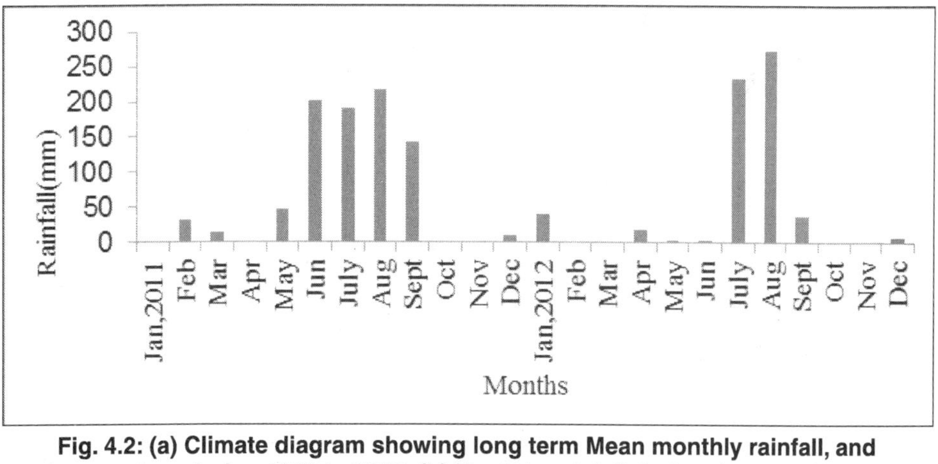

Fig. 4.2: (a) Climate diagram showing long term Mean monthly rainfall, and temperature during 1990 to 2000; (b) Monthly rainfall during January 2011 to December 2012 at Ambala, northern Haryana.

Soil of the study area is old alluvium which is sandy- loam in texture. In reduced tillage, the organic carbon content of the soil varied from 0.43 to 0.24% (0 to 30cm soil depth) and in convention tillage, the organic carbon content of the soil varied from 0.35 to 0.18% (0 to 30cm soil depth) (Table. 1). Soil organic carbon concentrations showed significant decrease with increase in soil depth in both tillage systems. The bulk density varied from 1.50 to 1.61 and soil pH was 8.10 to 8.34 (Table. 4.1).

Table 4.1: Some soil characteristics of the rice-wheat cropping system under Reduced-tillage (RT) and Conventional tillage (CT) upto 0-30cm soil depth at Bhanokheri, Ambala.

Tillage system/ Soil depth (cm)	pH (1:2)	Bulk density (gm-3)	Organic carbon (%)	Total N (%)
RT				
0-7.5	8.13±0.09a	1.51±0.01	0.43±0.01a	0.04±0.001a
7.5-15	8.25±0.09ab	1.52±0.01	0.31±0.02b	0.029±10.97a
15-30	8.34±0.08ab	1.56±0.006	0.24±0.01c	0.021±0.001b
CT				
0-7.5	8.10±0.11a	1.50±0.01	0.35±0.01a	0.04±0.0006a
7.5-15	8.18±0.12a	1.58±0.01	0.26±0.02b	0.03±0.22a
15-30	8.23±0.12a	1.61±0.02	0.18±0.03c	0.02±0.002ab

2.3 Analysis of Crop Biomass and Nitrogen Uptake

The rice crop was grown as per conventional practices in both zero-tillage and conventional tillage plots. The rate of nitrogen fertilization was 150 kg N ha^{-1}for both wheat and rice crop. The basal dose of DAP (at the rate of 50 kg N ha$^{-1)}$ was given at the time of sowing of rice and wheat crop. Afterwards, nitrogen fertilization (at the rate of 150 kg N ha^{-1}) was in the three equal split doses at the time of tillering, flowering and seeding of wheat crop. In the field, five replicated plots of 10 x 20m were selected for analyzing crop yield, nitrogen uptake, and soil aggregate analysis. Crop biomass and grain yield were recorded from agricultural field site-I under rice- wheat crop rotation during April 2011 to October 2012. The root biomass upto 20cm depth was estimated using soil core method.

Nitrogen uptake by rice and wheat crop was calculated by multiplying the crop dry matter weight with average nitrogen concentration in respective plant components. At crop harvest, the weight of straw, roots and grains was multiplied by their nutrient concentration to obtain values of total N uptake.

2.4 Carbon and Nitrogen Analysis of Plant Samples

At crop maturity, the above ground parts of harvested samples were separated into grains and straw and dried. The plant samples of rice and wheat were oven dried at 40°C and powered in a Wiley mill equipped with 2mm sieve. Organic carbon in plant samples was analyzed following the method of Kalembasa and Jenkinson (1973). Total nitrogen concentration in roots, straw and grain were estimated after

Kjeldahl digestion and titration using Kelplus Nitrogen Estimation System based on micro-Kjeldahl method (Bremner, 1965).

2.5 Soil Nitrogen Pool

The amount of nitrogen in soil was estimated from bulk density, soil depth and nitrogen concentrations using the expression; Soil N (ha^{-1}) = ppm x bulk density x soil depth/10; where ppm= % N concentration; bulk density = 1.20 to 1.39; soil depth =0 to 100cm.

2.6 Statistical Analysis

One-way ANOVA was used to evaluate the effect of tillage practices on soil carbon, soil nitrogen, crop biomass, crop accumulation and nitrogen uptake in the rice-wheat systems. A significance level of $P<0.05$ was used for all tests. All analyses were conducted following SPSS, ver. 16.0.

3. Result and Discussion

3.1 Plant Biomass and Crop Yield of Rice-Wheat System

The shoot and root biomass of rice during the two years of study are given in (Table 4.2). The plant biomass in rice crop increased from transplantation to heading stage. Grain yield, straw production and root biomass were not significantly different between conventional and reduced tillage. At the time of crop harvest of rice, the root biomass varied from 951.61 to 959.63 kg ha^{-1}. The grain yield of rice ranged from 4793.45 to 5104.30 kg ha^{-1}. In the two tillage systems, the straw production was 4622 to 5568.60 kg ha^{-1}. Crop Yield in wheat system was found to be influenced by reduced tillage (Table. 2). At the time of crop harvest, the root biomass of the wheat crop in the two-tillage system was 862.11 to 867.83 kg ha^{-1} (CT) and 918.80 to 924.00 kg ha^{-1}. (RT). Yield of wheat and rice was found to be higher in reduced-tillage as compared to the conventional system. It has been reported that surface placement of residue in the no-till system increased spring wheat yield compared to residue incorporation in the conventional tillage system (Halverson *et al.*, 2002, Sainju *et al.*, 2009).

The root biomass of wheat in soils in surface 0.30 m depth was greater under reduced tillage as compared to conventional tillage at all the sites. Most of the roots biomass was concentrated in 0.15 m soil depth under conventional tillage, whereas same was observed to a depth of 0.30 m in under zero tillage. This is conformity to the reports by that root biomass density of wheat was significantly higher under RT as compared to CT at various experimental sites in north-western India (Singh *et al.*, 2014).

There are many factors influencing the results of direct sowing, including soil, climate and the duration of RT implementation (Martinez *et al.*, 2008). Higher plant yields from RT than from CT fields have been reported from dry climate or in years with less rainfall (Arshad *et al.*, 1999; Bonfil *et al.*, 1999; De Vita *et al.*, 2007; Wang *et al.*, 2007) because in a dry year, plants are less vulnerable to yield loss under RT rather than under CT method. In the moderate climate of West Poland, with mean annual precipitation of 550 mm and mean annual temperature of 8.50 C, zero tillage has not been found to generate higher yields than conventional soil cultivation methods.

The recent comprehensive meta-analysis of 5463 paired yield observations from 610 studies suggests that no-till could result in a yield reduction of around 10% overall (Pittelkow *et al.*, 2015a). The reduction in yield was greatest for cereal crops; however, oilseeds, cotton, and legumes gave comparable yields under no-till and conventional tillage (Pittelkow *et al.*, 2015b). The negative effects of no-till could be minimized when combined with mulching and crop rotation, characterizing the systems to be truly CA (Pittelkow *et al.*, 2015a). Yield benefits of CA under such conditions are mostly due to timelier sowing, early crop establishment, and summer fallow weed control.

Table 4.2: Biomass production of rice-wheat system in conventional tillage (CT) and zero-tillage (ZT) practices during April 2011 to October, 2011 (I-yr) and April 2012 to October, 2012(II-yr) at Bhanokheri, Ambala.

Tillage systems	Year/crop	Biomass (kg ha-1)		
		Grain	Straw	Roots
CT	I-yr Rice	4793.45±98.25a	4622.05±112.56a	951.61±10.82a
CT	II-yr Rice	5104.30±75.23a	5099.27±85.85b	959.63±22.02a
RT	I-yr Rice	4855.20±116.30b	5194.40±76.90b	1025.40±11.03b
RT	II-yr Rice	5436.00±79.85.07b	5568.60±109.96b	1029.80±24.70b
LSD(p<0.05)		9.681	15.953	5.221
F df -3,16		281.292	292.368	54.744
CT	I-yr Wheat	3857±132.42a	3671.84±106.5 a	867.83±17.01a
CT	II-yr Wheat	4574.29±112.70a	4204.46±145.5b	862.11±16.03a
RT	I-yr Wheat	3891.60 ±125.68b	4138.8±136.83b	924.00±10.26b
RT	II-yr Wheat	4876.40±112.67b	4496.0±176.71b	918.80±10.05b
LSD(p<0.05)		17.478	5.652	5.693
F		363.250	430.388	41.126
df -3,16				

3.2 Carbon and Nitrogen Concentration in Wheat and Rice

The concentration of nitrogen in straw, grain and roots was higher under zero-tillage (0.88 to 2.06 %) as compared to that of the conventional tillage (0.69 to 1.88 %) wheat system (Table. 4.3). The concentration of nitrogen in straw and roots of wheat were found to be higher by (9 to 30 %) in zero-tillage as compared to that of the conventional tillage wheat system. At the time of crop maturity, nitrogen concentration was more in grains (1.88 to 2.06 %) as compared to that of straw (0.69 to 0.90 %) in all the two tillage systems (Table. 3). Nitrogen concentration in wheat straw varied from 0.69 to 0.88 %. In roots, nitrogen concentration was 0.82 to 0.88 %. Organic carbon concentration in wheat straw and roots varied from 39.40 to 41.78 % (straw) and in roots and straw of rice, the carbon concentration varied from 37.28 to 40.75 % (root) (Table. 3). Rice grain showed slightly higher N concentration in zero-tillage treatment as compared to conventional tillage. In the case of rice, nitrogen concentration was: 1.19 to 1.25 % grains; 0.55 to 0.60 % straw; 0.64 to 0.69 % roots (Table. 4.3).

Table 4.3: Nitrogen concentration in wheat grain, straw and roots under conventional tillage (CT) and reduced-tillage (RT) practices.

Crop	Tillage systems		Nitrogen (%)
Wheat	conventional tillage CT	Grain	1.88 ± 0.032
		Straw	0.69 ± 0.014
		Roots	0.82 ± 0.018
	reduced-tillage RT	Grain	2.06 ± 0.024
		Straw	0.90 ± 0.018
		Roots	0.88 ± 0.016
Rice	conventional tillage CT	Grain	1.19 ± 0.02
		Straw	0.55 ± 0.023
		Roots	0.69 ± 0.015
	reduced-tillage RT	Grain	1.25 ± 0.04
		Straw	0.60 ± 0.024
		Roots	0.64 ± 0.014

3.3 Nitrogen Uptake and Nitrogen use Efficiency

In the monitored period, the uptake of nitrogen was significantly higher at the maturity of plant growth. For two years of study, the average nitrogen uptake by wheat was (kg N ha^{-1}):124.53 - 149(RT); 104 - 122 (CT). In the case of rice, the values of nitrogen uptake varied from 89.03 kg N ha^{-1} (CT) to 107.95 kg N ha^{-1} (RT). Thus, nitrogen uptake was enhanced in the case of reduced-tillage as compared to that of conventional tillage. The average nitrogen uptake by wheat and rice crops is given in (Table 4.4). From the results, it is evident that N uptake was higher in grain as compared to straw and roots in case of both CT and RT. Uptake of nitrogen in wheat increased in reduced-tillage systems as compared to conventional tillage. From the results, it is evident that N uptake was higher in grain as compared to straw and roots in case of both CT and RT. Uptake of nitrogen in wheat increased by about 24% in reduced-tillage systems as compared to conventional tillage. In reduced-tillage rice, the increase in nitrogen uptake was 10 to 12%. A number of studies have shown that tillage and nitrogen fertilization can significantly affect carbon and nitrogen accumulation in different cropping systems (Crozier and Naderman 1999; Qin and Stamp, 2004; Merrilland Black *et al.*, 1996).

The efficiency of nitrogen use in rice crop was 44.52 to 53.98 % and that in wheat crop it varied from 52.49 to 74.5 % (Table. 4). In the rice system, the efficiency of nitrogen is often low due to loss of nitrogen from flooded soil or leaching of nitrogen into ground water (De Datta and Buresh, 1989). The differences in the biomass yields between CT and RT could have been caused by the differences in the nourishment of wheat plants during the growing season. An improved crop residue management, combined with application of fertilizer N or incorporation of legumes greatly improves the N economy of cereal cropping systems and enhances crop productivity in soils with a low N content on the short term (Pandiaraj *et al.*, 2015).

Table 4.4: Nitrogen uptake by rice-wheat system under conventional tillage (CT) and reduced-tillage (RT) treatments during April 2011 to October, 2011 (I-yr) and April 2012 to October, 2012 (II-yr) at Bhanokheri.

Tillage systems	Year/crop	Total nitrogen uptake	Nitrogen use efficiency(%)
CT	I yr- Rice	89.03±1.23a	44.52
	II yr- Rice	95.41±1.04b	47.71
RT	I yr- Rice	98.42±1.61b	49.21
	II yr- Rice	107.95±1.63c	53.98
LSD(p<0.05)		20.31	
F Df- 3,16		0.345	
CT	I yr- Wheat	104.98±3.22a	52.49
	II yr- Wheat	122.08±2.81b	61.04
RT	I yr- Wheat	124.53±4.18b	62.27
	II yr- Wheat	149.00±3.20c	74.5
LSD(p<0.05)		25.23	
F Df- 3,16		0.652	

3.4 Soil Nitrogen Pool

There was decrease in soil nitrogen pool with soil depth in both reduced-tillage and conventional tillage systems. The soil N pool was greatest in 0-30cm soil depth, the values being 1269.33 kg N ha^{-1}(RT) and 1205.99 kg N ha^{-1}(CT). In the rice growing area of Asia, the total soil nitrogen pool amounts to 2000 to 3000 kg N ha^{-1} in the plough layer (Kundu and Ladha, 1995) and mineralization and immobilization of plant residues N are the two important biological processes affecting plant nitrogen uptake (Jarvis *et al.*, 1996). Thus, nitrogen pool in plough layer of the studied systems was lower as compared to that of other rice growing areas (Kundu and Ladha, 1995). The total nitrogen pool up to 100cm soil depth was 2686.35 to 2735.6 kg N ha^{-1} in two tillage systems.

4. Conclusions

Carbon accumulation and nitrogen uptake increased under zero tillage wheat. The efficiency of nitrogen use was greater in wheat under zero tillage practice. Reduced tillage could lead increased yield of crop by increasing soil moisture and improving nitrogen availability. The root biomass of wheat in soils in surface 0.3 m depth was greater under reduced tillage as compared to conventional tillage at all the sites. Most of the roots biomass was concentrated in 0.15 m soil depth under conventional tillage, whereas same was observed to a depth of 0.305 m in under reduced tillage. According to Singh *et al.* (2014), the practice of RT was found to increase the grain yield of wheat significantly over CT practice in clay loam soil because of increase in the test weight of grains whereas in loam no such increase in wheat yield was found. Uptake of nitrogen in wheat increased by about 24% in reduced-tillage systems as compared to conventional tillage. In reduced-tillage rice, the increase in nitrogen uptake was 10 to 12%. A number of studies have shown that tillage and nitrogen fertilization can significantly affect carbon and nitrogen accumulation in different cropping systems. Carbon sequestration potential of conservation tillage systems are derived by combined information of above ground carbon storage, time averaged carbon stock in plant biomass and soil carbon accumulation. The potential of carbon sequestration in agricultural systems will be useful to develop suitable management practices for sustainable agriculture. Implementing conservation agriculture practices helps to maintain levels of soil organic matter and soil fertility and provide an opportunity for soil carbon sequestration.

Acknowledgements

The financial assistance to Pushpa from Rajiv Gandhi National Fellowship from UGC, New Delhi is gratefully acknowledged.

References

Anonymous. 2005. Atlas of Rice & World. Rice Statistics. http://www.irri.org/science/ricestat/index.asp.

Arshad, M.A., Franzluebbers, A.J. and Azooz, R.H. 1999. Components of surface soil structure under conventional and no-tillage in northwestern Canada. *Soil Tillage Res.* 53: 41-47.

Bonfil, D.J., Mufradi, I., Klitman, S. and Asido, S. 1999. Wheat grain yield and soil profile water distribution in a no-till arid environment. *Agronomy J.* 91: 368-373.

Bremner, J.M. 1965. Inorganic forms of nitrogen. In: Black C.A., Evans D.D., White J. L., Esminger L. L. and Clark F. E. (Editors) Methods of Soil Analysis. *American Soc. Agronomy, Madison,* USA. 2: 1179-1237.

Crozier C.R., Naderman, G.C., Tucker, M.R. and Sugg, R.E. 1999. Nutrient and pH stratification with conventional and no-till management. *J. Soil Science Plant Analysis,* 30: 65-74.

De Datta, S.K. and Buresh, R.J. 1989. Integrated nitrogen management in irrigated rice. *J. Advances Soil Science,* 10: 143-169.

De Vita P., Di Paolo E., Fecondo G., Di Fonzo N. and Pisante M. 2007. No-tillage and conventional tillage effects on durum wheat yield, grain quality and soil moisture content in southern Italy. *Soil Tillage Res.* 92: 69–78.

Halvorson, A.D., Weinhold, B.J. and Black, A.L. 2002. Tillage, nitrogen, and cropping system effects on soil carbon sequestration. *Journal of Soil Science Society of America,* 66: 906–912.

Kalembasa, S.J. and Jenkinson, D.S. 1973. A comparative study of titrimetric and gravimetric methods for determination of organic carbon in soil *J. Science Food Agric.*, 24: 1085-1090.

Kumar, K., and Goh, K.M. 2000. Crop Residue management: Effects on soil quality, soil nitrogen dynamics, crop yield and nitrogen recovery. *Advanced Agronomy*. 68: 197-319.

Kundu, D.K. and Ladha, J.K. 1995. Enhancing soil nitrogen use and biology nitrogen fixation in wetland rice. *Exp. Agric.*, 31: 261-277.

Martinez, E., Fuentes, J., Silva, P., Valle, S. and Acevedo, E. 2008. Soil physical properties and wheat root growth as affected by no-tillage and conventional tillage systems in a Mediterranean environment of Chile. *Soil Tillage Res.*, 99: 232–244.

Merrill, S.D., Black, A.L. and Bauer, A. 1996. Conservation tillage affects root growth of dryland spring wheat under drought. *J. Soil Sci. Soc. America*, 60: 575-583.

Oussible, M., Crookstan, R.P. and Larson, W.E. 1992. Subsurface compaction reduces the root and shoot growth and grain yield of wheat. *Agronomy J.* 84: 34–38.

Pandiaraj, T., Selvaraj, S. and Ramu, N. 2015. Effects of Crop Residue Management and Nitrogen Fertilizer on Soil Nitrogen and Carbon Content and Productivity of Wheat (*Triticum aestivum* L.) in Two cropping Systems. Article 22, Volume 17, Issue 1, Pp. 249-260.

Pittelkow, C.M. Liang, X., Linquist, B.A., Van Groenigen, K.V., Lee, J., Lundy, M.E., Van Gestel, N., Six, J., Venterea, R.T. and Van Kessel, C. 2015a. Productivity limits and Potentials of the principles of conservation Agriculture. *Nature*, 517: 365-368

Pittelkow, C.M., Linquist, B.A., Liang, X, Van Groenigen, K.J., Lee, J., Lundy, M.E., Van Gestel, N., Six, J., Venterea, R.T. and Van Kessel, C. 2015b. when does no yield more? A global metaanalysis. *Field Crops Res.*, 183: 156-168.

Qin, R., Stamp, P. and Richner, W. 2004. Impact of tillage on root system of winter wheat. *J. Soil Sci. Soc. America*, 96: 1523-1530.

Sainju, U.M., Lenssen, A.W., Caesar-TonThat, T. and Evans R.G. 2009. Dryland crop yields and soil organic matter as influenced by long term tillage and cropping sequence. *Agron. J.* 101: 243-251.

Sharma, P.K. and De Datta, S.K. 1986. Physical properties and processes of puddled rice soils. *Adv. Soil Sci.* 5: 139–178.

Singh, A., Phogat, V.K., Dahiya, R. and Batra. S.D. 2014. Impact of long-term zero till wheat on soil physical properties and wheat productivity under rice–wheat cropping system. *Soil Tillage Res.* 140: 98–105.

Tisdall, J.M. and Oades, J.M. 1982. Organic matter and water-stable aggregates in soils. *Soil Sci*, 33: 141-163.

Usman, K., Khan, E.A., Khan, M.A., Ghulam, S., Khan, S and Baloch, J. 2013. Effects of tillage and nitrogen on wheat production, economics and soil fertility in rice-wheat cropping system. *American J. Plant Sci.*, 4: 17-25.

Verma, T.S. and Bhagat, R.M. 1992. Impact of rice straw management practices on yield, nitrogen uptake and soil properties in a rice-wheat rotation in northern India. *Fertility Research.* 33: 97–106.

Wang, X.B., Cai, D.X., Hoogmoed, W.B., Oenema, O. and Perdok, U.D. 2007. Developments in conservation tillage in rainfed regions of North China. *Soil Tillage Research.* 93: 239-250.

Chapter 5

Plant Growth Promoting Potential of Bacterial Endophytes From Medicinal Plants

Ram Kumar Pundir1*, Sandeep Bhat[1], Navpreet Kaur1, Pranay Jain[2], Satish Kumar[1], Viraj K. Mishra[1] and Rajesh Kumar[3]

[1]*Biotechnology Engineering Department, Ambala College of Engineering and Applied Research, Ambala, Haryana, 133101, India*

[2]*Department of Biotechnology, University Institute of Engineering & Technology, Kurukshetra University, Kurukshetra, Haryana, India*

[3]*Department of Agriculture Botany, Gochar Mahavidhyalya, Rampur Maniharan, Saharanpur, Uttar Pradesh, India*

**Corresponding Author E-mail: drramkpundir@gmail.com*

Abstract

Endophytes are those microbes that have ability to promote plant growth through various mechanisms such as production of plant growth hormones and anti-microbial substances, as well as to provide soils with nutrients like inorganic phosphate and nitrogen and many more. The present study aimed to evaluate the potential of endophytic bacteria isolated from leaves and stems of jamun (Syzygium sp.*) and tulsi (Ocimum* sp.*) plants to produce substances involved in the promotion of plant growth, such as indole-3-acetic acid, phosphate solubilization capacity, nitrogen fixation, catalase, amylase, cellulose activity, and ammonia production. It was found that out of five bacteria isolated, four isolates were Gram negative and one isolate to Gram positive. All bacterial isolates were producing ammonia and catalase positive. Four bacterial isolates produced HCN. One bacterial isolate was found to produce indole-3-acetic acid. Three isolated endophytic bacteria had phosphate solubilizing activity and two had nitrogen fixation activity. Bacterial isolate designated as B_5 was found to be most efficient bacterial agent that showed HCN production, nitrogen fixation, catalase, ammonia production and phosphate solubilisation activity. Present finding suggested that potent bacterial isolate (B_5) may be used for further study regarding the plant growth* *promoting activities in the field conditions because the field condition may vary as compare to laboratory conditions.*

Keywords: *Endophytes, plant growth promoters, catalase test, phosphate solubilizing, nitrogen fixing, HCN production.*

1. Introduction

Endophytes are defined as "microbes that colonize living, internal tissues of plants without causing any immediate, overt negative effects" (Bacon, 2000). Common endophytes include a variety of bacteria, fungi and actinomycetes (Panchal and Ingale, 2011). It is noteworthy that, of nearly 300,000 plant species that exist on earth, each individual plant is host to one or more endophytes (Muzzamal *et al.*, 2012). However, only a handful of plants, mainly grass species, have been completely studied in relation to their endophytic biology as suggested by Strobel *et al.* (2004). Many economically important grasses carry fungal endophytes, some of which may enhance host growth and may improve the plant's ability to tolerate abiotic stresses, such as drought (dry condition), High salt concentration, Heavy metal Concentration, and resistance to insects and mammalian herbivores.

Endophytes also help to prevent host plant to pathogenic organisms from colonizing them. Extensive colonization of the plant tissue by endophytes creates a "barrier effect", where the local endophytes out-compete and prevent pathogenic organisms from taking hold. Endophytes may also produce chemicals which inhibit the growth of competitors, including pathogenic organisms. Endophytic bacteria can promote plant growth and yield and can act as bio-control agents. Endophytes can also be beneficial to their host by producing a range of natural products that could be harnessed for potential use in medicine, agriculture or industry. In addition, it has been shown that they have the potential to remove soil contaminants by enhancing phytoremediation and may play a role in soil fertility through phosphate solubilization and nitrogen fixation (Frank *et al.*, 2006).

Endophytic bacteria produce a wide range of phyto-hormones, such as auxins, cytokinins, and the gibberellins. plant hormones produced by endophytic bacteria seem to be necessary for bryophytes development. Various plant growth promoting attributes such as nitrogen fixing activity, phosphate solubilizing activity, plant hormone producing capacity, HCN production ability, siderophore production, bio-control ability, cellulose, amylase and protease activities.

On the basis of above justification, the present endeavour was to study the bio-prospecting potential of endophytic bacteria isolated from medicinal plant.

2. Materials and Methods

2.1 Samples Collection

For the isolation of endophytic bacteria and fungus from healthy and fresh leaves and stems of tulsi (*Ocimum* sp.) and jamun (*Syzygium* sp.) plants were collected from Ambala, Haryana. Each sample was tagged and placed in separate polythene bags and transported aseptically to the microbiology laboratory and processed within

24 hours of collection. Fresh plant materials were used for isolation to reduce the chance of contamination. (Gayatri *et al.*, 2010)

2.2 Pre-treatment and Isolation of Endophytes

For the pre-treatment of leaf and stem samples and isolation of endophytes, all leaf and stem samples were excised and subjected to a surface sterilization by using the protocol as described by Strobel *et al.* (2004). The procedure is shown below;

Leaves and Stem cut into 1 1 cm pieces with scissor

↓

Immersed in 70% ethanol (1-2 minutes)

↓

Sterilized with aqueous sodium hypo-chloride 4% (3-5 mins)

↓

Rinsed in 70% ethanol (10-20 seconds)

↓

Sample placed on Nutrient agar (NA) Petri plates

↓

Petri dishes were kept in incubator for 2 days for growth of endophytic bacteria

2.3 Purification and Maintenance of Endophytic Bacteria

Bacteria were purified by streak plate method on sterile NA and incubated at 30^{0}C for 24hrs and transferred to nutrient agar slants and then maintained in refrigerator at 4^{0}C till further analysis (Aneja, 2003).

2.4 Screening of Endophytic Bacterial Isolates for Plant Growth Attributes

2.4.1 Screening for Phosphate Solubilization

Bacterial isolates were streaked onto the Pikovskayas agar media (Pikovskaya, 1948) containing (g/l) yeast extract-0.50, dextrose-10, calcium phosphate-5.0, ammonium sulfate-0.5, potassium chloride-0.2, magnesium sulfate-0.1, manganese sulfate-0.0001, ferrous sulfate-0.0001, agar-15 and incubated at 37^{0}C for 2 days. After incubation, clear halo around the streak indicates the phosphate solubilizing activity of endophyte (Pandey *et al.*, 2008).

2.4.2 Screening for Nitrogen Fixation

Bacterial isolates were streaked onto the Jensen agar medium containing (g/l) sucrose-20, dipotassium phosphate-1.0, magnesium sulfate-0.5, sodium chloride-0.5, ferrous sulfate-0.1, sodium molybdate-0.005, calcium carbonate-2.0, agar-15 and incubated at 30^{0}C for 2 days. After incubation, growth formation on Jensen agar medium indicates the nitrogen fixation activity of bacterial endophyte.

2.4.3 Screening for Catalase Production

Small amount of bacterial colony transferred onto a glass slide and placed a drop of 3% H_2O_2. Check the plates for result, if there was rapid evolution of oxygen as evidenced by bubbling, the result was positive (Aneja, 2003).

2.4.4 Screening for Indole-3-acetic Acid (IAA) Production

A total of 500µl of 24hrs old bacterial cultures were inoculated to 50ml nutrient broth containing 0.1% tryptophan and incubated in incubator shaker at 30°C for 48hrs. After incubation 500µl kowaski reagent was added to resultant culture broth. Observed for the change in colour of culture broth from pale yellow to reddish pink, which indicates the presence of indole-3-acetic acid (Hamdali *et al.*, 2008).

2.4.5 Screening for HCN Production

Bacterial cultures were streaked on NA containing 4.4g/l glycine and a whattman filter paper No.1 soaked in 0.5% picric acid solution (in 2% sodium carbonate) was placed inside the lid of Petri plate. Petri plates were sealed with para-film and incubated at 30°C for 3 days. After incubation, development of light brown to dark brown colour of whattman filter paper No.1 indicates the presence of HCN.

2.4.6 Screening for Ammonia Production

Overnight grown cultures were inoculated to 10ml peptone broth and incubated at 30°C for 48hrs. After incubation 0.5 ml Nesslers reagent was added. The development of faint yellow to dark brown colour of broth indicated the production of ammonia (Singh *et al.*, 2014).

2.4.7 Screening of Bacterial Isolates for their Antimicrobial Activity

Nutrient broth (100ml) was prepared and isolated endophytic bacteria were inoculated to it and kept in incubator for 10 days. For extracting secondary metabolites, the bacterial broth was filtered through filter paper and to filtrate ethyl acetate was added and stir continuously for 5 minutes, left undisturbed for 5 minutes. After 5 minutes the upper layer was extracted out using pipette and stored as secondary metabolite. Then the test pathogenic microorganisms were inoculated to NA Petri plates and 6mm diameter wells are made which can easily hold 50-70µl of metabolite. The secondary metabolites are added to well and kept in incubator for 2 days. The formation of zone of inhibition ensures the positive result (Sun *et al.*, 2013).

2.5 Identification of Endophytic Bacteria and Fungus

For the identification of endophytic bacterial and fungal isolates the method described by Cruickshank *et al.* (1975) was followed which includes morphological, cultural, biochemical tests.

3. Results and Discussion

3.1 Isolation of Endophytic Bacteria and Fungus

Endophytic bacteria and fungus were isolated from Tulsi and Jamun plants. The isolates were differentiated on the basis of their colony morphology; two-three distinct colonies were observed and selected from each sample. A total of five bacterial isolates and five fungal isolates were finally selected, purified and maintained for further analysis.

3.2 Activity of Endophytic Bacterial Isolates

It was observed that out of 5 bacterial isolates 3 (60%) were found positive for phosphate solubilisation, 2 (40%) were positive for nitrogen fixation, all isolates (100%) were catalase positive, one (20%) was positive for IAA production, all isolates (100%) were positive for ammonia production and 3 (60%) isolates were found positive for HCN production as shown in (Table 5.1). A research was done by Kumar *et al.* (2012) on French bean in which 30 bacterial were isolated out of which 5 (17%) isolates were positive for phosphate solubilisation, 10 (33%) were positive for IAA production, 20 (67%) were positive for ammonia production, 25 (83%) were positive for catalase production.

Table. 5.1: Results of screening of bacterial Endophytes.

S. No.	NFB test	PSB test	Anti-microbial test	Catalase Test	H C N test	Ammonia production test	IAA production test
B1	+VE	+VE	-VE	+VE	+VE	+VE	-VE
B2	-VE	-VE	-VE	+VE	+VE	+VE	-VE
B3	-VE	+VE	-VE	+VE	+VE	+VE	-VE
B4	-VE	-VE	-VE	+VE	-VE	+VE	+VE
B5	+VE	+VE	-VE	+VE	+VE	+VE	-VE

An important feature that can influence plant growth is the production of indolic compounds as phytohormones. These compounds stimulate root growth and increase root length, resulting in a larger root surface area that enables the plant to access more nutrients from the soil (Souza *et al.* 2013). The bacterial strain showed excellent IAA producing abilities and was able to produce IAA (4.74 μg ml^{-1}). IAA production was found to be more pronounced in the presence of tryptophan and varying levels of IAA production were recorded. The isolate showed increase in the amount of IAA production (7.52, 18.31, 28.81, 35.09 and 38.16 μg ml^{-1}) with the increased concentrations of l-tryptophan (50, 150, 200, 400 and 500 μg ml^{-1}). Majeed *et al.* (2015) and Iqbal *et al.* (2016) researchers also reported increased auxin production in response to l-TRP application. Iqbal *et al.* (2016) also carried out the research under saline field conditions found that the IAA producing rhizobacteria significantly enhanced the growth, physiology and yield of Maize plants.

In the present study, clear halo around the bacterial colonies on PVK agar plates indicated phosphate solubilizing activity of endophytes as shown in Fig. 5.1(a). Phosphate is growth limiting element for plants. Phosphorus has many important functions in plant, the first one being storage and transfer of energy through the plant. Plants are able to utilize phosphorous from soil only in the free available form. Micro-organisms offer a biological rescue system capable of solubilizing the insoluble inorganic phosphorus of soil and make it available to the plants and promoting growth of the plants.

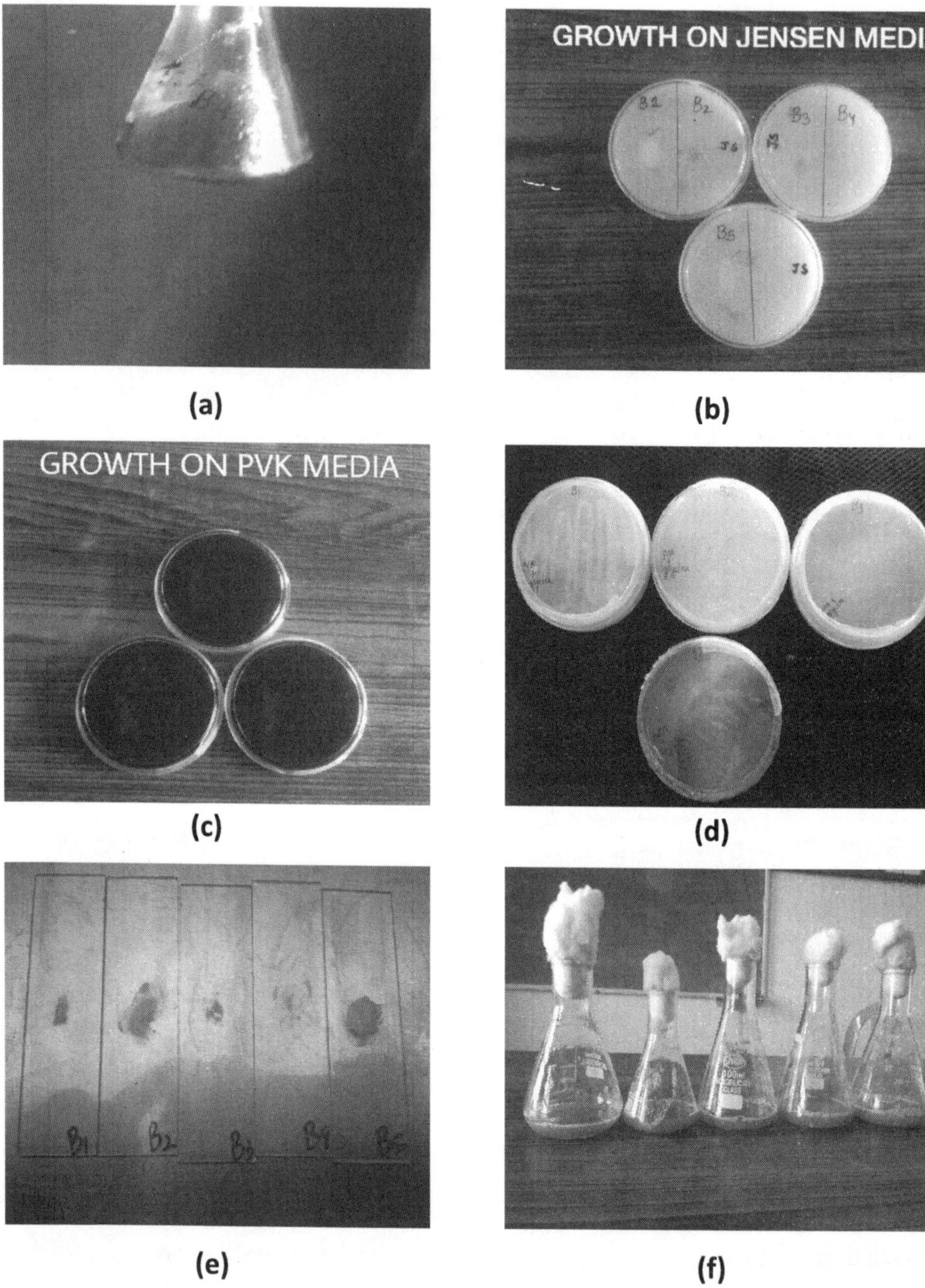

Fig. 5.1: (a) Clear halo around the streak on PVK media indicates the phosphate solubilizing activity (b) Growth formation on Jensen agar medium indicates the nitrogen fixation activity of endophytes (c) Dark pink colour of culture broth showing the production of indole-3-acetic acid (d) Dark brown colour of culture broth indicates the production of ammonia (e) Rapid bubbles formation on slide indicates catalase test positive (f) Dark brown colour of whattman filter paper No.1 inside the lid of Petri plates indicates the production of HCN.

Growth of bacterial colonies (Fig.5.1, b) on Jensen media indicated nitrogen fixation ability of endophytes. Nitrogen fixing micro-organisms fix atmospheric nitrogen gas for their cell protein synthesis, this cell protein is then mineralized in soil after the death of cells thereby contributing towards the nitrogen availability of the crop plants. The organism breaks down the complex nitrogenous material and releases ammonia in soil which is taken up by plant as a nutrient source. Accumulation of ammonia in the soil also creates the alkaline conditions which suppresses the growth of certain fungi (Jha *et al.*, 2012).

Pinkish colour of culture broth (Fig. 5.1, c) indicated the production of indole -3-acetic acid (IAA). IAA is naturally occurring, plant hormone of the auxin class. It induces cell elongation and cell division with all subsequent result for plant growth and development.

The development of faint yellow to dark brown colour of culture (Fig.5.1, d) indicated the production of ammonia. Ammonia is a key ingredient to fertilizer. Ammonia is a source of nitrogen for growing plants. Fertilizers are costly, thus the solution is inoculating the endophyte that have the capability to produce ammonia with the plants. They do not pollute the environment and easily produce ammonia from the nitrogen present in the soil or atmosphere and provide to the plant for its high yield.

Rapid bubbles formation on slide (Fig. 5.1, e) indicated the catalase test positive. Catalase is a common enzyme found in nearly all living organisms exposed to oxygen (such as bacteria, plants, and animals). It catalyzes the decomposition of hydrogen peroxide to water and oxygen. It is very important enzyme in protecting the cell from oxidative damage by reactive oxygen species. Dark brown colour of whattman filter paper No.1 (Fig. 5.1, f) indicated the production of hydrogen cyanide]. Hydrogen cyanide is a colour-less, extremely poisonous organic compound. It is the precursor to sodium cyanide and potassium cyanide, which are used mainly in gold and silver mining and for the electroplating of those metals.

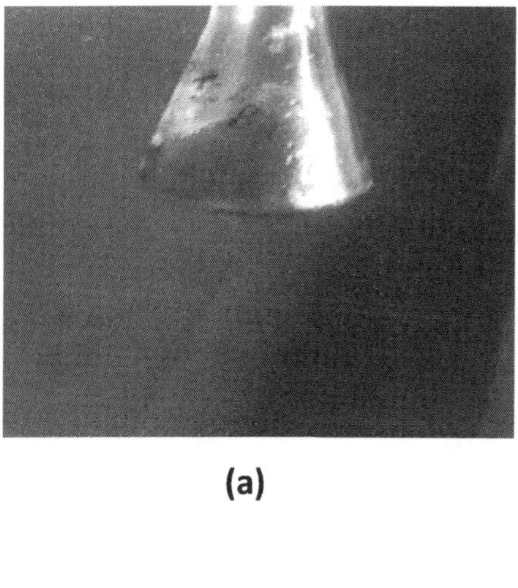

(a)

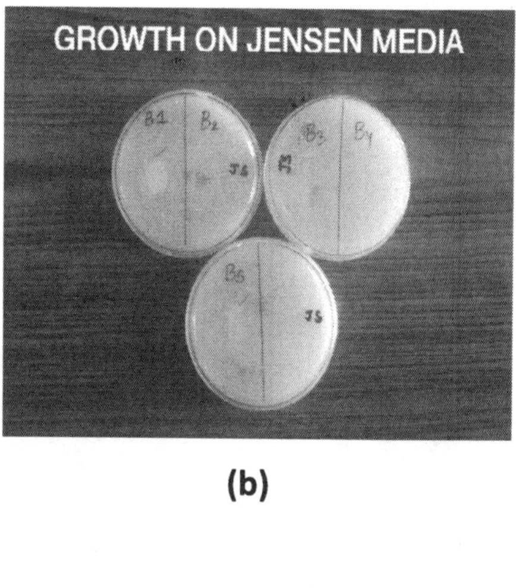

(b)

(c)

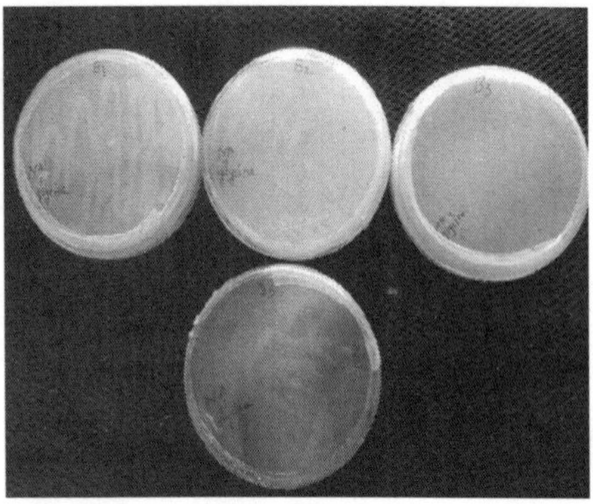

(d)

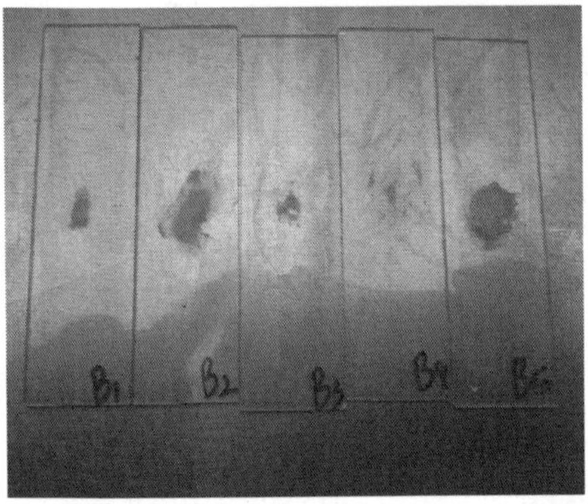

(e)

(f)

Fig. 5.2: (a) Clear halo around the streak on PVK media indicates the phosphate solubilizing activity (b) Growth formation on Jensen agar medium indicates the nitrogen fixation activity of endophytes (c) Dark pink colour of culture broth showing the production of indole-3-acetic acid (d) Dark brown colour of culture broth indicates the production of ammonia (e) Rapid bubbles formation on slide indicates catalase test positive (f) Dark brown colour of whattman filter paper No.1 inside the lid of Petri plates indicates the production of HCN.

3.3 Antimicrobial Activity of Endophytic Bacterial Isolates

None of bacterial isolates were found positive for antimicrobial activity against pathogens. Anti-microbial test was done by extracting the metabolites from both bacterial and fungal endophytes but no zone of inhibition was seen in both the cases. It may be due to difference in protocol or it can vary among different species and strain, also influenced by culture conditions. Singh *et al.* (2017) studied on isolation, morphological identification and in vitro antibacterial activity of endophytic bacteria isolated from *Azadirachta indica* (neem) leaves. The microscopic examination revealed that maximum isolates were Gram-positive and rod shaped. Good antibacterial activity was observed against *Staphylococcus aureus, Streptococcus pyogenes, E. coli,*

Salmonella Typhimurium, and *K. pneumoniae*. Recently, many known, as well as new endophytic bioactive metabolites, possessing a wide variety of biological activities as antibiotic, antiviral, anticancer, anti-inflammatory, antioxidant, etc., have been identified (Strobe and Daisy, 2003).

3.4 Identification of Endophytic Bacterial Isolates

For the morphological and biochemical characterization of bacterial isolates, different tests were performed. The morphological tests namely gram's staining, endospore staining and biochemical tests such as catalase test, indole acetic acid production test, MR test, VP test, citrate utilization test, H_2S test, lactose fermentation test and urease test were performed for identification of endophytic bacteria as shown in (Table 5.2).

Table 5.2: Morphological and biochemical characteristics of bacterial isolates.

Bacterial Isolates	B_1	B_2	B_3	B_4	B_5
Gram's Staining	–VE	–VE	–VE	+VE	–VE
Endospore Staining	–VE	–VE	–VE	+VE	–VE
IAA TEST	–VE	–VE	–VE	+VE	–VE
Methyl red Test	–VE	–VE	+VE	+VE	+VE
Catalase Test	+VE	+VE	+VE	+VE	+VE
Citrate Utilization Test	+VE	+VE	–VE	+VE	+VE
H2S Test	+VE	–VE	–VE	–VE	–VE
VP Test	+VE	–VE	–VE	–VE	–VE
Urease Test	+VE	+VE	–VE	–VE	+VE
Lactose Fermentation Test	+VE	+VE	+VE	+VE	+VE
Suspected Bacterial identification	*Proteus* sp.	*Klebsiella* sp.	*Shigella* sp.	*Bacillus* Sp.	*Proteus* sp.

Endophytic bacterial isolates namely B_1, B_2, B_3, B_5, were found to be Gram negative and B_4 was found to be Gram positive; isolate B_4 was endospore forming and B_1, B_2, B_3, B_5, were non endospore forming; isolate B_4 was positive for IAA production and isolate B_1, B_2, B_3, B_5, were negative; bacterial isolates B_3, B_4, B_5 were found positive for MR test (fig. 2.a) and B_1, B_2, were found negative for MR test ; isolate B_1, was found positive for VP test (fig. 2.b) and isolates B_2, B_3, B_4, B_5 were found negative for VP test ; isolates B_1, B_2, B_4, B_5 were found positive (fig. 2c) and B_3 was found negative for citrate utilization test ; all isolates were found positive for lactose fermentation (fig. 2d); isolates B_1, B_2, B_5 were found positive for urease activity (fig. 2e) and B_3, B_4, were found negative for urease activity . On the basis of these test B_5 was identified as *Proteus* sp., B_3 were identified as *Shigella* sp. Yu and Lee (2013) observed that *Proteus vulgaris* JBLS202 stimulates the seedling growth of Chinese cabbage through indole emission. Barthakur and Bezbaruah (1999) studied on two strains of *Proteus* isolated from tea plantation soil were tested for their ability to colonise the roots of gram (*Cicer arietinum*), bean (*Phaseolus radiatus*) and mung (*Phaseolus mungo*) using a gnotobiotic system. Seeds bacterized with the two strains grew faster and showed significant increase in root and shoot enlargement of the plants tested. According to Nehra *et al.* (2016), the *Brevibacillus brevis* isolate was found positive for many of the PGP attributes like IAA, ARA, anti-fungal activity and ammonia production. No phospahate solubilizing activity was observed by them.

Hung and Anapurna (2004) identified endophytic bacteria on the basis of morphological and physiological test including Gram's staining, capsule staining, cellulose, pectinase, Motility test, indole acetic acid production, fluorescent pigmentation test and BIOLOG system identification test. Isolation of endophytic staphylococcus was reported by Carrim *et al.*, 2006). Muzzamal *et al.* (2012) identified various bacterial species belonging to genera *Bacillus, Pseudomonas, Serratia, Stenotrophomonas* and *Micromonospora.*

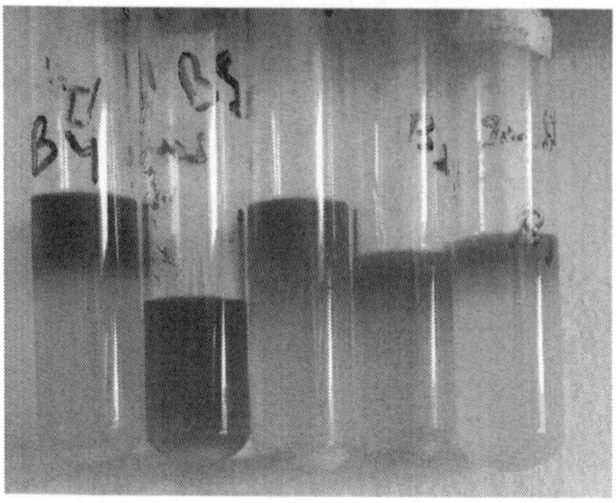

(a)

(b)

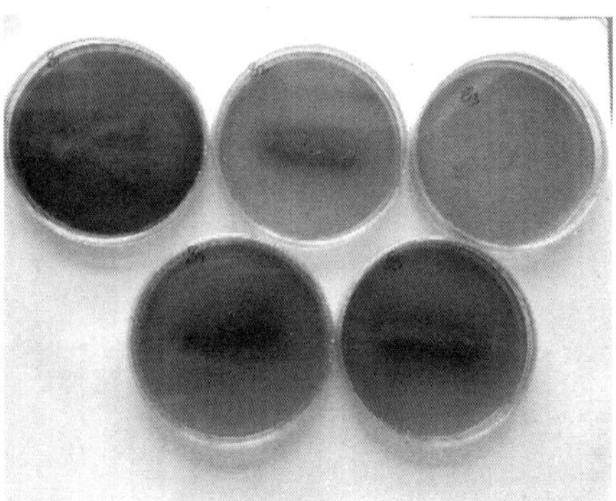

(c)

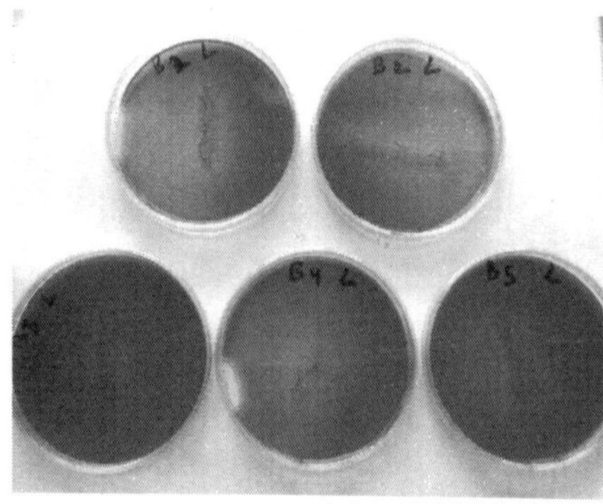

(d)

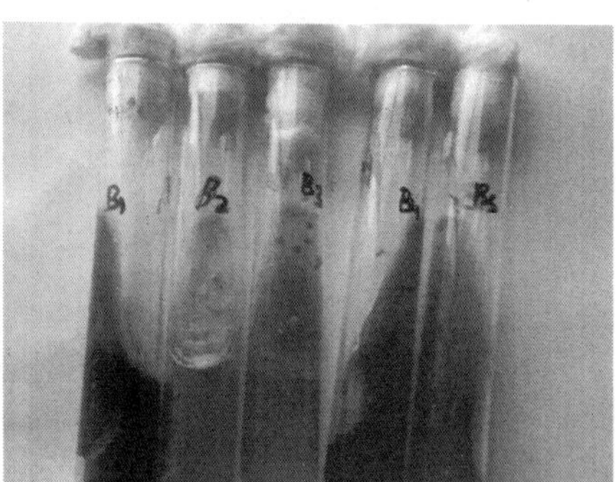

(e)

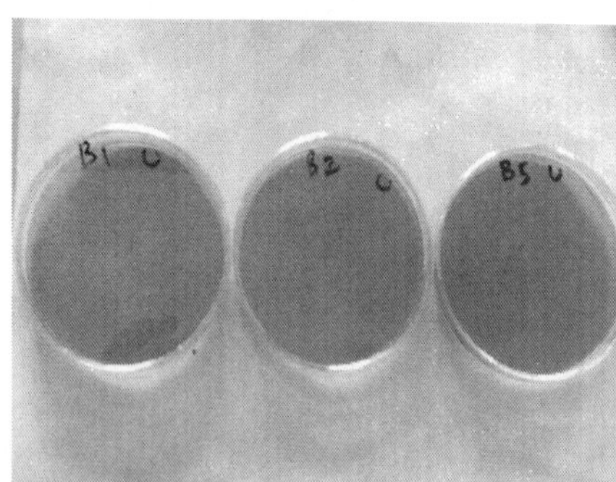

(f)

Fig. 5.3: (a) Blood red colour indicates MR test positive (b) Red colour indicates the VP test positive (c) Blue colour around bacterial colonies indicates positive citrate utilization test (d) shows the lactose fermentation (e) pink colour shows urease activity (f) Shows H_2S production, glucose and lactose fermentation and gas production.

4. Conclusion

Endophytes are a promising source for production of essential bioactive compounds. They are poorly investigated group of micro-organisms that represent an abundant and dependable source of bioactive and chemical novel compounds with potential for exploitation in a wide variety of agricultural, industrial and medicinal areas. The mechanisms through which endophytes exist and respond to their surroundings must be better understood in order to be more predictive about which higher plants to seek, study, and spend time isolating micro-floral components. This may facilitate the product discovery processes.

Acknowledgements

The authors are grateful to the Management of Ambala College of Engineering and Applied research (ACE), Devsthali, PO-Sambhalkha, Ambala, Haryana (India) for providing laboratory facilities in Department of Biotechnology Engineering and also encouraging us to write this article.

References

Aneja, K. R. 2003. Experiments in Microbiology, Plant Pathology and Biotechnology. New Age International (P) Publishers, New Delhi.

Bacon, C.W. and White, J.F. 2000. Microbial endophytes. Marcel Dekker Inc., New York, N.Y.

Barthakur, M. and Bezbaruah, B. 1999. Plant beneficial effect of two strains of *Proteus vulgaris* isolated from tea plantations. *Ind. J. Exp. Biol.* 37: 919-24.

Carrim, A. J.I., Barbosa, E.C. and Vieira, J.D.G. 2006. Enzymatic activity of endophytic bacterial isolates of *Jacaranda decurrens* cham. *Braz. Arch. Biol. Tech.* 49: 353-359.

Cruickshank, R., Duguid, J.P., Marimon, B.P. and Swian, R.A.H. 1975. Medical microbiology. 2nd edition. Vol. 1 church hill Livingstone, Edinburgh.

Franks, A., Ryan, P.R., Abbas, A., Mark, G.L. and O'Gara, F. 2006. Molecular Tools for Studying Plant Growth-Promoting Rhizobacteria (PGPR): Molecular Techniques for Soil and Rhizosphere Microorganisms. CABI Publishing, Wallingford, Oxfordshire, UK.

Gayatri, S., Saravanan, D., Radhakrishnan, M., Balagurunthan, R. and Kathiresan, K.2010. Bio-prospecting potential of fast growing endophytic bacteria from leaves of mangrove and salt-marsh plant species. *Ind. J. Biotechnol.* 9: 397-402.

Germaine, K., Keogh, E., Borremans, B. *et al.* 2004. Colonisation of poplar trees by gfp expressing bacterial endophytes. *FEMS. Microbiol. Ecol.* 48: 109–118.

Hamdali, H.M., Virolle, M.J. and Achdouch, Y., 2008. Rock-phosphate solubilizing actinomycetes: screening for plant growth promoting activities. *World J. Microbial. Biotechnol.* 24: 2565-2575.

Hung, P.Q. and Annapurna, K. 2004. Isolation and characterization of endophytic bacteria in soybean (*Glycine* sp.). *Omonrice.*12: 92-101.

Iqbal, M.A., Khalid, M., Zahir, Z.A., Ahmad, R. 2016. Auxin producing plant growth promoting rhizobacteria improve growth, physiology and yield of maize under saline field conditions. *Int. J. Agric. Biol.* 18: 37–45.

Jha, C.K., Patel, B. and Sarf, M. 2012. Stimulation of the growth of *Jatropha curcas* by the plant growth bacterium Enterobacter cancerogenus MSA2. *World J. Microbiol. Biotechnol.*, 28: 891-899.

Nehra, V., Saharan, B.S. and Choudhary, M. 2016. Evaluation of *Brevibacillus brevis* as a potential plant growth promoting rhizobacteria for cotton (*Gossypium hirsutum*) crop. *Springerplus*. 5:948.

Majeed, A., Abbasi, M.K., Hameed, S., Imran, A., Rahim, N. 2015. Isolation and characterization of plant growth-promoting rhizobacteria from wheat rhizosphere and their effect on plant growth promotion. *Front Microbiol*. 6: 198.

Muzzamal, H., Sarwar, R.0., Sajid, I. and Hasnain, S. 2012. Isolation, identification and screening of endophytic bacteria Antagonistic to Biofilm formers. *Pak. J. Zool.* 44: 249-257.

Panchal, H. and Ingale, S. 2011 Isolation and Characterisatation of Endophytes from root of medicinal Plant C*hlorophylum borivilianum* ((Safed musli) J. Adv. Dev. Res. 2: 205-209.

Pandey, A., Das, N., Kumar, B., Rinu, K. and Trivedi, P. 2008. Phosphate solubilization by *Penicillium* spp. Isolated from soil samples of Indian Himalayan region, *World J. Microbial. Biotechnol.* 24: 97-102.

Pikovskaya, R.I. 1948. Mobilization of phosphorous in soil in connection with vital activity of some microbial species. *Mikrobiologiya.*17: 362–370.

Singh, A.K., Sharma, R.K., Sharma, V., Singh, T., Kumar, R., Kumari, D. 2017. Isolation, morphological identification and in vitro antibacterial activity of endophytic bacteria isolated from *Azadirachta indica* (neem) leaves. *Veterin. World.* 10: 510-516.

Singh, P., Kumar, V. Agrawal, S. 2014. Evaluation of phytase producing bacteria for their plant growth promoting activities, Int J Microbiol, 426-483.

Souza, D.R., Beneduzi, A., Ambrosini, A., Costa, P.B., Meyer, J., Vargas, L.K., Schoenfeld, R., Passaglia, L.M.P. 2013. The effect of plant growth-promoting rhizobacteria on the growth of rice (*Oryza sativa* L.) cropped in southern Brazilian fields. *Plant Soil*. 366:585–603.

Strobel, G., Daisy, B., Castillo, U. and Herper, J. 2004. Natural Products from endophytic microorganism. *J. Nat. Prod.* 67: 257-268.

Sun, H., He, Y., Xiao, Q., Ye, R. and Tian, Y. 2013. Isolation, characterization, and antimicrobial activity of endophytic bacteria from *Polygonum cuspidatum*. *Afr. J. Microbiol. Res.* 7(16): 1496-1504. doi:10.5897/AJMR12.899.

Tan, Rx. and Zou, W.X. 2001. Endophytes: a rich source of functional metabolites. *N. Prod. Rep.*. 18: 448-459.

Yu, S. M. and Lee, Y. H. 2013. Plant growth promoting rhizobacterium *Proteus vulgaris* JBLS202 stimulates the seedling growth of Chinese cabbage through indole emission. Plant and Soil. 370: 485-49.

Chapter 6

Cytological Study on two Species of *Cymbopogon* Spreng.- A Grass with Medicinal Importance

Akshita Dhaliwal[1*], Rajdeep Singh Dhaliwal[1] & R.C. Gupta[2]

[1]Post Graduate Department of Agriculture, G.S.S.D.G.S. Khalsa College, Patiala.

[2]Department of Botany, Punjabi University, Patiala.

**Corresponding author E-mail- akshi2081@gmail.com*

Abstract

Presently, two species of Cymbopogon have been studied cytologically i.e. C. jwarancusa (Jones) Schult. and C. martinii (Roxb.) Wats. C. jwarancusa commonly found in Shiwaliks and in sandy habitats. This is the common scented grass. In all the populations, during meiosis unequivocally 10 II have been counted at M-I indicating that the worked out plants are diploids. C. martinii is common in sandy soil near humid habitats. It is cultivated for its volatile oil (Palmarosa oil), which contains the active compound geraniol and is valued for its scent and for a number of traditional medicinal uses. The Palmarosa oil has been shown to be an effective insect repellant, an antihelmintic, an antifungal and mosquito repellent. The meiosis of the species also shows the presence of 10 bivalents at M-I in all the populations.

Keywords *- Cytology, Cymbopogon, Meiosis.*

1. Introduction

Cymbopogon genus belonging to the family Poaceae is an Herb. The 140 species are known to occur mostly in tropics and sub-tropics of the world (Airy Shaw, 1973). In all, 45 species are reported from the Indian sub-continent (Bor, 1960). Presently, two species *C. jwarancusa* and *C. martinii* are discussed cytologically on population basis. These species are known worldwide for their high essential oil content which has been used for cosmetics, pharmaceuticals and perfumery applications (Khanuja *et al.*, 2005). The commercial value of some *Cymbopogon* species is further enhanced by their ability to grow in moderate and extremely harsh climatic conditions (Padalia *et al.*, 2011).

2. Material and Methods

Materials for male meiotic studies were collected from wild plants growing in various localities of Haryana district with adjoining Shiwalik Hills. The localities were visited two to four times during the months of August to February. For meiotic studies, unopened and young floral buds/inflorescences from healthy plants were fixed in freshly prepared Carnoy's fixative (1part glacial acetic acid: 3 parts chloroform: 6 parts absolute alcohol) for 24 h. The materials were subsequently transferred to 70% alcohol and kept at 4 C in the refrigerator until analysis. Meiotic studies were carried out through standard acetocarmine squash technique for which developing anthers were squashed in 1% acetocarmine.

A number of freshly prepared slides were examined for meiotic analysis in each accession and chromosome counts were made. Pollen fertility in each species/taxa was estimated through stainability tests for which the mature anthers from different flowers were squashed in glyceroacetocarmine (1:1) mixture. Well-filled pollen grains with stained nuclei and cytoplasm were scored as apparently fertile while grains that were shrivelled and showed unstained/poorly stained cytoplasm as sterile. Pollen grain size was measured using an occulomicrometre. Photomicrographs of well-spread meiotic chromosome counts, sporads and pollen grains were made from the freshly prepared/temporary slides using a Nikon Eclipse 80*i* microscope fitted with a Digital Imaging System. Voucher specimens of the meiotically worked out specimens were deposited in the Herbarium, Department of Botany, Punjabi University (PUN), Patiala and accession numbers are given in the Table 6.1.

Table 6.1: Data on Population number, locality with altitude, accession number, meiotic chromosome number, pollen fertility (%), pollen size (μm) and ploidy level of two species of *Cymbopogon*.

Population Number	Locality with altitude in meters	Accession Number (PUN)	Meiotic Chromosome Number (n)	Pollen Fertility (%)	Pollen Size (μm)	Ploidy Level	Remarks
***Cymbopogon jwarancusa* (Jones) Schult.**							
P-1	Barog, Solan (1350 m)	50218	10	92.8	16.9-18.8 x 16.9-18.8	2x	N
P-2	Jabli, Shiwaliks (900 m)	50219	10	100	15.0-16.9 x 15.0-16.9	2x	N
P-3	Dharampur, Solan (1350 m)	50220	10	96.7	16.9-18.8 x 16.9	2x	N
P-4	Kalka, Shiwaliks (624 m)	50221	10	97.3	16.9 x 16.9	2x	N
***C. martinii* (Roxb.) Wats**							
P-1	Barog, Solan (1350 m)	50187	10	98.7	16.9-18.8 x 16.9-18.8	2x	N

Population Number	Locality with altitude in meters	Accession Number (PUN)	Meiotic Chromosome Number (n)	Pollen Fertility (%)	Pollen Size (μm)	Ploidy Level	Remarks
P-2	Jabli, Shiwaliks (900 m)	50188	10	95.4	15.0-16.9 x 16.9	2x	N
P-3	Dharampur, Solan (1350 m)	50189	10	100	15.0-16.9 x 15.0-16.9	2x	N
P-4	Kalka, Shiwaliks (624 m)	50190	10	99.2	16.9-18.8 x 16.9	2x	N

3. Results and Discussion

In *C. jwarancusa* (Jones) Schult. lower glume of the sessile spikelet was smooth and glabrous; pedicel of the lowest pedicelled spikelet not enlarged; basal sheaths not wooly; pedicelled spikelet glabrous whereas in *C. martinii.* lower glume of sessile spikelets with a narrow slit-like groove in the lower half; sessile spikelets long; leaf-blades linear lanceolate, cordate, amplexicaul at the base.

3.1 *C. jwarancusa* (Jones) Schult.

The species is commonly found in Shiwaliks and in sandy habitats. The common name is "Iwarancusa grass". This is a common scented grass. In all the four populations, during meiosis unequivocally 10_{II} have been counted at M-I (Fig. 6.1) indicating that the worked out plants were diploids. Normal meiosis leads to high pollen fertility with some variation in pollen grain size (Table 6.1). The chromosome number 2n=20 is well established as is clear from the available data. The tetraploid cytotype (2n=40) is reported by Mehra and Kohli (1966); Mehra *et al.* (1968) and Sahni and Bir (1985) from India whereas Baquar and Saeed (1969) recorded tetraploid cytotype from Pakistan. The presence of 0-8 B-chromosomes recorded by Sobti *et al.* (1979) and Verma and Sobti (1985) in the diploid cytoytpe. Therefore, the species embraces 2x and 4x cytotypes.

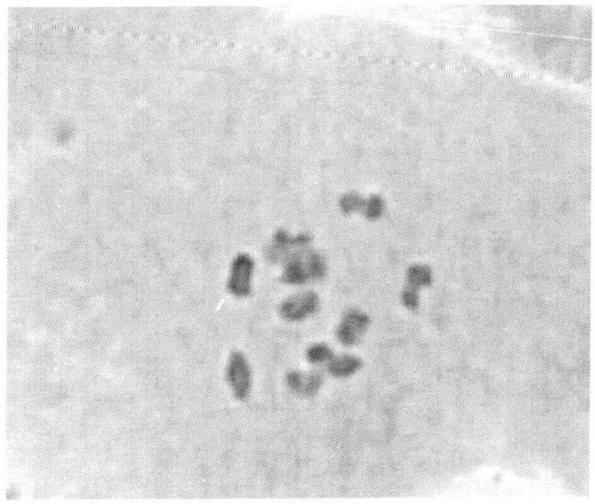

Fig. 6.1: *C. jwarancusa*- M-I showing 10_{II}

3.2 *C. martinii* (Roxb.) Wats

C. martinii is distributed in sandy soil near humid habitats. The species is commonly known as "Palmarosa", "Indian geranium" and "Rosha grass". It is cultivated for its volatile oil (Palmarosa oil), which contains the active compound geraniol and is valued for its scent and for a number of traditional medicinal uses.

The Palmarosa oil has been shown to be an effective insect repellant (Kumar *et al.*, 2007), an antihelmintic (Kumaran, 2003) and an antifungal and mosquito repellent (Duke and DuCellier, 1993). The meiosis of the species shows the presence of 10 bivalents at diakinesis stage (Fig. 6.2) in all the populations. The meiosis is quite normal with high pollen fertility (95.4-100.0 per cent) (Table 6.1). Both the diploid (2n=20) and tetraploid (2n=40) cytotypes were already well known from India. Gupta, (1965) reported the presence of one B-chromosome in some PMCs of the diploid cytotype. Thus, the intraspecific polyploidy (2x, 4x) is well represented in the species.

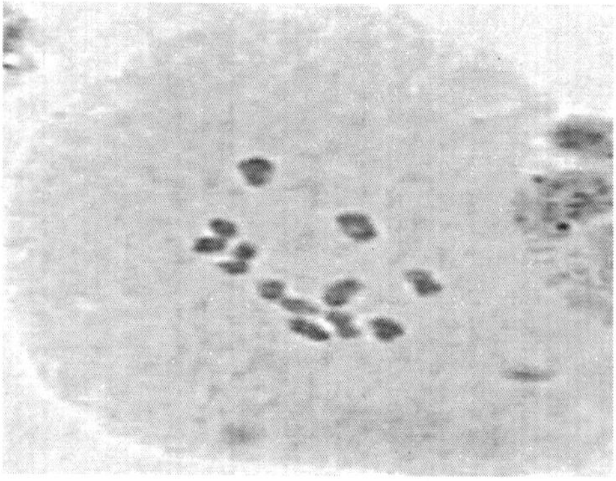

Fig. 6.2: ***C. martinii*- Diakinesis showing 10_{II}**

The genus *Cymbopogon* has immense medicinal importance to human beings but very little attention is given to the exploration of grasses of the region. So keeping this in mind, efforts were made to explore for genetic diversity of the important medicinal grasses which can be further subjected to morphological, cytological and chemical characterization on population basis from different areas.

Acknowledgements

The authors are grateful to the Department of Botany, Punjabi University, Patiala for providing help during research work. Thanks are also due to Principal, GSSDGS Khalsa College, Patiala for necessary laboratory and library facilities.

References

Airy Shaw, H.K. 1973. *A dictionary of Flowering plants and Ferns* (Revised 8th ed. of the late J.C. Willis), Univ. Press Cambridge.

Baquar, S.R. and Saeed, M. 1969. Chromosome studies and polyploidy analysis in grasses of West Pakistan I. *Caryologia* 22: 103-111.

Bor, N.L. 1960. *The Grasses of Burma, Ceylon, India and Pakistan.* Pergamon Press.

Duke, J.A. and DuCellier, J. 1993. *CRC Handbook of Alternative Cash Crops*. Boca Raton: CRC Press. 214.

Gupta, B.K. 1965. Chromosome numbers in some species of *Cymbopogon*. *Proc. Ind. Acad. Sci.*, Sect. B., 62, 3: 155-157.

Khanuja, S.P.S., Shasany, A.K., Pawar, A., Lal, R.K., Darokar, M.P., Naqvi, A.A., Rajkumar, S., Sundaresan, V., Lal, N. and Kumar, S. 2005. Essential oil constituents and RAPD markers to establish species relationship in *Cymbopogon* Spreng. (Poaceae). *Biochem. Syst. Ecol.*, *33*, 171–186.

Kumar, R., Srivastava, M. and Dubey, N.K. 2007. Evaluation of *Cymbopogon martinii* oil extract for control of postharvest insect deterioration in cereals and legumes. *Journal of Food Protection* 70(1): 172-178.

Kumaran, A.M. 2003. Geraniol, the putative anthelmintic principle of *Cymbopogon martinii. Phytotherapy Res.* 17(8): 957.

Mehra, P.N. and Kohli, B.L. 1966. Cytological studies in North Indian grasses. *Proc. 53th Ind. Sci. Congr.*, Part 3: 277-278.

Mehra, P.N., Khosla, P.K., Kohli, B.L. and Koonar, J.S. 1968. Cytological studies in North Indian grasses. I. *Res. Bull. Panjab Univ., Sci.* 19: 157-230.

Padalia, R.C., Verma, R.S., Chanotiya, C.S. and Yadav, A. 2011. Chemical fingerprinting of the fragrant volatiles of nineteen indian cultivars of *Cymbopogon* Spreng (Poaceae). *Rec. Nat. Prod. 5*, 290–299.

Sahni, M. and Bir, S.S. 1985. SOCGI plant Chromosome number reports – III. *J. Cytol. Genet.* 20: 205-206.

Sobti, S.N., Verma, V., Rao, B.L. and Pushpangadan, P. 1979. *In* IOPB Chromosome number reports LXV. *Taxon* 28: 627.

Verma, V. and Sobti, S.N. 1985. Cytogenetical studies of accessory chromosomes in the genus *Cymbopogon* Spreng. *Cytologia* 50: 671-682.

Chapter 7

Agro-biochemical Traits Variation in Indian Wheat Genotypes Grown Under Irrigated and Water Stress Field Conditions

Vandita Mittal[1,2], Sonia Sheoran[1] and Narender Singh[2,*]

[1]*ICAR – Indian Institute of Wheat and Barley Research, Karnal, India.*

[2]*Department of Botany, Kurukshetra University, Kurukshetra, India.*

**Corresponding author E- mail: nsheorankukbot11@gmail.com*

Abstract

Water stress has become a serious constraint to wheat productivity as the climate change has created unpredictability in the timing and amount of rain fall worldwide. In the present study, thirteen wheat cultivars with a wide range of genetic variation for drought were selected from different agro-climatic zones related to irrigated and rainfed sowing conditions. The cultivars were grown in field under two water conditions i.e. normal water supply condition in open field (IR) and water stress condition in rain out shelter (ROS). Plants were sampled at heading (HD) and anthesis stage (AN) for several agro-biochemical traits i.e. 12 morpho-physiological and 7 yield traits, evaluation. Our findings indicated that water stress significantly reduced biomass/plant (50.1%), tiller number/plant (38.4%) but increased proline content i.e. 75%. Antioxidant enzyme activity also showed a significant percent increment under water stress condition as compared to control condition. Water stress also decreased relative water content, chlorophyll content, normalized vegetation index (NDVI), plant height (PH) while electrical conductivity, canopy temperature, catalase enzyme activity and total phenolic content increased more significantly at anthesis stage of growth. In the present study positive and significant correlation of SOD enzyme activity has been found with POX, NDVI (AN), PH and total phenolic content while a significant and negative correlation was found with canopy temperature (AN) under water stress. Principle component analysis (PCA) of these genotypes identified four principle components that explained 78.4% of maximum varaition at anthesis stage. Remarkebly, physiological traits at anthesis stage, antioxidant enzyme activity, proline content, total phenolic content and yield parameters could provide a meaningful tool for futher improvement and future breeding programmes for drought tolerance.

Keywords: *ROS, water stress, proline content, antioxidant enzyme activity, PCA, anthesis stage.*

1. Introduction

Climate change increases the negative impact of abiotic stresses- drought on wheat production by initializing different physiological responses as senescence, disturbance of photosynthesis, lower transpiration rate, storage of carbon products, early ground cover, osmotic adjustment (OA), decrease of evaporation etc. (Monneveux *et al.*, 2012). In India water deficit stress limits crop production about 67% of net sown area and in the developing world yield reduced to 50-90% of their irrigated potential (Devi *et al.*, 2011). Plant response to drought is complex as it involves a number of physio-biochemical processes at the cellular level and different interacting component traits with different responses at the whole plant level (Kadam *et al.*, 2012). Decline in relative water content and chlorophyll content are the common and primary changes under osmotic stress (Lascano *et al.*, 2001; Bayoumi *et al.*, 2008). Water potential is the major factor showing diverse relation with other physiological traits, affecting growth and yield potential (Nezhadahmadi *et al.*, 2013; Saqib *et al.*, 2013). One of the main effects of dehydration is the production of ROS (reactive oxygen species) such as singlet oxygen, superoxide anion radicals, hydroxyl radicals and hydrogen peroxide (Devi *et al.*, 2011). Generation of ROS leads to the lipid peroxidation that disrupts the lipid bilayer structure and reduces the function of membrane bound enzymes. Antioxidant defense systems including activation of the antioxidant enzymes superoxide dismutase (SOD), ascorbate peroxidase (APX), catalase (CAT) and peroxidase (POD), production of osmolytes etc., are well known for scavenging ROS (Lascano *et al.*, 2001; Marcinska *et al.*, 2013). Drought effects nearly all the plant growth processes; however the stress response depends upon the stage of crop growth (Nezhadahmadi *et al.*, 2013). In wheat transition from vegetative to reproductive stage is important. Recent studies indicate that drought stress has a major influence on yield at anthesis stage of wheat (Saqib *et al.*, 2013). Around the time of anthesis vegetative biomass tends to decline due to remobilization of assimilates and translocation to the developing grains (Ferris *et al.*, 1998). Significant increase in the enzyme activity of Catalase and superoxide dismutase was observed in some Indian wheat genotypes in a pot experiment (Chakraborty and Pradhan, 2012) and at seedling stage (Devi *et al.*, 2011). Osmotic adjustment in plants may occur through accumulation of some organic solutes. Proline accumulation is found to be osmotically active and contributing to the membrane stability in plants (Marcinska *et al.*, 2013). Protein and total phenols are also considered to be affected under drought stress condition (Narwal *et al.*, 2014). Considerable attention has been given by plant breeders and biotechnologists to increase the crop productivity and to minimize the losses caused by adverse environmental conditions. Different wheat genotypes differ for drought resistance and leaf water relation, show a positive correlation with the yield under water stress condition developed in PVC pipes (Kadam *et al.*, 2012). The variation in the level of resistance to drought stress is known to exist amongst the genotypes of plant species e.g. in maize (Lorens *et al.*, 1987) and wheat (Winter *et al.*, 1988). So, the identification of plant varieties, resistant to drought is a necessity. In this review, complete drought stress was applied under field condition and anthesis stage of growth was focused to study the response of selected drought tolerant, susceptible and moderate tolerant Indian wheat genotypes. This study will also help to find out the direct practical measure to select tolerant genotypes for breeding of wheat cultivars under drought prone areas.

2. Material and Methods

2.1 Plant Material

Thirteen wheat cultivars with a wide range of drought genetic variation were selected from different agro-climatic zones related to irrigated and rainfed sowing conditions. The cultivars used in this study, were provided from Indian Wheat and Barley Research Institute (IWBR), Karnal (Table 7.1) including AKAW3717, C306, DBW17, DL153-2, DPW621-50, GW322, HD2781, HD2987, NI5439, PBW175, PBW343, PBW396 and PBW596. The plants were grown in field under two water conditions i.e. normal water supply condition in open field (IR) and water stress condition (DT) in rain out shelter (ROS). In ROS, irrigation was restricted after sowing of seeds. Plant samples were sown in alpha lattice design in three replications i.e. 32 plants of each genotype in one block as one replication, during two consecutive years 2012-13 and 2013-14. Sowing was done in the first week of November in both years. Observations were made from randomly selected five plants from each block for the following parameters:

Table 7.1: Varieties of bread wheat used for the experiment.

S.No.	Genotype	Pedigree	Zone	Sowing time
1.	AKAW3717	HW 2035/NI54349	PZ	RTS
2.	C306	RGN/CSK3 //2*C591/3/C217/N14 //C281	NWPZ/ NEPZ	RTS
3.	DBW17	CMH79A.95/3*CNO79//RAJ3777	NWPZ	ITS
4.	DL153-2	TANORI71/NP890	NWPZ	RTS
5.	DPW621-50	KAUZ//ALTAR84/AOS/3/MILAN/KAUZ/4/ HUITES	NWPZ	ITS
6.	GW322	GW 173/GW 196	PZ/CZ	ITS
7.	HD2781	BOW /C 306 //C591/HW 2004	PZ	RTS
8.	HD2987	HI 1011/HD 2348//MENDOS//IWP 72/DL 153-2	NWPZ	ITS
9.	NI5439	REMP 80 /3* NP 710	PZ	RTS, ITS
10.	PBW175	HD 2160 /WG 1025	NWPZ	RTS
11.	PBW343	ND/VG9144//KAL/BB/3/YCO"S" /4/VEE5	NWPZ/ NEPZ	ITS
12.	PBW396	CNO67/MFD //MON"S"/3/SERI	NWPZ	RTS
13.	PBW596	PBW 343/DHARWAR DRY//PBW 343	PZ	RTS

* ITS Irrigated timely sown, RTS Rainfed timely sown

2.2 Physiological Parameters

Plants were sampled at heading and anthesis stage for different physiological parameters. Relative water content (RWC) was estimated by using RWC % = (FW-DW)/ (TW-DW)x100 (Barrs and Weatherley 1962). For this, middle portion of leaf 10cm long leaf was excised, fresh weight (FW) was recorded and the leaf segment was immediately immersed in double distilled water (DDW) to record turgid weight (TW). Then, the samples were placed in a paper bag and dried in a hot air oven at 80 C for dry weight (DW). Chlorophyll content index (CCI) was measured by using chlorophyll meter (atLEAF$^+$, FT GREEN LLC; Wilmington, DE, 19801, USA). Electrical conductivity (EC) was estimated by using EC% = (Fresh EC/Final EC) x 100. For this leaf samples were excised and immediately put in 25 ml DDW in tarson tubes for fresh EC and final EC was measured after autoclaving the samples by using EC-TDS analyzer (EUTECH PC700). Canopy temperature (CT) was measured during 1200 hrs and 1400 hrs with a handheld infrared thermometer HTC MT4 (Pask *et al.*, 2012). Normalized vegetation index (NDVI) was recorded using an optical handheld GreenSeeker sensor (Trimble industries, Inc.).

2.3 Yield Parameters

The number of the productive tillers per plant (TN/P) were recorded at physiological maturity and plant height (PH) was measured in centimeters (cm) from the ground to the tip of spike from five randomly selected plants from each plot before harvesting. Biomass/plant (BM/P), thousand grain weight (TGW), total grain number/spike (GN/S) and total grain weight/spike (GW/S) were recorded by using seed counter (WAVER, Daidex Co. Ltd.) and weighed on electric balance and expressed in grams (g). Days to heading (DTH) were counted as the number of days from sowing to the day of heading, when 75% of the heads emerged fully from the boot of flag leaf in each plot.

2.4 Biochemical Parameters

To study the biochemical changes occurred under stress, proline concentration, MDA content, total phenols concentration, protein content and the antioxidant defence components (SOD, CAT, POX enzyme activities) were recorded. The plants were sampled around midday between 11:00 and 12:00 h from fully emerged young leaves from both field (irrigated) and ROS (drought) condition. Samples were transferred immediately in liquid nitrogen and then stored at -80 C for further experiment. The Proline concentration (PC) was determined according to the method of Bates *et al.* (1973) and the amount of proline was calculated against pure proline standard and expressed in μMg^{-1}fresh weight (FW). Measurement of lipid peroxidation (Malondialdehyde, MDA) was taken according to Heath and Packer (1968) and concentration was expressed in $\mu M\ g^{-1}$ FW using a molar extinction coefficient equal to 155 105 $mM^{-1}\ cm^{-1}$. Phenolic content was measured spectrophotometrically according to Singleton et al (1999) by and expressed as the µg Gallic acid equivalent (GAE) per gram leaf sample. SOD activity was determined by using the method SPCYO01 of Sigma (Mccord and Fridovich, 1969) and expressed in units/mg protein CAT enzyme activity was calculated by using extinction coefficient of H_2O_2 39.4 mMcm-1(Rao *et al.*, 1996) and expressed in µMoles/min/mg proteins. POX activity was calculated by using extinction coefficient of 26.6 $mM^{-1}cm^{-1}$(Jebara *et al.*, 2005) and expressed in units/mg protein. Protein content was measured by using Bradford's method (Bradford, 1976). Bovine serum albumin (BSA) was used for preparation of standard curve to estimate protein content present in leaf samples.

2.5 Statistical Analysis

The experiment was performed in completely randomized design. Standard errors of means were calculated for all parameters. Data was analyzed using ANOVA with IRRI Crop Stat (version 7.2.3) and SAS 9.3 software. Principal component analysis (PCA) was done by using JMP 9.0 software to find out the principal component for drought tolerance accounting maximum variation and loading plot and scree plot were also made. Values are presented in the graph as the mean ± standard error (SE).

3. Results

3.1 Physiological Parameters

Table 7.2 summarizes the mean value, standard deviation, minimum, maximum and coefficients of variation (CVs) for all the traits studied in two water conditions for all genotypes. Performance of individual genotype was showed graphically for all the traits. The germination percentage (GER%) under water stress condition is affected in almost all genotypes. The mean GER% under stressed and irrigated conditions was 85% and 90% respectively. Maximum reduction in germination percentage was reported in DPW621-50 (16.9%) followed by GW322 (14.7%) while minimum reduction was found in PBW175 i.e. 1.7%. Maximum reduction in relative water content (RWC) at heading stage was found in PBW343 (11.4%) while C306 showed minimum reduction in RWC (0.8%) under water stress condition which was increased at anthesis stage with maximum reduction value 15% in PBW343 followed by HD2987 i.e. 13% and minimum reduction 3.3% in C306 (Fig. 7.1, a, b). Under irrigated conditions, RWC was observed to be lower at anthesis stage in all genotypes with the mean value 86% as compared to 97% at heading stage. At heading stage maximum reduction in chlorophyll content index (CCI) was observed in PBW343 (11.4%) followed by HD2987 (10.2%) and DPW621-50 (9.6%) (Fig. 7., 1c). At anthesis stage, lowest reduction in CCI was noticed in C306 (0.2%) while the highest reduction was noticed in HD2987 (18%) followed by DBW17 (16%) (Fig. 7.1, d). In this study all genotypes recorded up regulation of canopy temperature under drought stress. At heading, minimum percentage increment in canopy temperature (CT) under drought condition was recorded in HD2781 (-30.5%) while the highest %age value of CT increment was observed in DBW17 i.e. -58.4% (Fig. 7.1, e). Similarly at anthesis stage, minimum increment in CT was recorded in HD2781

Table 7.2: Conditional phenotypic values for different traits observed under two water regimes drought (DT) and irrigated (IR) at heading (HD) and anthesis (AN) stage.

	Traits	Sowing Condition	Mean	SD	MAX	MIN	CV%	%age change due to stress
PHYSIOLOGICAL	Germination %	DT	84.73	5.41	91.67	76.67	6.38	6.10
		IR	90.23	3.58	95.93	84.54	3.97	
TRAITS	Relative water content (HD)	DT	92.80	3.14	96.69	85.88	3.39	3.87
		IR	96.54	1.33	98.20	94.01	1.37	
	Chlorophyll content index (HD)	DT	49.05	1.91	53.12	45.30	3.89	4.61
		IR	51.42	1.77	53.80	48.65	3.45	
	Canopy temperature (HD)	DT	18.82	1.17	21.60	16.87	6.24	-43.23
		IR	13.14	1.25	16.55	11.42	9.53	
	Electrical conductivity (HD)	DT	14.61	1.52	17.21	12.21	10.43	-19.27
		IR	12.25	0.85	13.80	10.91	6.90	
	Normalized vegetation index (HD)	DT	0.57	0.04	0.63	0.51	6.93	12.31
		IR	0.65	0.03	0.69	0.60	3.94	
	Relative water content (AN)	DT	79.39	5.66	85.90	66.12	7.68	7.77
		IR	86.08	7.68	94.06	77.38	7.53	
	Chlorophyll content index (AN)	DT	49.52	3.58	53.55	42.50	7.23	8.30
		IR	54.00	2.85	57.75	48.00	5.30	
	Canopy temperature (AN)	DT	27.75	1.71	31.33	25.58	6.09	-11.00
		IR	25.00	1.90	29.27	22.00	7.50	
	Electrical conductivity (AN)	DT	19.60	2.20	24.47	16.50	11.93	-27.36
		IR	15.39	2.08	18.89	12.44	12.89	
	Normalized vegetation index (AN)	DT	0.34	0.07	044	0.23	21.06	12.82
		IR	0.39	0.08	0.55	0.31	18.95	
	Plant height maturity	DT	88.24	9.52	108.56	69.89	10.79	11.59
		IR	99.81	11.00	121.22	83.56	11.02	

	Traits	Sowing Condition	Mean	SD	MAX	MIN	CV%	%age change due to stress
YIELD TRAITS	Tiller number/ Plant	DT	4.20	1.07	5.88	2.71	25.43	38.36
		IR	6.83	1.16	9.60	5.21	17.01	
	Days to heading	DT	81.38	4.75	89.00	70.33	5.84	15.50
		IR	96.31	2.48	99.67	90.67	2.58	
	Biomass/Plant (gm.)	DT	12.22	2.11	16.14	8.91	17.29	50.1
		IR	24.47	3.59	30.42	15.19	14.68	
	Thousand grain weight (gm.)	DT	37.81	4.87	45.92	27.58	12.89	13.87
		IR	43.90	5.42	55.37	30.85	12.35	
	Grain no/ spike	DT	34.02	4.55	44.26	27.64	13.37	11.87
		IR	38.60	5.79	52.16	29.94	15.01	
	Grain weight/ spike	DT	1.38	0.22	2.02	1.12	16.13	14.55
		IR	1.62	0.34	2.20	1.14	21.35	
BIOCHEMICAL TRAITS	Superoxidase dis mutase (units/mg protein)	DT	227.37	41.30	399.50	120.94	41.19	-72.15
		IR	132.08	35.35	190.27	55.94	26.77	
	Peroxidase activity (Moles/Min/mg)	DT	8.41	2.41	13.01	5.28	28.63	-52.32
		IR	5.52	1.26	7.25	3.35	22.75	
	Catalase activity ((Moles/Min/mg)	DT	0.0001	3.6E-05	0.00028	0.00007	-	-66.18
		IR	0.0002	6.02E05	0.00017	0.00006	-	
	MDA content (M g-1 FW)	DT	51.04	6.20	59.47	42.26	12.10	-19.59
		IR	42.68	8.23	53.76	25.90	19.62	
	Protein content(mg)	DT	0.69	0.33	1.24	0.10	52.85	38.19
		IR	1.12	0.31	1.64	0.60	27.25	
	Proline content (Mg-1fwt)	DT	5.37	1.44	8.07	2.19	26.86	-75.49
		IR	3.06	1.33	5.49	1.25	46.18	
	Total phenolic content (g GAE / g FW)	DT	4956.12	1588.00	8252.27	3104.55	32.04	-19.91
		IR	4133.39	1527.64	7320.45	2604.55	36.96	

Negative numbers indicate value higher than the control

(-0.8%) and the maximum for PBW343 (-27.5%) (Fig. 7.,1, f). Experimental plants displayed higher canopy temperature at anthesis stage as compared to heading stage which is approximately doubled with the progressive stage of growth. In this study, percentage increment in electrical conductivity (EC) was recorded to be minimum in C306 (-5.4%) followed by DL153-2 (-6.2%) at heading stage and NI5439 (-3.4%) followed by DL153-2 (-4.9%) at anthesis stage (Fig.1g-h). The range of EC under stress condition was increased from heading to the anthesis stage i.e. 12-15 to 15-20 respectively. Under stress condition maximum value for normalized vegetation index (NDVI) was recorded for C306 (0.63) at heading stage and DL153-2 (0.44) at anthesis stage while the minimum value for NDVI was observed for DBW17 at both heading (0.51) and anthesis stage (0.23) (Fig. 7.1, i, j). An observable difference in plant height (PH) was recorded in genotypes grown under irrigated and drought condition. The mean plant heights under stressed and irrigated conditions were 88.24cm and 99.81cm, respectively. Under stressed conditions, the shortest genotype was DBW17 (69.89cm) while the tallest was C306 (108.56cm) (Fig. 7.1, k). Table 3 summarizes the results from analysis of variance for all the studied physiological, yield and biochemical traits. According to the analysis of variance highly significant difference was observed for treatments and genotypes in maximum all the physiological traits.

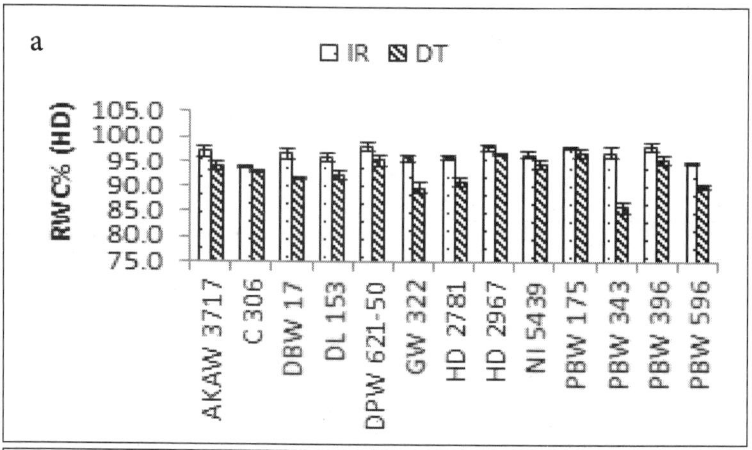

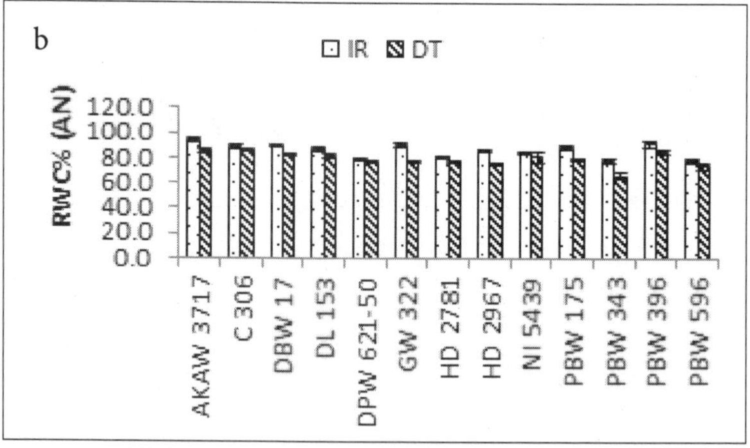

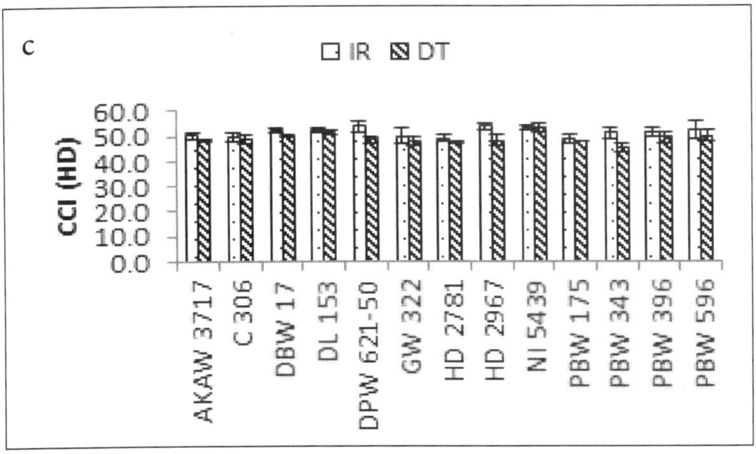
c
IR
DT
CCI (HD)
60.0
50.0
40.0
30.0
20.0
10.0
0.0
AKAW 3717
C 306
DBW 17
DL 153
DPW 621-50
GW 322
HD 2781
HD 2967
NI 5439
PBW 175
PBW 343
PBW 396
PBW 596

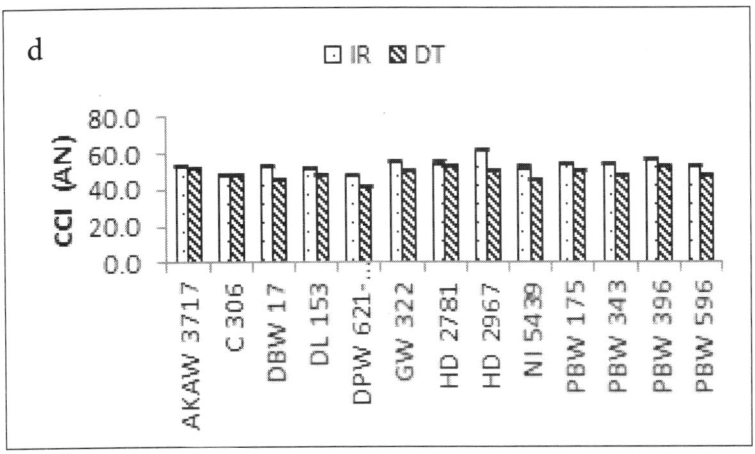
d
IR
DT
CCI (AN)
80.0
60.0
40.0
20.0
0.0
AKAW 3717
C 306
DBW 17
DL 153
DPW 621-...
GW 322
HD 2781
HD 2967
NI 5439
PBW 175
PBW 343
PBW 396
PBW 596

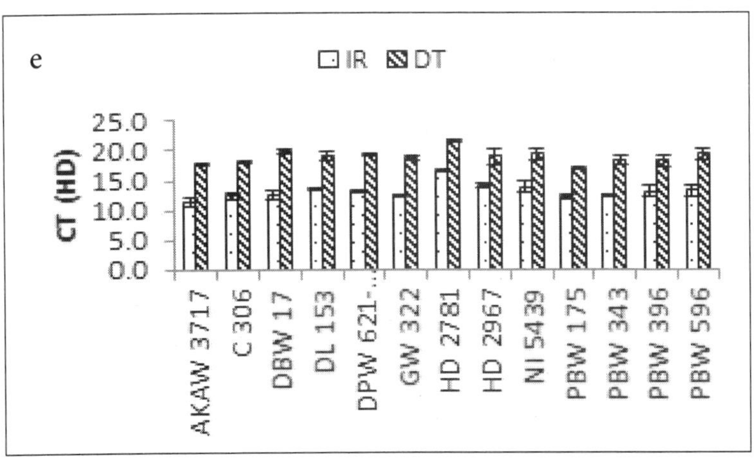
e
IR
DT
CT (HD)
25.0
20.0
15.0
10.0
5.0
0.0
AKAW 3717
C 306
DBW 17
DL 153
DPW 621-...
GW 322
HD 2781
HD 2967
NI 5439
PBW 175
PBW 343
PBW 396
PBW 596

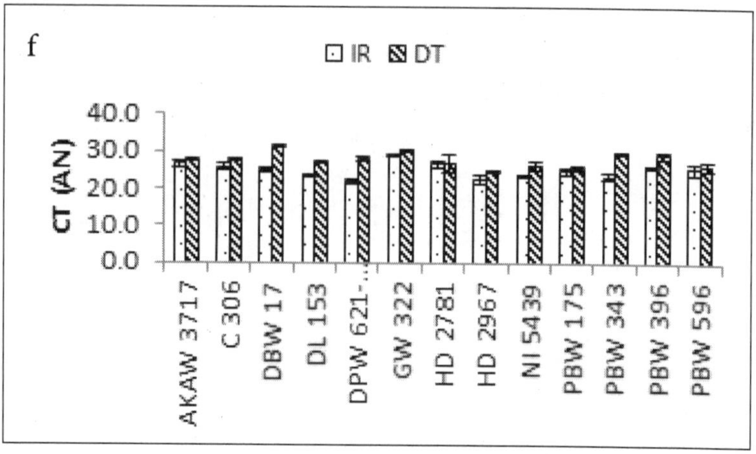
f
IR DT
CT (AN)
40.0
30.0
20.0
10.0
0.0
AKAW 3717
C 306
DBW 17
DL 153
DPW 621-...
GW 322
HD 2781
HD 2967
NI 5439
PBW 175
PBW 343
PBW 396
PBW 596

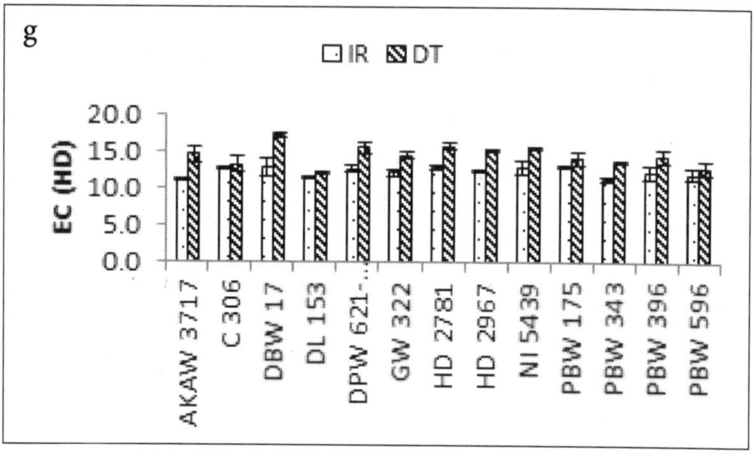
g
IR DT
EC (HD)
20.0
15.0
10.0
5.0
0.0
AKAW 3717
C 306
DBW 17
DL 153
DPW 621-...
GW 322
HD 2781
HD 2967
NI 5439
PBW 175
PBW 343
PBW 396
PBW 596

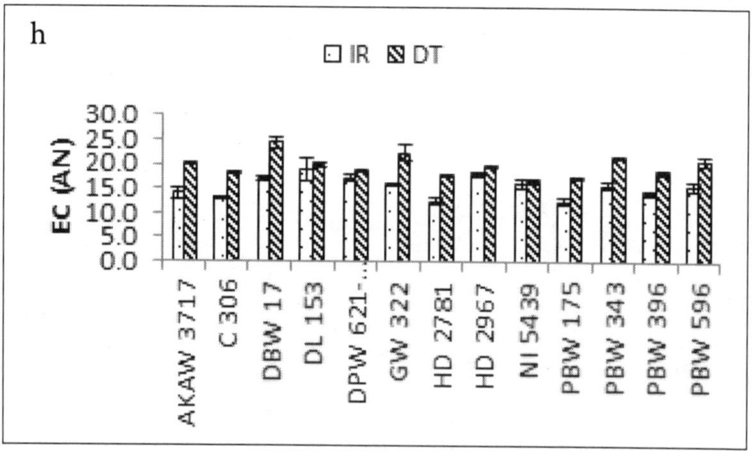
h
IR DT
EC (AN)
30.0
25.0
20.0
15.0
10.0
5.0
0.0
AKAW 3717
C 306
DBW 17
DL 153
DPW 621-...
GW 322
HD 2781
HD 2967
NI 5439
PBW 175
PBW 343
PBW 396
PBW 596

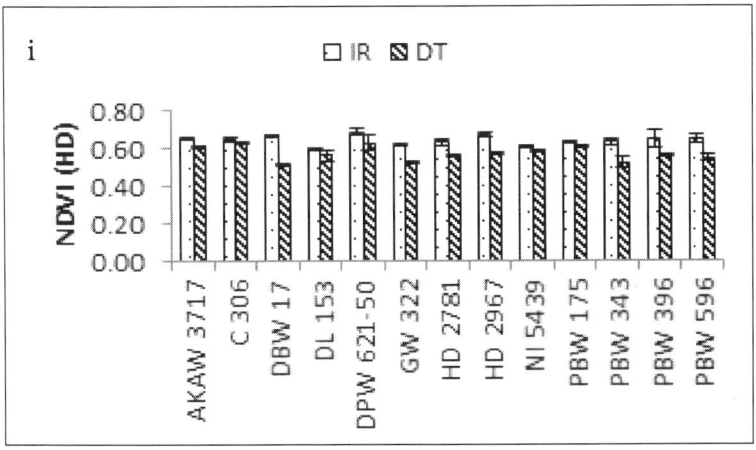
i
IR
DT
NDVI (HD)
0.80
0.60
0.40
0.20
0.00
AKAW 3717
C 306
DBW 17
DL 153
DPW 621-50
GW 322
HD 2781
HD 2967
NI 5439
PBW 175
PBW 343
PBW 396
PBW 596

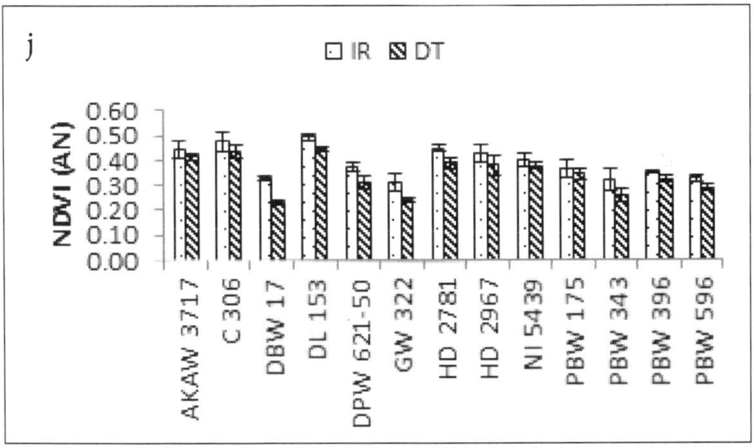
j
IR
DT
NDVI (AN)
0.60
0.50
0.40
0.30
0.20
0.10
0.00
AKAW 3717
C 306
DBW 17
DL 153
DPW 621-50
GW 322
HD 2781
HD 2967
NI 5439
PBW 175
PBW 343
PBW 396
PBW 596

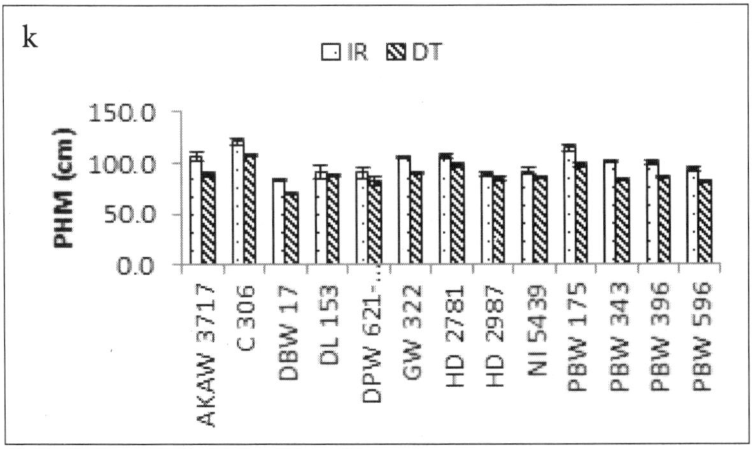
k
IR
DT
PHM (cm)
150.0
100.0
50.0
0.0
AKAW 3717
C 306
DBW 17
DL 153
DPW 621-...
GW 322
HD 2781
HD 2987
NI 5439
PBW 175
PBW 343
PBW 396
PBW 596

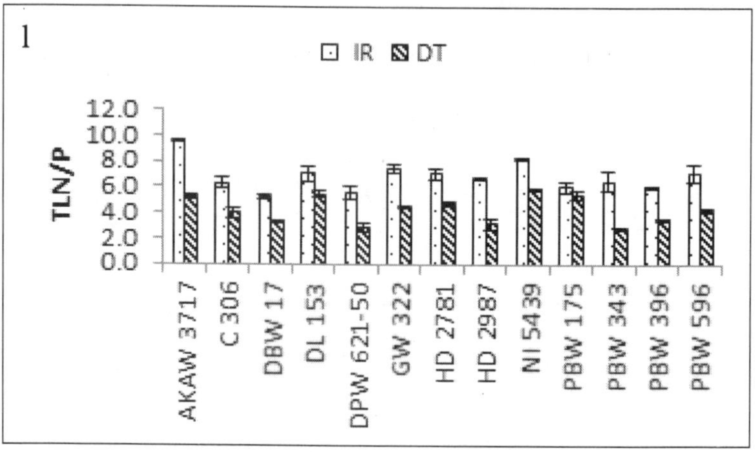
l
IR
DT
TLN/P
12.0
10.0
8.0
6.0
4.0
2.0
0.0
AKAW 3717
C 306
DBW 17
DL 153
DPW 621-50
GW 322
HD 2781
HD 2987
NI 5439
PBW 175
PBW 343
PBW 396
PBW 596

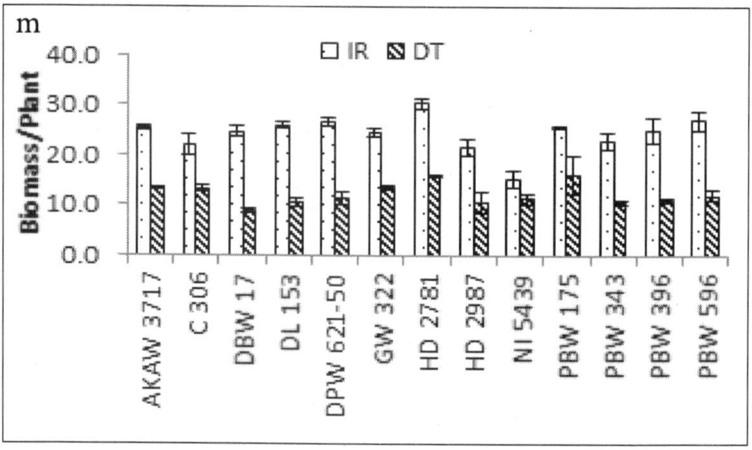
m
IR
DT
Biomass/Plant
40.0
30.0
20.0
10.0
0.0
AKAW 3717
C 306
DBW 17
DL 153
DPW 621-50
GW 322
HD 2781
HD 2987
NI 5439
PBW 175
PBW 343
PBW 396
PBW 596

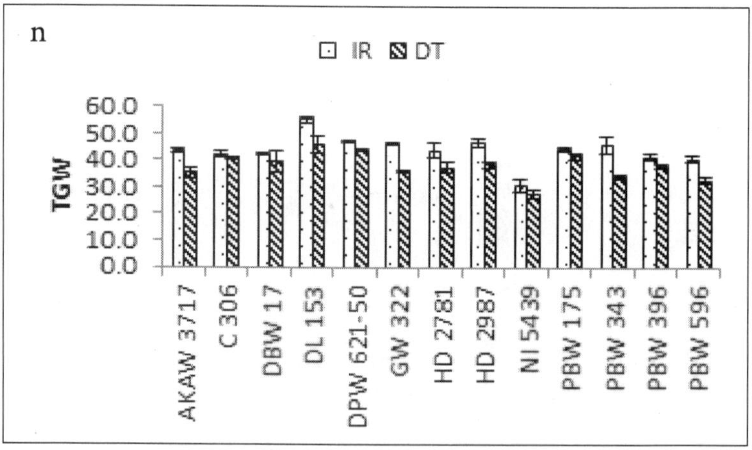
n
IR
DT
TGW
60.0
50.0
40.0
30.0
20.0
10.0
0.0
AKAW 3717
C 306
DBW 17
DL 153
DPW 621-50
GW 322
HD 2781
HD 2987
NI 5439
PBW 175
PBW 343
PBW 396
PBW 596

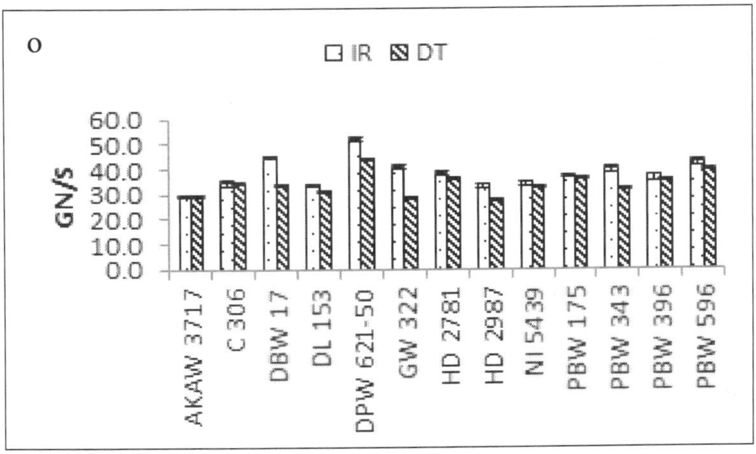

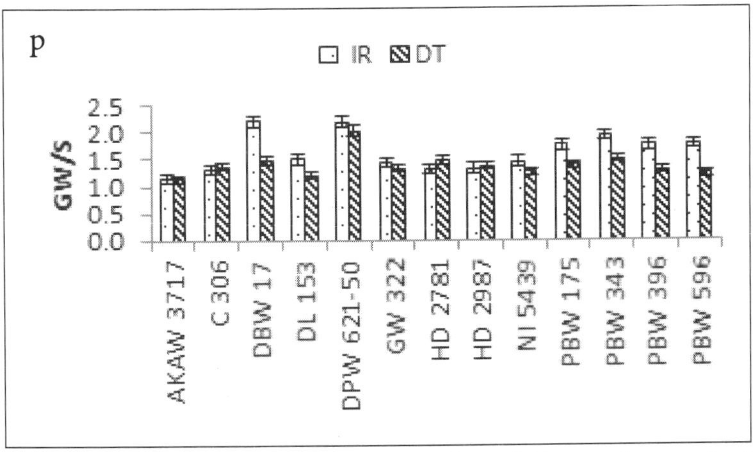

Fig. 7.1: Variation in differént parameters a, b- relative water content (RWC); c, d- chlorophyll content index (CCI); e, f- canopy temperature (CT); g, h- electrical conductivity (EC); i, j- normalized vegetation index (NDVI) at heading (HD) and anthesis (AN) stage respectively; k- plant height at maturity (PH); l- tiller number/ plant (TN/P); m- biomass/ plant (BM/P); n- thousand grain weight (TGW); o- grain number/ spike (GN/S); p- grain weight/ spike.

3.2 Yield and Yield Components

We found that osmotic stress significantly affected the different yield components as tiller number per plant (TN/P), days to heading (DTH), biomass per plant (BM/P), thousand grain weight (TGW), total grain number per spike (GN/S) and total grain weight per spike (GW/S) (Table 7.3). A highly significant effect on genotypes and treatments was observed in case of all these parameters. Mean values of GN/S, GW/S, TGW, BM/P were observed slightly lower under stressed condition as compared to irrigated conditions as shown in Table 2. Under stress condition, a reduction in average tiller number was observed from 6.83 to 4.21. Genotypes NI5439 and PBW175 developed the highest number of tillers, 5.88 and 5.43, under stressed condition respectively (Fig.1l). Average DTH was observed slightly lower under stressed condition (81.38) than irrigated condition (96.31). Under irrigated condition BM/P was observed highest for HD2781 i.e. 30.4g while under stressed condition highest BM/P was observed for PBW175 i.e. 16.1g (Fig. 7.1, m). TGW decreased 14% as a mean under stressed condition as compared to control and the maximum value of TGW obtained in DL153-2 i.e. 46g under stressed condition (Fig. 7.1, n). Reduction in GNS and GWS under stressed condition has also been observed in different genotypes. Under stressed condition maximum reduction in GNS and GWS was obtained in GW322 and DBW17 respectively (Fig. 7.1, o-p). Analysis of variance showed significant variation for genotypes and treatment for almost all the traits studied except GW/S (Table. 7.3).

Table 7.3: Analysis of variance showing mean square values for different traits observed under two water regimes drought (DT) and irrigated (IR) at heading (HD) and anthesis (AN) stage.

Traits	Between genotypes	Between treatments	Interaction
Germination %	31.04**	195.75***	11.01
Relative water content % (HD)	8.02*	90.88***	3.67
Chlorophyll Content Index (HD)	4.98**	36.70***	1.78
Canopy Temperature (HD)	2.7***	209.67***	0.27
Electrical Conductivity (HD)	1.96*	36.01***	0.59
Normalized vegetation index (HD)	0.003	0.03***	0.001
Relative water content %(AN)	53.74**	336.45***	15.47
Chlorophyll Content Index (AN)	16.55**	117.97***	4.39
Canopy Temperature (AN)	4.14*	49.62***	2.39
Electrical Conductivity (AN)	5.83	168.86***	3.80
Normalized vegetation index (AN)	0.01***	0.02***	0.001
Plant height at maturity (cm)(PH)	110.91	724.93**	83.61
Tiller no / plant (TN/P)	2.04**	38.70***	0.53
Days to heading (DTH)	24.78***	1447.69***	3.94
Biomass per plant (BM/P)	12.64**	949.01***	5.42
1000 grain weight (gm)(TGW)	47.31***	241.38***	5.82
Grain number per spike (GN/S)	45.86**	136.07***	8.58
Grain weight per spike (GW/S)	0.27	0.01	0.17
Super oxidase dismutase (SOD)	6335.52	59031.4**	3687.15
Peroxidase (POX)	4.6	54.25***	2.77
Catalase (CAT)	0.004*	0.03***	0.001
Lipid peroxidation (MDA)	72.99*	568.83***	33.15
Protein content	0.11	1.16***	0.05
Proline content	2.61	34.60***	1.24
Total phenolic content (TPC)	4604790***	4399720***	250634

*Significant at the $p \leq 0.05$; **Significant at the $p \leq 0.01$; ***Significant at the $p \leq 0.001$

3.3 Biochemical Parameters

In this study proline content increased in all genotypes significantly. In C306 5.16 μMg^{-1}fwt proline content was found under drought condition as compared to 1.25 μMg^{-1}fwt proline content under irrigated condition with a maximum percentage increment while the minimum increase in proline content was recorded in HD2987 i.e. 14% and DBW17 i.e. 20% (Fig. 7.2, a).

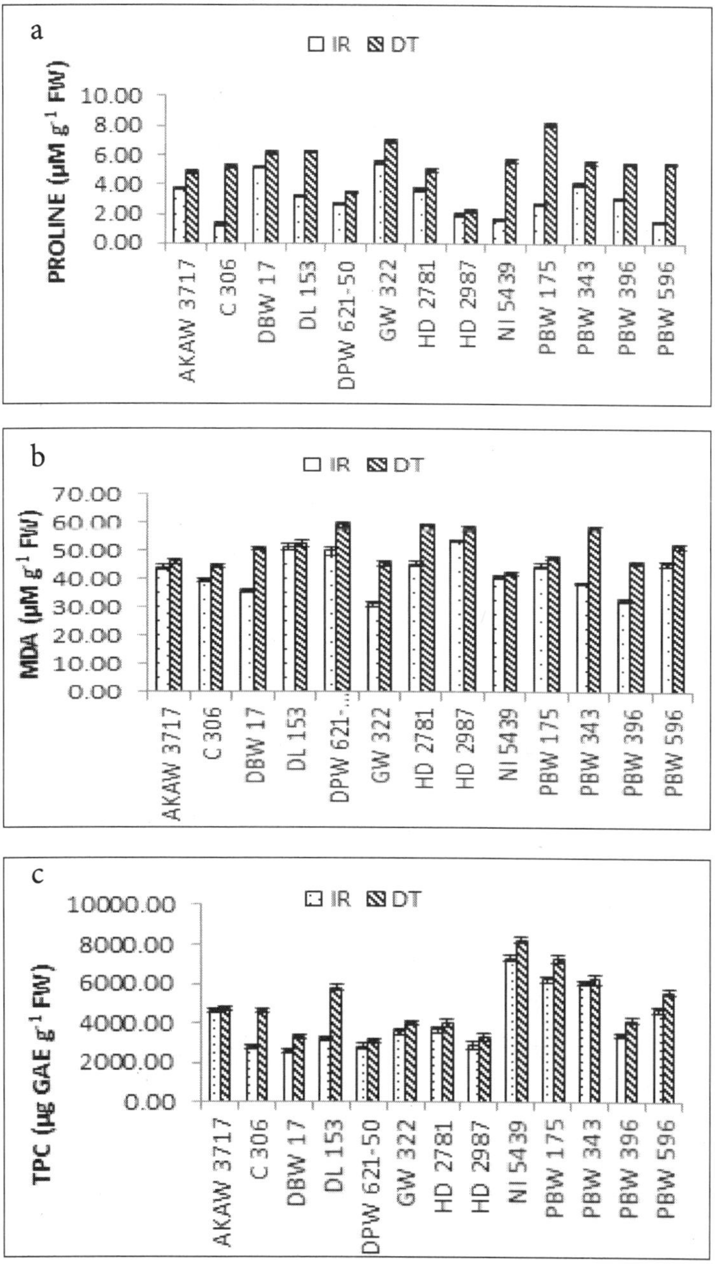

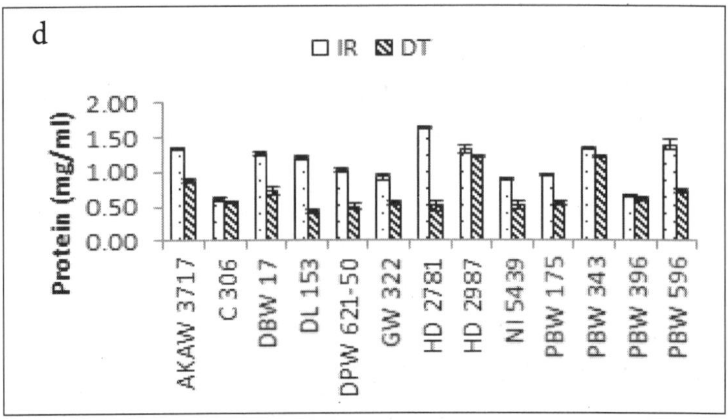

Fig. 7.2: Variation in different biochemical parameters (a) Proline content (PC- μM g^{-1} FW), (b) MDA content (μM g^{-1} FW), (c) Total phenolic content (TPC- μg GAE /g FW) and (d) Protein content (mg/ml) in wheat genotypes grown under irrigated (IR) and drought (DT) condition.

The highest value for proline content was obtained for PBW175 i.e. 8.07 μMgm^{-1}fwt under drought condition. Maximum increase of MDA content was noticed in PBW343 (-50.83%) and DBW17 (-41.69%) while the minimum increase in MDA content was noticed in DL153-2 i.e. 2.12% followed by NI5439 (3.15%) over non-stressed genotypes (Fig. 7.2, b). A wide variation was observed in total phenolic content ranged from 2605 μg GAE/g in DBW17 under irrigated condition to 8252 μg GAE/g in NI5439 under drought condition (Fig 7.2, c). A wide variation was observed under drought and irrigated condition in protein content also in different genotypes (Fig. 7.2, d). Minimum protein content was found in C306 i.e. 0.60 mg/ml under irrigated condition and the maximum protein content was calculated in HD2781 i.e.1.64 mg/ml.

3.4 Antioxidant Enzyme Activity

In the present study, activity of SOD showed a significant increase (P< 0.005) in DL153-2, PBW596, GW322, NI5439, C306 and PBW175 wheat genotypes grown under water stress condition as compared to irrigated one (Fig. 7.3, a). The highest activity of the enzyme was reported in DL153-2 i.e. 399 units/mg protein showing its drought tolerance behavior. Compared to control, increase in CAT activity was noted during water stress in almost all genotypes excepting GW322 and HD2987. The maximum enzyme activity was reported in DL153-2 i.e. 0.00028 μMoles/Min/mg proteins (Fig. 7.3, b). In all genotypes water deficit leads to the up-regulation of POX activity in leaves with a mean 5.52 units/mg protein under irrigated condition and 8.41 units/mg proteins under drought condition. Maximum POX activity was reported in PBW175 i.e. 13 units/mg protein followed by DL153-2 (12.48 units/mg protein) under drought condition (Fig. 7.3, c).

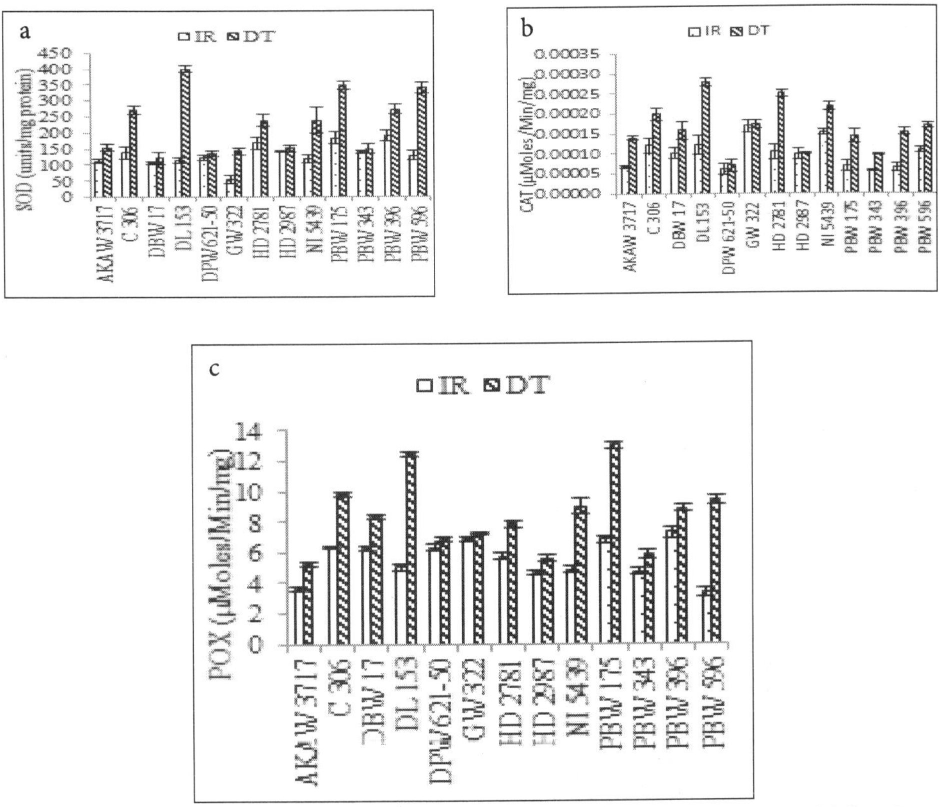

Fig. 7.3: Activity of antioxidant enzymes: (a) Superoxide dismutase (SOD), (b) Catalase (CAT) and (c) Peroxidase (POX) in wheat genotypes grown under irrigated (IR) and drought (DT) condition.

3.5 Means of Correlation Coefficient

Under drought condition a highly significant and positive correlation of SOD antioxidant enzyme activity was found with POX, TPC, NDVI (AN) and PH while a significant and negative correlation was found with CT(AN) under drought. CAT showed a positive and significant correlation with POX, NDVI and PH while a negative and significant correlation with protein content under drought condition. A significant positive correlation between POX and proline content and TPC with proline content has also been found under drought. Under irrigated condition less correlation between antioxidant enzyme activity and other biochemical components was observed (Table 7.4). The results showed the role of biochemical components in gaining drought tolerance under water stress condition.

Table 7.4: Correlation analysis of different biochemical and morpho-physiological traits.

	SOD	CAT	POX	MDA	TPC	PROLINE	PROTEIN	CT (AN)	NDVI (AN)	EC (AN)	R W C (AN)	PH
SOD		-0.66**	0.18	0.21	0.16	-0.42	-0.06	-0.2	0.21	-0.59*	-0.07	0.3
CAT	0.6		-0.05	-0.12	0.02	0.03	-0.07	0.33	0.13	0.24	0.05	-0.03
POX	0.86***	0.55*		-0.47	-0.3	0.3	-0.62*	0.13	-0.15	-0.17	0.24	0.19
MDA	-0.28	-0.29	-0.33		0.03	-0.57*	0.34	-0.3	0.52*	0.01	-0.17	0.1
TPC	0.53*	0.29	0.49	-0.44		-0.2	0.01	-0.28	-0.23	-0.17	-0.32	0.07
PROLINE	0.43	0.38	0.66**	-0.49	0.51*		0.24	0.33	-0.29	0.23	0.16	-0.2.
PROTEIN	-0.46	-0.59*	-0.64**	0.35	-0.14	-0.48		0	0.22	0.09	-0.32	-0.26
CT (AN)	-0.66**	-0.3	-0.45	0.12	-0.57*	-0.02	0.3		0.03	-0.45	0.59*	0.46
NDVI (AN)	0.50*	0.53*	0.32	-0.26	0.25	-0.12	-0.32	-0.61*		-0.26	0.13	0.31
EC (AN)	-0.22	-0.25	-0.12	-0.01	-0.4	0.23	0.27	0.73**	-0.64*		-0.25	-0.8***
RWC (AN)	-0.03	0.3	0.1	-0.62*	-0.23	0.14	-0.48	0.13	0.27	0.13		0.3
PH	0.73**	0.54*	0.46	0.08	0.17	-0.09	-0.39	-0.66**	0.69**	-0.52*	-0.1	

SOD- Superoxide dismutase, CAT- Catalase, POX- Peroxidase, MDA- Malondialdehyde content, TPC- Total phenolic content, proline, protein, CT (AN) Canopy temperature at anthesis, NDVI (AN) Normalized vegetation index at anthesis, EC (AN) Electrical conductivity at anthesis, RWC (AN)- Relative water content at anthesis, PH- Plant height.

*Significant at the $p \leq 0.05$; **Significant at the $p \leq 0.01$; ***Significant at the $p \leq 0.001$

Note: Shaded area is under drought

3.6 Principle Component Analysis (PCA) for Biochemical Traits

At anthesis stage principal component analysis for biochemical traits and some physiological traits under irrigated and drought condition accounted maximum varaition (78.4%) upto four principal components shown by scree plot (Fig. 7.4, a). First principle component (PC1) accounted 37.9% maximum variation for electrical conductivity at anthesis (EC-AN), proline, catalase (CAT) and peroxidase (POX). Second principal component (PC2) showed maximum effect for superoxidase dis mutase (SOD) and plant height (PH) and normalized vegetation index (NDVI-AN) and total phenolic content (TPC). PC3 accounted maximum varaition for relative water content (RWC-AN) and chlorophyll content index at antheis (CCI-AN). PC4 accounted maximum varation canopy temperature (CT-AN) (Table 7.5). Biochemical components were superimposed in the plot showing their positive correaltion with each other and their significant effect on the diversity of genotypes. Loading plot showed strong negative association of CCI, NDVI, RWC and PH with antioxidant enzyme activity, MDA content and EC (Fig. 7.4, b). Principle component biplot analysis showed smaller angles between dimension vectors in same direction indicated higher correlation between the traits.

Table 7.5: Principle component analysis of biochemical and physiological traits on the basis of pooled data under irrigated and drought conditions.

	PC1	PC2	PC3	PC4
SOD	0.291	0.457	-0.079	0.027
CAT	0.285	0.303	0.164	0.219
POX	0.345	0.348	0.143	0.112
MDA	0.219	-0.106	-0.588	0.237
TPC	0.154	0.329	-0.038	-0.639
PROLINE	0.373	0.077	0.372	-0.043
CT (AN)	0.248	-0.251	0.299	0.361
CCI (AN)	-0.276	0.066	0.356	-0.255
NDVI (AN)	-0.240	0.384	-0.165	0.329
EC (AN)	0.397	-0.258	0.039	0.076
RWC (AN)	-0.298	0.058	0.452	0.313
PH	-0.253	0.409	-0.104	0.253

SOD- Superoxidee dismutase, CAT- Catalase, POX- Peroxidase, MDA- Malondialdehyde content, TPC- Total phenolic content, proline, CT (AN) Canopy temperature at anthesis, CCI (AN) Chlorophyll content index, NDVI (AN) Normalized vegetation index at anthesis, EC (AN) Electrical conductivity at anthesis, RWC (AN)- Relative water content at anthesis, PH- Plant height

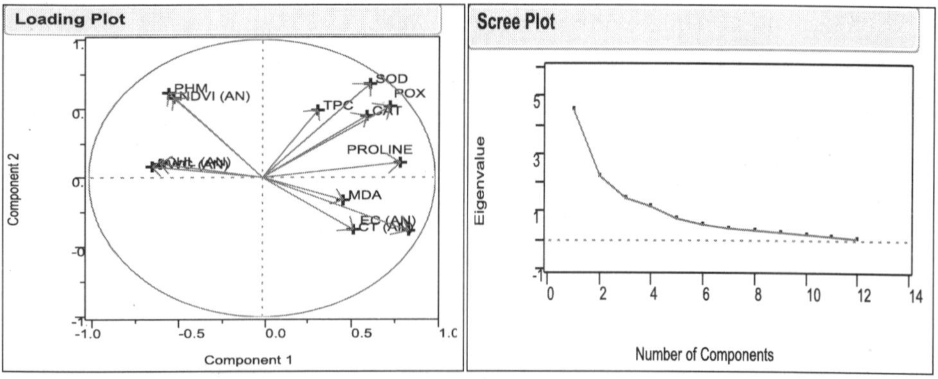

Fig. 7.4: Principal component analysis (a) scree plot, (b) loading plot of biochemical and physiological traits at anthesis, under irrigated and drought condition.

4. Discussion

Among different abiotic stresses, water stress is the most devastating abiotic stress affecting plant growth and development. Water stress imposes major limits on wheat yield which is further affected by different factors as plant species, plant age, phases of growth and development, level and duration of drought and physical parameters. Development of drought tolerant wheat genotypes is the main objective of wheat breeders due to impending climate change. Effective germ-plasm screening for drought tolerance under field drought condition is very effective in selecting material for advanced breeding. Therefore, we analyzed 13 Indian wheat genotypes for their physiological, biochemical and yield effect under drought and irrigated conditions applied in field. The result of analysis of variance for different traits showed the presence of significant genotypic variations under both water conditions.

Under stress condition an external osmotic potential limits the absorption of water by the seeds and the roots and resulted in loss of germination %age, consistent with the earlier reports (Sayar et *al.*, 2010; Guo *et al.*, 2013). RWC depicts the present water status of plant. Under water stress condition gradual reduction in RWC was observed in all the genotypes at heading and anthesis stages. Less reduction in RWC in some genotypes showing that some genotypes are drought resistant as compared to others (Arjenaki *et al.*, 2012). This variation may be due to the difference in their efficiency to absorb water from the soil and less stomatal conductance behavior (Keyvan, 2010). Maintenance of relative water content is vital for the development of drought resistance in plants (Khanna-Chopra and Selote, 2007). Chlorophyll content is a promising trait to depict photosynthetic ability of the plant which is reduced under stress environment (Nezhadahmadi *et al.*, 2013). Lower chlorophyll content in drought sensitive genotypes in comparison to drought tolerant has been reported in a large number of studies earlier (Arjenaki *et al.*, 2012). Under drought significant damage of Photosynthetic apparatus in drought sensitive genotypes as compared to tolerant one has been reported in barley (Fischer *et al.*, 1998). Canopy temperature reflects the interaction among plants, soil and atmosphere and is also related to the water stress and provides a good indication of plant water potential

(Jackson *et al.*, 1981). Lower canopy temperature is associated with high stomatal conductance and photosynthetic rate (Fischer *et al.*, 1998). Under drought condition, low canopy temperature wheat could maintain higher superiority with high canopy temperature wheat and can be used as a useful tool for wheat breeding for drought tolerance (Feng *et al.*, 2004). Cell membrane damage due to water stress is a basic physiological change has been studied in wheat (Sairam *et al.*, 1998). Under stress high electrolyte leakage can be related to the higher content of H_2O_2 and MDA causing membrane damage in wheat (Sanchez-Rodriguez *et. al.*, 2010). Drought stress causes reduction in plant growth due to cell elongation inhibition (Blum 1998).

Drought imposed significant reduction in different yield traits as GN/S in all genotypes except C306, AKAW3717 and PBW175. The stem water soluble carbohydrates and the current photosynthesis are very important for grain development in wheat. This material is translocated from leaves, culm and head to the grain at the grain filling stage (Blum, 1998). Damaged root system and impaired water and nutrient uptake results reduced no of tillers per plant and grain weight (Saqib *et al.*, 2013) which finally results in lower grain yield in drought susceptible genotypes. The difference in the grain yield among genotypes and between treatments (ANOVA) was found to be highly significant. Water stress during anthesis causes low yield can be contribute by pollen sterility or abortion of embryos or premature end of grain filling in plants (Passioura, 2006). Drought tolerant genotypes which can maintain a high plant water status even in water stress contribute to a high yield potential phenotype (Blum, 2005).

Plants possess efficient systems for scavenging ROS that protect them from destructive oxidative damages (Feng *et al.*, 2004). In this study, SOD, POX, CAT, MDA content, total phenols and proline content are taken as important indices for evaluating the status of wheat redox. During osmotic stress plant cell accumulate organic solutes such as proline which act as a mediator of osmotic adjustment and eliminator for free radicals (Vendruscolo *et al.*, 2007). These results are in accordance with the result obtained under heat stress in wheat seedlings (Wang *et al.*, 2014). Higher proline synthesis in stressed plants at field condition could favour a better recovery of these plants (Vendruscolo *et al.*, 2007). In wheat seedlings increase in proline content with increased PEG concentration under laboratory has also been reported (Guo *et al.*, 2013). Content of the MDA has been considered as an indicator of the oxidative damage and a suitable marker of membrane lipid peroxidation (Khan and Naqvi, 2013). Higher MDA content in drought susceptible wheat varieties than tolerant varieties has been observed under lab drought condition (Chakraborty and Pradhan, 2012). The dramatic increase in MDA content was related to reductions in SOD activity (Shahbazi *et al.*, 2012). In present study PBW343 showed highest accumulation of MDA content i.e. 60μMg^{-1}FW, expressing the worst membrane completeness at anthesis stage of growth. Osmolyte accumulation is reported to be a key mechanism for increasing crop yield under drought stress contributed by maintenance of root development to reach water that may available deep in soil (Serraj and Sinclair, 2002). Another metabolic process that has been related to the responses of the water stress in plants involves total phenols. Varieties with higher TPC including phenolic acids, flavonoids and proanthocyanidins, contributed to higher antioxidant activity and plays important role in scavenging free radicals under stress condition (Narwal *et al.*, 2014; Radi *et al.*, 2013). Higher increase in total phenols has been confirmed in plants acclimated to drought stress and higher reduction is essential to eliminate ROS in cells (Sanchez-Rodriguez et al 2010). Present study showed the clear differences in the participation of antioxidant enzyme components in drought tolerance mechanism. Within a cell, the SODs constitute the first line of defense against ROS. The enhancement in the SOD activity in drought-stressed leaves was also in accordance with previous findings (Chakraborty and Pradhan, 2012). Lower MDA content and higher POX activity has been observed in potted wheat genotypes (Khan and Naqvi, 2013). Higher POX activity under drought condition is associated with their drought tolerant behavior (Devi *et al.*, 2011).There are many reports on the increase, decrease or no effect of water stress on CAT activity (Feng *et al.*, 2004 and Sanchez-Rodriguez *et al.*, 2010). Increased catalase activity under drought stress can be due to higher production of H_2O_2 (Shahbazi *et al.*, 2012) and decrease in activity can be due to light inactivation of this enzyme and prevention of its synthesis in the dark (Zhang *et al.*, 1994). The genotypes PBW175, C306, HD2781 are significantly higher yielder under both conditions and also perform better in other drought related physiological and biochemical traits studied in this experiment. Correlation among different traits showed the moderate to highly significant and positive effect of antioxidant enzyme activity, proline and TPC on TN/S depicts their direct contribution of these components to yield. Negative and significant correlation of CT and NDVI and MDA content and RWC also found to be considered as important target traits during selection.

Principle component analysis indicated that yield traits- DTH, BM/P and TN/P first have more contribution in selecting genotypes followed by biochemical traits- SOD, CAT, POX enzyme activity and TPC and after that physiological traits- CT and RWC of anthesis stage. This indicate that the genotypes with higher RWC, lower CT showed more effect on the accumulation of higher biomasss and tiller number. It also indicates the role of antioxidant enzyme activity and TPC on higher yield production.

Our results indicated that the water stress has a different effect on different genotypes. Some genotypes showing higher RWC, CCI, TN/P and BM/P depict drought tolerance behaviour as compared to other genotypes. Physiological traits observed at anthesis stage also showing importance as compared to heading stage. Our results also indicated higher antioxidant enzyme activity in some genotypes as compared to others. Different response of different genotypes showing their genotypes difference in stress tolerance. In this study, proline content is also showing its major contribution in attaining drought tolerance. This system can be used in understanding the mechanism of drought tolerance and improving the genotypes.

Acknowledgement

This work is a part of PhD and authors greatly appreciate the facilities provided by the Indian Institute of Wheat and Barley Research (IIWBR), Karnal, for conducting the experiment.

References

Arjenaki, F.G., Jabbari, R. and Morshedi, A. 2012. Evaluation of drought stress on relative water content, chlorophyll content and mineral elements of wheat (*Triticum aestivum* L.) varieties. *Int. J. Agric.Crop Sci.*, 4(11): 726-729.

Barrs, H.D. and Weatherley, P.E. 1962. A re-examination of relative turgidity technique for estimating water deficit in leaves. *Aust. J. Biol. Sci.*, 15: 413-428.

Bates, L.S., Waldeen, R.P. and Teare, I.D. 1973. Rapid determination of free proline for water stress. *Plant Soil*, 39: 205-207.

Bayoumi, T.Y., Eid, M.H. and Metwali, E.M. 2008. Application of physiological and biochemical indices as a screening technique for drought tolerance in wheat genotypes. *Afr. J. Biotechnol.*, 7(14): 2341-2352.

Blum, A. 2005. Drought resistance, water-use efficiency, and yield potential- are they compatible, dissonant, or mutually exclusive? *Aust. J. Agric. res.*, 56: 1159-1168.

Blum. 1998. Improving wheat grain filling under stress by stem reserve mobilization. *Euphytica*, 100: 77-83.

Bradford, M.M. 1976. A rapid and sensitive method for the quantification of microgram quantities of protein utilizing the principle of protein-dye binding. *Anal. Biochem.*, 72: 248-254.

Chakraborty, U. and Pradhan, B. 2012. Oxidative stress in five wheat varieties (*Triticum aestivum* L.) exposed to water stress and study of their antioxidant enzyme defense system, water stress responsive metabolites and H_2O_2 accumulation. *Braz. J. Plant Physiol.*, 24(2): 117-130.

Devi, R., Kaur, N. and Gupta, A.K. 2011. Potential of antioxidant enzymes in depicting drought tolerance of wheat (*Triticum aestivum* L.). *Ind. J. Biochem. Biophysics*, 49(4): 257-265.

Feng, Z., Jin-Kui, G., Ying-Li, Y., Wen-Liang, H. and Li-Xin, Z. 2004. Changes in the pattern of antioxidant enzymes in wheat exposed to water deficit and rewatering. *Acta Physiol. Plant.*, 26(3): 345-352.

Ferris, R., Ellis, R. H., Wheeler, T. R. and Hadley, P. 1998. Effect of high temperature stress at anthesis on grain yield and biomass of field-grown crops of wheat. *Ann. Bot.* 82: 631-639.

Fischer, R.A., Rees, D., Sayre, K.D., Lu, Z.M., Condon, A.G. and Saavedra, A.L. 1998. Wheat yield progress associated with higher stamatal conductance and photosynthetic rate, and cooler canopies. *Crop Sci.*, 38: 1467-1475.

Guo, R., Hao, W.P., Gong, D.Z., Zhong, X.L. and Gu, F.X. 2013. Effects of water stress on germination and growth of wheat, photosynthetic efficiency and accumulation of metabolites. Intech: Soil Processes and Current Trends in Quality Assessment, China, pp. 367-380.

Heath, R.L. and Packer, L. 1968. Photoperoxidation in isolated chloroplasts: I. Kinetics and stoichiometry of fatty acid peroxidation. *Arch. Biochem. Biophys.*, 125: 180-198.

Jackson, R.D., Idso, S.B., Reginato, R.J. and Pinter, P.J. 1981. Canopy temperature as a crop water stress indicator. *Water Resour. Res.*, 17: 1133-1138.

Jebara, S., Jebara, M., Limam, F. and Aouani, M.E. 2005. Changes in ascorbate peroxidase, catalase, guaicol peroxidase and super-oxide dismiutase activities in common bean nodules under salt stress. *J. Plant Physiol.*, 162: 929-936.

Kadam, S., Singh, K., Shukla, S., Goel, S., Vikram, P., Pawar, V., Gaikwad, K., Chopra, R.K. and Singh. N.K. 2012. Genomic associations for drought tolerance on the short arm of wheat chromosome 4B. *Funct. Integr. Genomics,* 12: 447-464.

Keyvan. 2010. The effect of drought stress on yield , relative water content, proline, soluble carbohydrate and chlorophyll of bread wheat cultivars. *J. Anim. Plant Sci.*, 3: 1051-1060.

Khan, N. and Naqvi, F.N. 2013. Agro-biochemical traits of wheat genotypes under irrigated and non-irrigated conditions. *Cereal Res. Commun.*, 41(12): 243-254.

Khanna-Chopra, R. and Selote, D.S. 2007. Acclimation to drought stress generates oxidative drought-resistant than susceptible wheat cultivar under field conditions. *Environ. Exper. Bot.*, 60: 276-283.

Lascano, H.R., Antonicelli, G.E., Luna, C.M., Melchiorre, M.N., Gomez, L.D., Racca, R.W., Trippi, V.S. and Casano, L.M. 2001. Antioxidant system response of different wheat cultivars under drought: field and in vitro studies. *Aust. J. Plant Physiol.*, 28: 1095-1102.

Lorens, G.F., Bennett, J.M., Loggale, L.B. 1987. Differences in drought resistance between two corn hybrids. II. Component analysis and growth rates. *Agron. J.*, 79: 808-813.

Marcinska, I., Czyczyło-Mysza, I., Skrzypek, E., Filek, M., Grzesiak, S., Grzesiak, M.T., Janowiak, F., Hura, T., Dziurka, M., Dziurka, K., Nowakowska, A. and Quarrie, S.A. 2013. Impact of osmotic stress on physiological and biochemical characteristics in drought-susceptible and drought-resistant wheat genotypes. *Acta Physiol. Plant.*, 35: 451-461.

Mccord, J.M. and Fridovich, I. 1969. Superoxide Dismutase: An enzymic function for erythrocuprein (Hemocuprein). *J. Boil. Chem.*, 244: 6049-6055.

Monneveux, P., Jing, R. and Misra, S.C. 2012. Phenotyping for drought adaptation in wheat using physiological traits. *Front. Physiol.*, 3: 1-12.

Narwal, .S, Thakur, V., Sheoran, S., Dahiya, S., Jaswal, S. and Gupta, R.K. 2014. Antioxidant activity and phenolic content of the Indian wheat varieties. *J. Plant Biochem. Biotechnol.*, 23(1): 11-17.

Nezhadahmadi, A., Prodhan, Z.H., Faruq, G. 2013. Drought tolerance in Wheat (Review article). *Sci. World J.*, 2013: 1-12.

Pask, A., Pietragalla, J. and Mullan, D. 2012. Physiological breeding II: A field guide to wheat phenotyping. Cimmyt, internation maize and wheat improvement center, Mexico, DF (Mexico).

Passioura.2006. Increasing crop productivity when water is scarce – from breeding to field management. *Agric. Water Manag.*, 80(1–3): 176-196.

Radi, A.A. , Farghaly, F.A. and Hamada, A.M. 2013. Physiological and biochemical responses of salt-tolerant and salt-sensitive wheat and bean cultivars to salinity. *J. Biol. Earth Sci.*, 3(1): B72-B88.

Rao, M.V., Paliyath, G. and Ormrod, D.P. 1996. UV-B and Ozone induced biochemical changes in antioxidant enzymes of *Arabidopsis thaliana*. *Plant Physiol.*, 110: 125-136.

Sairam, R.K., Deshmukh, P.S. and Saxena, D.C. 1998. Role of antioxidant system in wheat genotypes tolerance to water stress. *Biol. Plant.*, 41(3): 387-394.

Sanchez-Rodriguez, E., Rubio-Wilhelmi, M., Cervilla, L.M., Blasco, B., Rios, J.J., Rosales, M.A., Romero, L. and Ruiz, J.M. 2010. Genotypic differences in some physiological parameters symptomatic for oxidative stress under moderate drought in tomato plants. *Plant Sci.*, 178: 30-40.

Saqib, M., Akhtar, J., Abbas, G. and Nasim, M. 2013. Salinity and drought interaction in wheat (*Triticum aestivum* L.) is affected by the genotype and plant growth stage. *Acta Physiol. Plant.*, 35: 2761-2768.

Sayar, R., Bchini, H., Mosbahi, M. and Ezzine, M. 2010. Effects of salt and drought stresses on germination, emergence and seedling growth of Durum wheat (*Triticum durum* Desf.). *J. Agric. Res.*, 5(15): 2008-2016.

Serraj, R. and Sinclair, T.R. 2002. Osmolyte accumulation: can it really help increase crop yield under drought conditions? *Plant, Cell Environ.*, 25: 333-341.

Shahbazi, H., Aalii, E. and Parchehbaf, R.S.A. 2012. Physiological traits related to yield of wheat under drought stress in early, mid and late stages of grain filling. *Ann. Biol. Res.*, 3(6): 2947-2952.

Singleton, V.L., Orthofer, R. and Lamuela-Raventos, R.M. 1999. Analysis of total phenols and other oxidation substrates and antioxidants by means of Folin-Ciocalteu reagent. *Methods Enzymol.*, 299: 152-178.

Vendruscolo, E.C.G., Schuster, I., Pileggi, M., Scapim, C.A., Molinan, H.B.C., Marur, C.J. and Vieira, L.G.E. 2007. Stress induced synthesis of proline confers tolerance to water deficit in transgenic wheat. *J. Plant Physiol.*, 164: 1367-1376.

Wang, C., Wen, D., Sun, A., Han, X., Zhang, J., Wang, Z. 2014. Differential activity and expression of antioxidant enzymes and alteration in osmolyte accumulation under high temperature stress in wheat seedlings. *J. Cereal Sci.*, 60: 653-659.

Winter, S.R., Musick, J.T. and Porter, K.B. 1988. Evaluation of screening techniques for breeding drought resistant winter wheat. *Crop Sci.*, 28: 512-516.

Zhang, J. and Kirkham, M.B. 1994. Drought-stress induced changes in activities of superoxide dismutase, catalase and peroxidases in wheat leaves. *Plant Cell Physiol.*, 35: 785-791.

Chapter 8

Efficacy of Different Concentrations of Refinery Waste along with *Acaulospora laevis* on Growth Response and Mycorrhization of Maize, Barley and Wheat

Neetu[1,*], Esha Jangra[2], Anju Tanwar[3] and Ashok Aggarwal[2]

[1]*Department of Botany, Govt. PG College for Women, Panchkula, Haryana, India*

[2]*Department of Botany, Kurukshetra University, Kurukshetra, Haryana, India*

[3]*Department of Botany, Govt. P. G. College, Ambala Cantt., Haryana*

**Corresponding Author E-mail: aggarwal_vibha@rediffmail.com; dr.neetu28@gmail.com*

Abstract

A pot experiment was conducted to see the impact of arbuscular mycorrhizal (AM) fungi, i.e., Acaulospora laevis on maize (Zea mays L.), barley (Hordeum vulgare L.) and wheat (Triticum aestivum L.) with different concentrations of refinery waste i.e. without substrate, 25 g/pot, 50 g/pot, 75 g/pot, 100 g/pot and 200 g/pot with five replicates resulted in an effective plant height, shoot and root biomass, root length, per cent root colonization and AM spore number. Results pertaining to inoculum production of Acaulospora laevis with refinery waste as a substrate depicted that 25 g concentration of waste increased all growth parameters and mycorrhization of selected host plants. It was found that increasing in concentration of refinery waste decreases the plant growth and mycorrhization. Higher concentration (200g) of refinery waste was found to be inhibitory for plant growth and mycorrhization. All the inoculated host plants produced AM fungal spores and colonized roots, extramatrical and intraradical hyphae, arbuscules and vesicles. Formation of arbuscules facilitated the nutrient exchange between host plants and fungal symbionts.

Keywords: *AM fungi, refinery waste, mycorrhization, growth response.*

1. Introduction

Nowadays, the extensive use of conventional fertilizers has significantly increased the crop biomass and productivity. However, use of chemical fertilizers has deleterious impact on the environment due to high level of run-off resulting in contamination of groundwater and streams. An attractive, economically suitable alternative is the use of Arbuscular Mycorrhizal Fungi (AMF), a symbiotic association which have been known to improve plant nutrient uptake (Fattah, 2013), growth (Rangel *et al.*, 2014), productivity and yield (Habte, 2006; Chen *et al.*, 2007). AM fungi play a pivotal role in agriculture and land reclamation also (Klironomos and Hart, 2002). Hence, it would be more feasible to manipulate the AM fungi for increased agricultural productivity and plant health assurance; so that the end users may be satisfied by the producers and both will be benefited. But, the mass production of inoculum is one of the hindrances in the large scale application of AM fungi as these are obligate symbionts (Mangla *et al.*, 2012) and production of AM fungal inocula of the best quality is still challenging (Kuppusamy and Kumutha, 2012). Various methods have been proposed for inoculum production of AM fungi such as aeroponics, hydroponics, nutrient film technique, *in vitro* root organ culture and axenic culture (Hung and Sylvia, 1988; Dugassa *et al.*, 1995; Mosse and Thompson, 1984). Among all these techniques, traditional pot-culture method is most commonly used for inoculum production of AM fungi where fungi are multiplied in conjunction with suitable host plant roots.

Type of substrate for mass multiplication of AM fungi is an important factor for the propagation of AM fungal propagules. Many organic wastes which are rich in nutrients and their positive effect on AM root colonization have been proposed by many workers (Gaur and Adholeya, 2002; Perner *et al.*, 2006; Tanwar *et al.*, 2013a). Azcón- Aguilar and Barea (1997) suggested that the growth substrate that favours the formation and functioning of AM fungi should be selected. A well aerated substrate low in nutrients is generally said to be the best for growth and proliferation of AM fungi (Gaur and Adholeya, 2000). Selection of an appropriate host plant for the mass production of AM inoculum is also needful as host plant plays an important role in AM fungal propagation (Tanwar *et al.*, 2013b). Plant tolerance to a wide range of temperature, low levels of nutrients such as phosphorus and inherent resistant to diseases and insects are important determinants for mass multiplication of AM fungi (Millner and Kitt, 1992).

Gianinazzi and Vosatka (2004) reported that production of AM inoculum is a complex process which involves not only the development of the necessary biotechnological expertise, but also the ability to respond to the especially related legal, ethical, educational and commercial requirements. The inoculum obtained after the process should be first screened in pots and later under field condition for various crops as it is very important that inoculum should be able to colonize plant roots effectively, multiply in the substrate and would be able to enhance the plant growth response.

In general, culturing of AM fungi under pot conditions is considered the most convenient method for large scale production with changes in the nutrient contents of soil by addition of substrates affecting the AM fungal propagules (Douds and Schenck, 1990a). AM fungi could be able to colonize the host plants by many sources including spores, colonized or infected roots and extraradical mycelia (Smith and Read, 1997). AM spore number formation depends on the various factors such as seasonality (Bever *et al.*, 2001), physiological changes in host plant (Sylvia and Schenck, 1983), fungal species (Bever, 2002) and nutrient levels (Douds and Schenck, 1990b).

The most important objective of any attempt at commercial production of AM fungal inoculum is the production of pure, pathogen free AM spores on large scale, having maximum capability of colonizing host. Inocula containing one or more AM fungi are available commercially and are produced in a variety of ways (Ijdo *et al.*, 2011). In present investigation, maize (*Zea mays* L.), barley (*Hordeum vulgare* L.) and wheat (*Triticum aestivum* L.) were used as host plants for multiplication of *Acaulospora laevis* in the presence of different concentrations of substrate i.e. refinery waste with an aim to see a) the possible role of substrate in enhancing AM fungal sporulation and root colonization, b) whether addition of substrate affects the contribution of AM fungi to host plant growth improvement and, c) to find out the best host-substrate combination that would maximize AM fungal spore number and root colonization, and thereby plant growth.

2. Materials and Methods

This experiment was undertaken to standardize the spore density for production of AM inoculum using refinery waste as substrate and maize, barley and wheat as host plants.

2.1 Experimental Design

This experiment was conducted in polyhouse at Botany Department, Kurukshetra University, Kurukshetra. In polyhouse, light was provided by cool white fluorescent lamps (8000 lux) under a 16-h photoperiod. The polyhouse also received natural sunlight having controlled temperature (25° ± 5°C) and humidity (50-70%). The experiment was a 3 6 factorial for refinery waste in a completely randomized design employing three hosts (maize, barley and wheat) and six different concentrations (without substrate, 25, 50, 75, 100 and 200 g pot^{-1}) of substrate used. Each treatment has five replicates.

2.2 Physico-chemical Characteristics of Experimental Soil

The soil used in the present investigation have the pH - 6.8, sand - 64.2%, silt - 21.81%, clay - 3.90%, phosphorus content - 7.30 Kg/acre, total nitrogen content - 0.042%, organic carbon - 0.06%, potassium - 88Kg/acre, sulphur - 14.80 ppm and electrical conductivity (EC) of 0.25 dS/m.

2.3 Preparation of Starter Inoculum

Acaulospora laevis was found to be the dominant AM fungal strains. These spores were mass multiplied with maize for a period of three months using the funnel technique (Menge and Timmer, 1982). For this, the selected isolated spores were first surface sterilized with 2% (w/v) Chloramine-T for about 15 minutes followed by washing with sterilized deionized water. Viability of these spores was checked by using Thionin stain. Ten healthy and viable spores were then used for the starter inoculum by using maize as a host plant. Sterilized Borosil funnel of 250 ml capacity was filled with sterilized soil: sand (3:1) mixture, inoculated with the selected spores. Five properly disinfected seeds of maize were sown in each funnel. After 60 days of growth, plant roots were analyzed for mycorrhizal colonization and rhizospheric soil was tested for quantification of AM spores. This inoculum was further used for the multiplication of AM fungi first under earthen funnels followed by the bigger earthen pots.

2.4 Selection of Substrate

Refinery waste was used as a substrate. Refinery waste was taken from Karnal refinery, Haryana, India.

2.5 Selection of Host Plants

Three different host plants were tested for each substrate. These were maize, barley and wheat. All these plants are monocots and members of family Poaceae.

2.6 Experimental Setup

Top soil (0-3 cm) from Botanical Garden was first air dried, pulverized and then sieved through 2 mm sieve and this soil was mixed with the sand to form sand: soil ratio 1:3. This soil was sterilized in autoclave at 121°C for two consecutive days at 15 lb pressure for 15 minutes. Different concentrations of each substrate (without substrate, 25, 50, 75, 100 and 200g pot^{-1}) were added to earthen pots (25 25 cm), and thoroughly mixed with sand: soil mixture to make final volume of 2 kg. To this, 200g of AM inoculum (chopped AM colonized root pieces of maize along with soil containing 400-450 AM spores 100g^{-1}) was added. Healthy seeds of maize, wheat and barley were surface sterilized with 0.5% sodium hypochlorite for 10 min, and subsequently washed with sterilized deionized water. Five seeds were sown in each pot above the inoculum. Plants were watered daily. After regular interval of 15 days, Hoagland nutrient solution (without KH_2PO_4) (Hoagland and Arnon, 1950) was also added to each pot.

2.7 Harvest and Analysis

After 90 days, vegetative growth response was assessed by manually uprooting the whole plant. Plant height (cm) and root length (cm) were recorded followed by washing of plants with running tap water. Roots and shoots were separated to determine their fresh weight (g) and then placed in an oven for their dry weight at 70°C.

2.8 Isolation of AM Spores

Isolation of AM spores was done by using 'Wet sieving and Decanting technique' of Gerdmann and Nicolson (1963). In this technique, 50 g of dried, sieved rhizospheric soil was thoroughly mixed in 500 ml of water taken in a beaker, stirred well on magnetic stirrer for about 30 minutes and allowed to settle overnight. The suspension of the sample was decanted through a series of sieves placed in descending order ranging from 150 μm, 120 μm, 90 μm, 63 μm and 45 μm from top to bottom, on which spores were trapped. The trapped spores were transferred by repeated washing with water to Whatman's filter paper No. 1 which was divided in many small compartments by marking with ball point pen.

2.9 Quantification of AM Spores

Quantification of isolated AM spores was done by "Grid Line Intersect Method" proposed by Gaur and Adholeya (1994). In this technique, filter paper was spread on inverted Petri dish and seen under a stereo binocular microscope at 60X magnification. All the intact spores were counted by adding the number of spores present in each sector. These spores were then transferred on glass slide having polyvinyl lactic acid (PVLA) by using a hypodermic needle.

2.10 Identification of AM Fungi

Mounted AM spores were examined under high power research microscope (Suswox Optic, Sudheer Scientific Works-133001, India) for their identification. Identification of AM spores was done on the basis of their morphological features, namely, colour, size, shape, wall structure, surface ornamentation, nature and size of subtending hyphae, bulbous suspensor, and the number and arrangement of spores in sporocarp. These AM spores were identified by using identification manual of Walker (1983), Schenck and Perez (1990), Morton and Benny (1990), Mukerji (1996) and Aggarwal *et al.* (2012).

2.11 AM Root Colonization

Mycorrhizal root colonization was done by "Rapid Clearing and Staining Method" of Phillips and Hayman (1970). In this technique, collected roots were washed several times in tap water and cut into small pieces of 1 cm length segments and then 15-30 segments were selected randomly. These roots were cleared in 10% KOH (Potassium hydroxide) solution overnight at room temperature. Decanted this KOH solution, and rinsed the roots with tap water followed by acidification of roots with 1% HCl for 5-10 minutes. HCl was poured out and root segments were again washed with tap water, stained with 0.5% trypan blue overnight. After this, these roots were destained with lactophenol to remove excess of stain.

Evaluation of root colonization was done by Root slide Technique of Giovannetti and Mosse (1980). Twelve to sixteen stained roots were mounted under two glass slides with lactic acid: glycerol (1:1). These stained roots were then examined under a light microscope at 100X for the presence of AM mycorrhizal root colonization and different AM structures. Total number of colonized root segments was noted down. Per cent root colonization was calculated by the following formula:

$$\text{Percentage of AM root colonization} = \frac{\text{No. of root segments with infection}}{\text{Total number of root segments studied}} \times 100$$

2.12 Statistical Analysis

The experimental data was subjected to analysis of variance and means were separated with least significant difference test using the Statistical Package for Social Sciences (ver. 11.5, Chicago (IL), USA).

3. Results and Discussion

For inoculum production of *Acaulospora laevis*, plant height (81.5±2.50 cm), fresh shoot weight (3.03±0.06 g), dry shoot weight (0.403±0.008 g), fresh root weight (1.017±0.040 g), dry root weight (0.204±0.008 g), root length (26.80±0.86 cm), percent root colonization (78.29±1.29 %) and AM spore number (78.8±2.77) were found to be the best at 25 g concentration of refinery waste when we used maize as a host plant (Table-1 and 2). Likewise, 25 g concentration of refinery waste also increased the plant growth and mycorrhization in barley plants. All the growth parameters such as plant height (52.6±2.36 cm), fresh shoot weight (2.14±0.07 g), dry shoot weight (0.325±0.015 g), fresh root weight (0.783±0.003 g), dry root weight (0.186±0.005 g) and root length (19.54±0.60 cm) were found to be increased at 25 g concentration of refinery waste. Same concentration of waste also increased the percent root colonization (71.43±0.06 %) and AM spore number (66.8±6.61) in barley plants. In case of wheat, plant height (73.1±2.15 cm), fresh shoot weight (3.16±0.05 g), dry shoot weight (0.473±0.020 g), fresh root weight (0.961±0.047 g), dry root weight (0.225±0.015 g), root length (19.52±1.16 cm), percent root colonization (77.72±2.33 %) and AM spore number (76.4±4.77) were found to be highest at 25 g concentration of waste (Table. 8.1 & 8.2).

Overall results showed that 25 g concentration of refinery waste increased the growth performance and mycorrhization in all selected host plants for inoculum production of *Acaulospora laevis*. It is evident from the results that all the inoculated host plants produced extramatrical and intraradical hyphae, arbuscules and vesicles. For *Acaulospora laevis*, maize showed maximum mycorrhization and AM spore number; hence plant growth performance was better in maize in comparison to wheat and barley. Second most effective results were found in wheat plants in plant growth enhancement and mycorrhization.

Results of present investigation showed that AM fungi increased the plant height, biomass, root length, per cent root colonization and AM spore number in all inoculated host plants i.e. maize, barley and wheat. AM fungi also have been reported to be involved in improvement of plant growth and enhancing accumulation of plant nutrients through greater soil exploration by mycorrhizal hyphae. Our results vindicated the findings of Mangla *et al.* (2012) who observed a significant enhancement in plant growth, biomass and mycorrhization in AM inoculated wheat and barley plants when mustard seed waste was used as a substrate.

Table 8.1: Efficacy of refinery waste and their different concentrations along with Acaulospora laevis on growth response of maize, barley and wheat.

Type of host Plant	Parameters→ Concentration (g pot -1)↓	Plant height (cm)	Fresh shoot weight (g)	Dry shoot weight (g)	Fresh root weight (g)	Dry root weight (g)	Root length (cm)
Maize	0	34.6±1.73de‡	1.13±0.10fg	0.191±0.008de	0.186±0.009h	0.035±0.006h	12.14±0.61f
	25	81.5±2.50a	3.03±0.06a	0.403±0.008ab	1.017±0.040a	0.204±0.008ab	26.80±0.86a
	50	61.4±2.55c	2.43±0.13b	0.322±0.010b	0.808±0.012c	0.182±0.014b	22.24±0.81b
	75	50.9±2.88cd	2.03±0.05c	0.201±0.009d	0.597±0.014e	0.125±0.009d	18.68±0.80c
	100	41.5±2.62d	1.31±0.04f	0.248±0.009c	0.482±0.012f	0.083±0.009e	15.32±0.96d
	200	30.4±2.42e	1.00±0.09fg	0.140±0.006f	0.102±0.016i	0.021±0.003i	09.68±0.76g
Barley	0	23.8±1.89f	0.89±0.01g	0.122±0.014g	0.118±0.001i	0.017±0.001i	08.76±0.84h
	25	52.6±2.36cd	2.14±0.07bc	0.325±0.015b	0.783±0.003d	0.186±0.005b	19.54±0.60c
	50	41.3±2.42d	1.96±0.03d	0.201±0.016d	0.603±0.009e	0.164±0.001c	17.26±0.77cd
	75	34.5±2.40de	1.64±0.04e	0.178±0.011e	0.406±0.006f	0.136±0.004d	13.96±1.23e
	100	27.5±2.58e	1.53±0.03ef	0.145±0.014f	0.200±0.002g	0.057±0.008f	11.76±0.75f
	200	19.9±2.06g	0.77±0.02h	0.100±0.010h	0.099±0.004j	0.014±0.001j	06.48±0.51i
Wheat	0	28.8±2.95e	1.22±0.05f	0.161±0.010e	0.229±0.033g	0.041±0.014g	09.13±0.90g
	25	73.1±2.15b	3.16±0.05a	0.473±0.020a	0.961±0.047b	0.225±0.015a	19.52±1.16c
	50	58.8±3.04c	2.45±0.11b	0.395±0.015ab	0.844±0.051c	0.183±0.016b	15.64±1.26d
	75	50.4±2.69cd	1.95±0.05c	0.314±0.015b	0.896±0.078bc	0.195±0.025ab	16.28±1.25d
	100	38.4±2.26d	1.47±0.07ef	0.224±0.018d	0.660±0.048e	0.126±0.011d	12.57±1.49f
	200	25.0±2.31f	1.01±0.04fg	0.113±0.004gh	0.196±0.015h	0.031±0.010h	08.41±1.17h
A N O V A (F5,72)	H type	815.426	127.732	356.457	717.699	297.606	648.251
(F2,72)	S conc.	102.23	91.876	297.804	261.227	156.303	883.429
(F10, 72)	H type × S conc.	21.555	23.641	56.303	41.589	66.789	19.160

‡ - Each value is a mean of five replicates, ± - standard deviation, H- host, S- substrate, values in columns followed by the same alphabet are not significantly different, (P≤0.05), least significant difference test

Table 8.2: Efficacy of refinery waste and their different concentrations along with *Acaulospora laevis* on Inoculum production of maize, barley and wheat.

Type of host plant	Parameters→ Concentration (g pot -1)↓	AM Root colonization (%)	AM Spore number/10 g of soil	Mycelium	Vesicles	Arbuscules
Maize	0	27.19±2.15g‡	12.8±4.65i	+	-	-
	25	78.29±1.29a	78.8±2.77a	+ + +	+ + +	+ + +
	50	64.29±0.94c	63.8±4.14c	+ + +	+ +	+ +
	75	56.59±1.67d	48.6±4.44e	+ + +	+ +	+
	100	42.93±1.63e	35.6±5.12g	+ +	+	-
	200	14.17±0.96h	04.8±2.58j	+	-	-
Barley	0	21.42±0.02g	11.6±4.33i	+	-	-
	25	71.43±0.06b	66.8±6.61b	+ + +	+ + +	+ +
	50	64.28±0.01c	55.6±5.72d	+ +	+ +	+ +
	75	57.14±0.02d	46.0±6.08e	+ +	+	+
	100	42.80±0.16e	32.2±5.80h	+ +	+	+
	200	14.24±0.06h	06.6±3.36j	+	-	-
Wheat	0	35.61±2.26f	14.8±4.14i	+	-	-
	25	77.72±2.33a	76.4±4.77a	+ + +	+ + +	+ + +
	50	64.76±1.89c	67.6±6.69b	+ + +	+ +	+ +
	75	62.17±2.42c	42.8±8.52f	+ + +	+ +	+
	100	57.25±1.62d	32.8±5.63h	+ +	+	+
	200	33.92±1.94f	11.0±6.40i	+	+	-
ANOVA (F5,72)	H type	783.020	818.60			
(F2,72)	S conc.	905.63	654.47			
(F10, 72)	H type × S conc.	141.509	68.93			

‡ - Each value is a mean of five replicates, AM - Arbuscular Mycorrhiza, ± - standard deviation, H- host, S- substrate, (-) - absent, + - scanty, ++ - moderate, +++ - abundant, values in columns followed by the same alphabet are not significantly different, ($P \leq 0.05$), least significant difference test.

Likewise, Kuppusamy and Kumutha (2012) also found an increased root dry weight, root colonization and AM spore count in AM inoculated maize plants when raw vermiculite was used as a substrate. It is evident from the results that increasing in concentration of refinery waste decreases the plant growth and mycorrhization. Higher concentration (200g) of refinery waste was found to be inhibitory for plant growth. Our results are confirmed by the findings of Konwar and Jha (2010) who reported a significant reduction in plant height, shoot and root biomass, panicle length and grain weight of rice plants when treated with different doses of refinery waste effluents. Likewise, Tanwar *et al.* (2013a) found that higher concentration of sugarcane bagasse waste inhibits the growth of AM fungus in AM inoculated onion plants. The present findings are in conformity with the findings of Roy (2001), who also observed negative influence of refinery effluents on the yield of rice. The low root dry weight of the selected host plants affected with refinery waste may be due to the inhibitory effects of the refinery effluents on root growth. Singh (1996) reported that higher the pollution stress of a locality, the poorer is the development of the root system.

Addition of refinery waste changes the soil chemical and physical properties which became more conducive for the sporulation of AM fungi. Additionally, improved colonization with refinery waste could be attributed to the uniform spreading of AM fungal hyphae in the added substrate with high porosity and thus increases the chances of root contact with AM fungal structures and hence colonization. Higher number of root colonization may be attributed to higher number of secondary roots and their density, which favour active mycorrhizal per cent root colonization (Javaid and Riaz, 2008).

Jo *et al.* (2008) reported that discharge of petroleum refinery wastewater results in environmental and human health effects due to release of toxic contaminants such as hydrocarbons (benzene, toluene, ethyl benzene and xylene), phenols, dissolved minerals (inorganic compounds, anions and cations) and heavy metals (sulphide, ammonia, zinc, lead and copper) (Diya'uddeen *et al.*, 2011). Konwar and Jha (2010) reported a negative correlation between plant height and soil nitrogen content in rice plants when treated with different doses of refinery waste effluents. Similarly, Azcon *et al.* (2001) also reported that higher dose of nitrogen present in soil might have inhibited the development of free living bacteria in the rhizosphere which affected the plant height. It has been observed that Phosphorus content increased due to increased refinery effluents in contaminated soil which lead to reduced AM colonization (Khaliq and Sanders, 2000; Rubio *et al.*, 2003). Deka *et al.* (2000) found that the increased impermeability to water and air in the soil due to the accumulation of oil and grease may be one of the reason behind poor root growth in the contaminated fields resulting into poor yield of rice. Thabaraj *et al.* (1964) observed that excess amount of sodium and chloride present in soil inhibits the uptake of other elements like magnesium, potassium, and phosphorus by the plants. Similarly, Babich and Stotzky (1985) reported that heavy metals inhibited the primary productivity of the plants.

AM root colonization and spore number formation depends on the type of host as well as the duration of infection of these symbiotic microorganisms (Chaurasia and Khare, 2005). *Glomus mosseae* enhanced the AM spore production might be due to enhanced growth and spread of AM hyphae (Joner and Jakobsen, 1992). The existence of a varied range of colonization might be attributed to the soil factors which affected the number of vesicles per root and AM spores in the rhizospheric soil. Number of vesicles was found to be correlated with AM spore number per root (Gupta *et al.*, 2006). Vesicles are formed later and persist during the growing period (Van Duin *et al.*, 1989). Formation of arbuscules facilitated the nutrient exchange between host plant and fungal symbionts.

4. Conclusions

Results pertaining to inoculum production of *Acaulospora laevis* with refinery waste as a substrate depicted that 25 g concentration of waste increased all growth parameters and mycorrhization of selected host plants. It was found that increasing in concentration of refinery waste decreases the plant growth and mycorrhization. Higher concentration (200g) of refinery waste was found to be inhibitory for plant growth and mycorrhization. The present study might be highly significant as the goal of our research was to make an inexpensive, species specific and highly effective inoculum. And as the substrate used was a waste, it is an eco-friendly step too. Use of molecular tools can further strengthen our research to make the inoculum of these fungi on large scale.

Acknowledgements

Authors are grateful to Department of Botany, Kurukshetra University, Kurukshetra for providing necessary facilities to carry out research. Main author (Neetu) is also thankful to UGC for providing research fellowship in the form of RGNF.

References

Aggarwal, A., Tanwar, A., Neetu and Mehrotra, R.S. 2012. Taxonomy of arbuscular mycorrhizal fungi with special reference to *Glomus* species. In: Bagyaraj, D.J., Tilak, K.V.B.R. and Kehri, H.K. (Eds.), Microbial Diversity and Functions, NIPH, New Delhi, pp. 39-94.

Azcon, R., Ruiz-Rozano, J.M. and Rodriguez, R. 2001. Differential contribution of arbuscular mycorrhizal fungi to plant nitrate uptake (^{15}N) under increasing N supply to the soil. *Can. J. Bot.* 97: 1175-1180.

Azcón-Aguilar, C. and Barea, J.M. 1997. Applying mycorrhiza biotechnology to horticulture: Significance and potentials. *Sci. Hortic.* 68: 1-24.

Babich, H. and Stotzky, G. 1985. Heavy metal toxicity to microbe-mediated ecological processes: A review and potential application to regulatory polices. *Environ. Res.* 36: 111-137.

Bever, J.D. 2002. Host-specificity of AM fungal population growth rates can generate feedback on plant growth. *Plant Soil.* 244: 281-290.

Bever, J.D., Peggy, A., Schultz, Pringle, A. and Joseph, B.M. 2001. Arbuscular mycorrhizal fungi: More diverse than meets the eye, and the ecological tale of why. *Bioscience.* 51: 923-932.

Chaurasia, B. and Khare, P.K. 2005. *Hordeum vulgare*: A suitable host for mass production of arbuscular mycorrhizal fungi from natural soil. *Appl. Ecol. Environ. Res.* 4(1): 45-53.

Chen, B., Xiao, X., Zhu, Y.G., Smith, F.A., Xie, Z.M. and Smith, S.E. 2007. The arbuscular mycorrhizal fungus *Glomus mosseae* gives contradictory effects on phosphorus and arsenic acquisition by *Medicago sativa* Linn. *Sci. Total Environ.* 379: 226-234.

Deka, D.S., Barua, I.C. and Deka, J. 2000. Impact of crude oil pollution on survival of rice crop. *Ann. Agric. Biol. Res.* 5(1): 49- 51.

Diya'uddeen, B.H., W. Daud, W.M.A. and Abdul Aziz, D.R. 2011. Treatment technologies for petroleum refinery effluents: A review, Process *Saf. Environ. Protect.* 175: 11.

Douds, D.D. Jr. and Schenck, N.C. 1990a. Increased sporulation of vesicular arbuscular mycorrhizal fungi by manipulation of nutrient regimes. *Appl. Environ. Microbiol.* 56: 413-418.

Douds, D.D. Jr. and Schenck, N.C. 1990b. Relationship of colonization and sporulation by VA mycorrhizal fungi to plant nutrient and carbohydrate contents. *New Phytol.* 116: 621-627.

Dugassa, D.G., Grunewaldt-Stöcker, G. and Schönbeck, F. 1995. Growth of *Glomus intraradices* and its effect on linseed (*Linum usitatissimum* L.) in hydroponic culture. *Mycorrhiza.* 5: 279-282.

Fattah, O.A. 2013. Effect of mycorrhiza and phosphorus on micronutrients uptake by soybean plant grown in acid soil. *Intl. J. Agron. Plant Prod.* 4 (3): 429-437.

Gaur, A. and Adholeya, A. 1994. Estimation of VAM fungal spores in soil: A modified method. *Mycorrhiza News.* 6: 10-11.

Gaur, A. and Adholeya, A. 2000. Effects of the particle size of soil-less substrates upon AM fungus inoculum production. *Mycorrhiza.* 10(1): 43-48.

Gaur, A. and Adholeya, A. 2002. Arbuscular- mycorrhizal inoculation of five tropical fodder crops and inoculum production in marginal soil amended with organic matter. *Biol. Fert. Soils.* 35(3): 214-218.

Gerdemann, J.W. and Nicolson, Y.H. 1963. Spores of mycorrhiza *Endogone* species extracted from soil by wet sieving and decanting. *Trans. Brit. Mycol. Soc.* 46: 235-244.

Gianinazzi, S. and Vosátka, M. 2004. Inoculum of arbuscular mycorrhizal fungi for production systems: Science meets business. *Can. J. Bot.* 82: 1264-1271.

Giovannetti, M. and Mosse, B. 1980. An evaluation of techniques for measuring vesicular arbuscular infection in roots. *New Phytol.* 84: 489-500.

Gupta, N., Rautaray, S. and Basak, U.C. 2006. The growth and development of AMF and its effect on the growth of maize under different soil composition. *Mycorrhiza News.* 18(3): 15-23.

Habte, M. 2006. The role of arbuscular mycorrhiza in plant and soil health. In: Uphoff *et al.*, (Eds.). Biological Approaches to Sustainable Soil systems, Boca Raton, Florida, USA, pp. 129-148.

Hoagland, D.R. and Arnon, D.I. 1950. The water-culture method for growing plants without soil. *Calif. Agric. Exp. Stn. Circ.* 347: 1-32.

Hung, L.L.L. and Sylvia, D.M. 1988. Production of vesicular–arbuscular mycorrhizal fungus inoculum in aeroponic culture. *Appl. Environ. Microbiol.* 54: 353-357.

Ijdo, M., Cranenbrouck, S. and Declerck, S. 2011. Methods for large-scale production of AM fungi: Past, present, future. *Mycorrhiza.* 21: 1-16.

Javaid, A. and Riaz, T. 2008. Effect of application of green leaf manure on growth and mycorrhizal colonization of maize. *Allelopathy J.* 21(2): 339-347.

Jo, M.S., Rene, E.R., Kim, S.H. and Park, H.S. 2008. An analysis of synergistic and antagonistic behaviour during BTEX removal in batch system using response surface methodology. *J. Haz. Mat.* 152: 1276-1284.

Joner, E.J. and Jakobsen, I. 1992. Enhanced growth of external VA mycorrhizal hyphae in soil amended with straw. In: Read, D.J., Lewis, D.H., Fitter, A.H. and Alexander, I.J. (Eds.), Mycorrhizas in Ecosystems, CAB, Oxford, pp. 387.

Khaliq, A. and Sanders, F.E. 2000. Effect of vesicular-arbuscular mycorrhizal inoculation on the yield and phosphorus uptake of field grown barley. *Soil Biol. Biochem.* 32: 1691-1696.

Klironomos, J.N. and Hart, M.M. 2002. Colonization of roots by arbuscular mycorrhizal fungi using different sources of inoculum. *Mycorrhiza.* 12: 181-184.

Konwar, D. and Jha, D.K. 2010. Response of Rice (*Oryza sativa* L.) to contamination of soil with refinery effluents under natural conditions. *Assam Uni. J. Sci. Technol.* 5(1): 14-22.

Kuppusamy, S. and Kumutha, K. 2012. Standardization of the spore density of AM fungal inoculum for effective colonization. *Intl. J. Agric. Sci.* 4(2): 176-181.

Mangla, C., Kumar, A. and Aggarwal, A. 2012. Inoculum production of endophytic mycorrhiza using mustard seed waste as substrate. *J. New Biol. Rep.* 1(2): 61-66.

Menge, J.A. and Timmer, L.M. 1982. Procedure for inoculation of plants with VAM in the laboratory, greenhouse and field. In: Schenck, N.C. (Ed.). Methods and Principles of Mycorrhizal Research, *Amer. Phytopath. Soc.*, St. Paul, USA, pp. 59-68.

Millner, P.D. and Kitt, D.G. 1992. The Beltsville method for soil less production of vesicular-arbuscular fungi. *Mycorrhiza.* 2: 9-15.

Morton, J.B. and Benny, G.L. 1990. Revised classification of arbuscular mycorrhizal fungi (Zygomycetes). New order Glomales, two new sub orders Glomineae and Gigasporineae and two new families Acaulosporaceae and Gigasporaceae with emendation of Glomaceae. *Mycotaxon.* 37: 471-491.

Mosse, B. and Thompson, J.P. 1984. Vesicular-arbuscular endomycorrhizal inoculum production. I. Exploratory experiments with beans (*Phaseolus vulgaris*) in nutrient flow culture. *Can. J. Bot.* 62: 1523-1530.

Mukerji, K.G. 1996. Taxonomy of endomycorrhizal fungi. In: Mukerji, K.G., Mathur, B., Chamola, B.P. and Chitralekha, P. (Eds.). Advances in Botany, A.P.H. Publishing Corporation, New Delhi, pp. 211-221.

Perner, H., Schwarz, D. and George, E. 2006. Effects of mycorrhizal inoculation and compost supply on growth and nutrient uptake of young leek plants grown in peat-based substrates. *Hort. Sci.* 41: 628-632.

Phillips, J.M. and Hayman, D.S. 1970. Improved procedures for clearing roots and staining parasitic and VAM fungi for rapid assessment of infection. *Trans. Brit. Mycol. Soc.* 55: 158-161.

Rangel, W.D.M., Schneider, J., de Souza Costa, E.T., Soares, C.R.F.S., Guilherme, L.R.G. and de Souza Moreira, F.M. 2014. Phytoprotective effect of arbuscular mycorrhizal fungi species against arsenic toxicity in tropical leguminous species. *Int. J. Phytoremed.* 16: 840-858.

Roy, R. 2001. A study on the effect of BRPL effluents on growth and yield of certain economic plants with particular reference to rice (*Oryza sativa*). Ph.D. Thesis, Gauhati University, Assam, India.

Rubio, R., Borie, F., Schalchli, C., Castillo, C. and Azcon, R. 2003. Occurrence and effect of arbuscular mycorrhizal propagules in wheat as affected by the source and amount of phosphorus fertilizer and fungal inoculation. *Appl. Soil Ecol.* 23: 245- 255.

Schenck, N.C. and Perez, Y. 1990. Manual for the identification of VA mycorrhizal fungi. University of Florida, Gainesville, Florida, USA, pp. 241.

Singh, S. 1996. Effect of soil pollution on mycorrhizal development. *Mycorrhiza News.* 8(2): 1- 8.

Smith, S.E. and Read, D.J. 1997. Vesicular-arbuscular mycorrhizas. In: Smith, S.E. and Read, D.J. (Eds.). Mycorrhizal Symbiosis 2nd Edn., Academic Press, London, pp. 9-160.

Sylvia, D.M. and Schenck, N.C. 1983. Application of superphosphate to mycorrhizal plants stimulates sporulation of phosphorus-tolerant vesicular-arbuscular mycorrhizal fungi. *New Phytol.* 95: 655-661.

Tanwar, A., Aggarwal, A. and Parkash, V. 2013a. Sugarcane bagasse: A novel substrate for mass multiplication of *Funneliformis mosseae* with onion as host. *J. Cent. Eur. Agric.* 14(4): 1502-1511.

Tanwar, A., Aggarwal, A., Yadav, A. and Parkash, V. 2013b. Screening and selection of efficient host and sugarcane bagasse as substrate for mass multiplication of *Funneliformis mosseae. Biol. Agric. Hortic.* 29(2): 107-117.

Thabaraj, G.J., Bose, S.M. and Nayaidaumu. 1964. Utilization of tannery effluents for agricultural purpose. *Environ. Hlth.* 6: 18- 36.

Van Duin, W.E., Rozema, J. and Ernst, W.H.O. 1989. Seasonal and spatial variation in the occurrence of vesicular arbuscular mycorrhiza in salt marsh plants. *Agric. Ecosyst. Environ.* 29: 107-110.

Walker, C. 1983. Taxonomic concepts in the Endogonaceae spore wall characteristics in species description. *Mycotaxon.* 18: 443-445.

Chapter 9

Antibacterial Activity of Selected Bryophytes from Nainital, Uttarakhand, India

Geetanjali Mehra[1], Ruchi Singh[1], Neeraj[2] and Preeti Chaturvedi[1,*]

[1]*Department of Biological Sciences, CBSH, G. B. Pant University of Agriculture& Technology, Pantnagar, -263145(Uttarakhand).*

[1]*Department of Botany, Govt. P.G. College, Sec-1, Panchkula*

**Corresponding Author E-mail: an_priti@yahoo.co.in*

Abstract

In the present study, crude extracts (acetone and methanol) of selected bryophytes viz. Artichum undulatum, Bryum capillare, Conocephalum conicum, Marchantia polymorpha, Rhodobryum roseum and Trachypodopsis serrulata were screened for antimicrobial activity against gram negative bacteria namely, Escherichia coli, Salmonella enterica and Pseudomonas aeruginosa. The highest antimicrobial activity against E. coli, S. enterica and P. aeruginosa was shown by methanolic extract of T. Serrulata (ZI= 16.66 ± 0.33mm, AI= 0.56), A. undulatum (ZI= 18.66 ± 0.33mm, AI= 0.55) and M. polymorpha (16.00 ± 0.57mm, AI = 0.53) respectively. Overall, inhibitory activity of methanol extracts was higher as compared to acetone extracts. Among all the tested extracts against different bacteria, acetonic extract of C. conicum gave lowest MIC (12 µg/ml) against P. aeruginosa followed by methanolic extracts of C. conicum (MIC =20 µg/ml) for E. coli and S. enterica.

Keywords: *Artichum undulatum, Bryum capillare, Conocephalum conicum, Marchantia polymorpha, Rhodobryum roseum, Trachypodopsis serrulata, antimicrobial activity*

1. Introduction

Bryophytes are the non-vascular land plants that are distributed in almost every kind of habitat. Traditionally, bryophytes are used in herbal medicine in India, China and North America. In Chinese herbal medicine, about 40 bryophytes species were used in treating illnesses related to heart and circulatory system (Ding, 1982). Many bryophytes are used in treating liver ailments, jaundice, hepatitis, cardiovascular diseases, lung diseases, skin diseases and also possess anti-bacterial and anti-inflammatory properties (Miller and Miller, 1979; Hu, 1987; Wu, 1982; Ding, 1982; Pant *et al.*, 1986; Frahm, 2004). Interestingly, bryophytes are a remarkable repository of novel natural products and various secondary metabolites like polyphenolic acids, flavonoids and alkaloids which also exhibit anti-oxidant properties to resist various stresses (Singh *et al.*, 2007). Biologically active compounds like monoflavonoids (apigenin, kaempferol, luteolin, orobol derivatives) and biflavonoids (apigenin, luteolin, eriodictyol) that are usually found in mosses are of high pharmaceutical value (Zinsmeister and Mues, 1980; Markam, 1990).

Indiscriminate usage of commercial antimicrobial compounds in the present healthcare systems leads to increasing cases of resistance among pathogenic microorganisms. Hence, there is an earnest need to search for the potent antimicrobial agents from natural sources. Not only the higher plants but the bryophytes also possess the interesting antibiotic substances (Wallach *et al.*, 2010; Montenegero *et al.*, 2009). In the present study, six bryophyte species *viz., Atrichum undulatum* (Hedw.) P. Beauv, *Bryum capillare* Hedw., *Conocephalum conicum* (L.) Dumort., *Marchantia polymorpha* L., *Rhodobryum roseum*(Hedw.) Limpr. and *Trachypodopsis serrulata* (P. Beauv.) M. Fleisch. (collected from Nainital, Uttarakhand) were evaluated for antimicrobial activity against some selected pathogen strains *viz., Escherichia coli, Salmonella enterica* and *Pseudomonas aeruginosa*.

2. Materials & Methods

2.1 Plant Materials

Plant samples were collected from Nainital (2084 masl, 29.38 N 72.45 E) in the month of March 2014. Identification and authentication of the collected samples was done by consulting NBRI herbarium, Lucknow.

2.2 Preparation of the Extracts

The collected plant material was washed with continuous running tap water and air dried. Plant material was then grounded to fine powder. 2g of the powdered plant material was separately added to 20ml of methanol (80%) and acetone (80%) to prepare the methanol and acetone extract respectively. The extract was filtered and concentrated using rotary evaporator (Star Scientific). The residue was kept in refrigerator for further experimentation.

2.3 Test Pathogens

Escherichia coli and *Salmonella enterica* were procured from Public Health Laboratory, College of Veterinary Sciences, G.B.P.U.A. & T., Pantnagar and *Pseudomonas aeruginosa* (MTCC 424) was obtained from IMTECH , Chandigarh(Punjab).

2.4 Antimicrobial Assay

Antibacterial activity of the methanol and acetone extracts of different species was screened by disc diffusion method (Bauer *et al*, 1966). A loop full of bacterial strain was inoculated in 10 ml of nutrient broth in a test tube and incubated for 24 hrs to get active strain. 6 mm filter discs were kept in bryophyte extracts for some time.

Nutrient agar was poured into petri plates. After solidification, the plates were seeded with the test organism with the help of cotton swab. The experiment was performed under strict aseptic conditions. Petri plates inoculated with bacterial strains were incubated at 37°C for 24 hrs. Antibacterial activity was expressed in terms of zone of inhibition (ZI)of the bacteria.

2.5 Determination of Minimum Inhibitory Concentration (MIC)

The agar diffusion method (Murray *et al.*, 1995) was used to determine MIC. Bacteria was cultured over night at 37°C in nutrient broth and used as inoculum. Final inoculums, using 100µl of suspension containing 10^8 CFU/ml of bacteria was swapped on to the nutrient agar. Sterile filter paper discs (6 mm in dia) containing 2.5-1000 µg/disc of plant extracts were placed on the surface of the medium. Culture containing standard drugs (streptomycin and cephotaxim) were used as positive controls whereas cultures without standard drugs and extract were used as negative control. MIC was regarded as lowest concentration of the extract that was able to inhibit visible bacterial growth. The experiment was performed in triplicate.

3. Results and Discussion

Both methanolic and acetone extracts of the bryophytes exhibited dose dependent inhibitory activities against the test bacteria *viz.*, *E. coli*, *S. enterica* and *P. aeruginosa* (Table 1). Highest antibacterial activity against *E. coli* was shown by *T. serrulata* (16.66±0.33mm) followed by *C. conicum* (14.33±0.33mm) and *R. roseum* (14.0±0.57mm) in methanolic extract. Acetone extract of *M. polymorpha* (16.0±0.57mm) showed maximum antibacterial activity against *E. coli* followed by *A. undulatum* (14.33±0.33mm) and *R. roseum* (13.66±0.33mm). *B. capillare, C. conicum* and *T. serrulata* did not show antimicrobial activity against *E. coli* in acetone extract. Altuner *et al.* (2011) also reported negligible antibacterial activity of *B. capillare* against *E. coli*. Methanolic extract of *M. polymorpha* showed antibacterial activity against *E.coli* with ZI of 9 mm (Mewari and Kumar, 2008) whereas *A. undulatum* showed 0.08±1mm ZI (Vats and Alam, 2013). Ulka and Karadge (2010) also reported the antimicrobial activity of *Targonia hypophylla* against *E.coli* with ZI of 12.3 ± 2.89mm.

Methanolic extract of *A. undulatum* showed highest antibacterial activity(18.66±0.33mm) against *S. enterica* followed by *M. polymorpha* (13.0±*0.25mm), B. capillare*(12±0.577mm), *T. serrulata* (11±0mm) and *C. conicum* (10.66±0.33mm); whereas in acetone extract, *M. polymorpha* (ZI of 11.66±0.57mm) showed highest antibacterial activity followed by *A. undulatum* (11.37±0.66mm), *C. conicum* (11.33±0.57mm) and *T. serrulata* (9±0.33mm). Elibol *et al.* (2011) also reported antibacterial activity of *B. capillare* against *Salmonella sp.* (7±0.1 mm ZI). Negi and Chaturvedi (2016a, 2016b) also reported significant antibacterial activity of *Rhynchostegium vagans* (21.00 ± 0.57 mm ZI) and *Plagiothecium appendiculatum* (ZI= 27.66 ± 0.88) against *S. enterica.*

Table 9.1: Antimicrobial activity of methanol and acetone extracts of bryophytes against E. coli, S. enterica and P. aeruginosa expressed as inhibition zone (mm).

	Strains Nature of extract	*E. coli*				*S. enterica*				*P. aeruginosa*			
		100	400	700	1000	100	400	700	1000	100	400	700	1000
M	*A. undulatum*	7.0±0.35	8.25±0.57	10.0±0.66	12.0±0.66	7.66±0.88	11.40±0.3	13.66±0	18.66±0	-	-	-	-
E													
T	*B. capillare*	7.33±0.33	8.0±0	9.0±0.57	10.66±0.33	-	9.0±0.88	11.0±0.5	12.0±0.57	-	-	-	-
H	*C. conicum*	7.66±0.66	9.66±0.33	12.66±0.57	14.33±0.33	6.66±0.66	7.0±0	8.33±0.6	10.66±0.	-	6.66±0.33	8.5±0.33	9.5±0.28
A N	*M. polymorpha*	-	-	8.66±1.20	11.33±0.88	-	8.91±0.3	11.54±0.43	13.0±0.25	7.0±1	10.66±0.33	13.0±0.57	16.0±0.57
O	*R. roseum*	-	7.33±0.88	10.33±0.33	14.0±0.57	-	-	-	-	-	-	-	-
L	*T. serrulata*	-	8.0±0.57	12.0±0.57	16.66±0.33	6.33±0.33	7.33±0.33	9.66±0.33	11.0±0	-	-	-	-
A	*A. undulatum*	8.0±0.33	9.33±0.33	11.66±0.33	14.33±0.33	7.0±0.25	8.4±0.54	10.0±0.33	11.37±0.66	-	-	-	-
C	*B.capillare*	-	-	-	-	-	-	-	-	-	-	-	-
E	*C. conicum*	-	-	-	-	8.3±0.66	9.0±0.42	10.66±0.33	11.33±0.57	6.66±0.33	8.9±0.59	10.4±0.3	11.66±0.33
T	*M. polymorpha*	7.33±0.33	8.33±0.88	14.0±0.57	16.0±0.57	-	-	8.33±0.33	11.66±057	7.66±0.33	9.66±0.66	12.66±0	14.33±0.33
O N	*R. roseum*	7.0±0.57	9.33±0.5	11.66±0.3	13.66±0.3	-	-	-	-	-	-	-	-
E	*T. serrulata*	-	-	-	-	-	-	7.45±0.5	9.0±0.33	-	7.0±0.57	9.66±0.33	11.66±0.33
	Cepholotaxime	30.0				23.0				25.0			
	Streptomycin	19.0				28.0				30.0			
		Sem		CD		Sem		CD		Sem		CD	
	Concentration (C)	0.18		0.51		0.14		0.41		0.18		0.51	
	Plant extract (P)	0.25		0.75		0.19		0.55		0.20		0.58	
	CxP	0.51		1.45		0.15		0.39		0.41		1.16	
	cv	9.761				10.59				7.74			

Both the organic extracts of *M. polymorpha* and *C. conicum* showed significant antibacterial activity against *P. aeruginosa* with maximum ZI for methanol (16.0±0.57mm) and acetone extracts(14.33±0.33mm) exhibited by *M. polymorpha*. Among mosses, only *T. serrulata*, showed antibacterial activity. Singh *et al.* (2011) reported significant antimicrobial activity of methanolic extract of *C. conicum* against *P. aeruginosa* with 32.47% of zone of inhibition. Sharma *et al.* (2013) reported 6 mm ZI of *Reboulia hemispherica* against *P. aeruginosa* whereas Negi and Chaturvedi (2016a) reported 19.33 ± 0.33 mm ZI of *R. vagans* against *P. aeruginosa*. *S. enterica*, among the tested pathogens, is most sensitive.

MIC of different extracts against the bacterial pathogens ranged from 12.5 µg/ml – 580 µg/ml (Table 2). Lowest MIC against *E. coli* was showed by *C. conicum* (20 µg/ml) followed by *T. serrulate* (40µg/ml). Methanolic extract of *M. polymorpha* possessed 0.875 mg/ml MIC against *E. coli* (Mewari and Kumar 2008). However, in the present study, methanolic extract of *M. polymorpha* exhibited 150µg/ml MIC. Methanolic extract of *C. conicum* showed lowest MIC (20µg/ml) against *S. enterica* followed by *A. undulatum*, *M. polymorpha* and *R. roseum* (MIC=100µg/ml). Lowest MIC was showed by acetone extracts of *C. conicum*(12.5 µg/ml) against *P. aeruginosa* followed by *M. polymorpha* (20 *µg/ml*) for both the extracts. Singh *et al.* (2011) also reported the same value of MIC (20 µg/ml) of *C. conicum* against *P. aeruginosa*. Kandpal *et al.* (2016) reported the higher MIC (62.5 µg/ml) of *M. palmata* against *P. aeruginosa*.

Table 9.2: Minimum Inhibitory Concentration (MIC) of different extracts of bryophytes against different pathogens (µg ml-1).

Nature of extracts		MIC (µg/ml) E. coli	*S. enterica*	*P. aeruginosa*
Methanol	*Atrichum undulatum*	50	100	-
	Bryum capillare	400	200	-
	Conocephalum conicum	20	20	100
	Marchantia polymorpha	150	350	20
	Rhodobryum roseum	-	400	-
	Trachypodopsis serrulata	40	380	-
Acetone	*Atrichum undulatum*	50	-	-
	Bryum capillare	-	-	-
	Conocephalum conicum	300	-	12
	Marchantia polymorpha	530	100	20
	Rhodobryum roseum	580	100	-
	Trachypodopsis serrulata	-	-	200

Indicates that MIC was not determined.

4. Conclusion

The present study reports fairly good antibiotic activity of both the liverworts and the mosses against the test pathogens. *T. serrulata* showed maximum antibiotic activity against *E. coli*. Similarly, *A. undulatum* showed maximum activity against *S. enterica* and *M. polymorpha* against *P. aeruginosa*. However, further chemical analysis of these bryophytes is essential to detect the active principle and the lead molecules that can be further used in disease control programmes replacing harmful chemicals.

References

Altuner, E.M., Ceter, T., Bayar, E., Aydin, S., Arici, F., Suleymanoglu, G. and Edis, A. 2011. Investigation on antimicrobial effects of some moss species collected from Kastamonu Region. *Commun. Fac. Sci. Univ. Ank. Series C, 23: 33-43.*

Bauer, A.W., Kirby, W.M.M, Sherris, J. C. and Turck, M. 1966. Antibiotic susceptibility testing by standard single disc diffusion method. *Am. J. Clin. Path.*, 45: 493-496.

Ding, H. 1982. Medicinal spore-bearing plants of China. Shanghai Science and Technology Press, Shanghai, pp. 409.

Elibo, B., Ezer, T., Kara, R., *Çelik*, G.Y. and *Çolak*, E. 2011. Antifungal and antibacterial effects of some acrocarpic mosses. *Afr. J. Biotechnol., 10(6): 986-989.*

Frahm, J.P. 2004. Recent developments of commercial products from bryophytes. *Bryologist*, 107: 277-283.

Hu, R. 1987. Bryology. Higher Education Press, Beijing, China, pp. 465.

Kandpal, V., Chaturvedi, P., Negi, K., Gupta, S. and Sharma, A. 2016. Evaluation of antibiotic and biochemical potential of Bryophytes from Kumaun hills and Tarai belt of Himalayas. *Int. J. Pharm. Pharm. Sci.*, 8(6): 65-69.

Markham, K.R. 1990. Bryophytes, their chemistry and chemical taxonomy. Clarendon Press, Oxford.

Mewari, N. and Kumar, P. 2008. Antimicrobial Activity of Extracts of *Marchantia polymorpha*. *Pharm. Biol.*, 46: 819-822.

Miller, N.G. and Miller, H. 1979. Make ye the bryophytes. *Horticult.*, 57(1): 40-47.

Montenegro, G., Portaluppi, M., Salas, F. and Diaz, M. 2009. Biological properties of the Chilean native moss *Sphagnum megellanium*. *Biol. Res.*, 42: 233-237.

Murray, P.R., Baroon, E.J., Pfaller, M.A., Tenover, F.C. and Yolke, R.H. 1995. Manual of Clinical Microbiology. 6: 15-18.

Negi, K. and Chaturvedi, P. 2016a. *In vitro* antimicrobial efficacy of *Rhynchostegium vagans* A. Jaeger (moss) against commonly occurring pathogenic microbes of India sub-tropics.*Asian Pac. J. Trop. Dis.*, 6(1): 10-14.

Negi, K. and Chaturvedi, P. 2016b. Anti-bacterial potential of lesser known Himalayan medicinal plants viz. Bryophytes. *Bioscan*, 11(4): 2231-2235.

Pant, G., Tewari, S. D., Pargaien, M. C. and Bisht, L. S. 1986. Bryological activities in North-West Himalaya – II. A bryophyte foray in the Askot region of district Pithoragarh (Kumaun Himalayas). Bryol. Times, 39: 2-3.

Sharma, C., Sharma, A. and Katoch, M. 2013. Comparative evaluation of antimicrobial activity of methanolic extract and phenolic compounds of a liverwort, *Reboulia hemispherica*. *Archive Bryol.*, 192: 1-6.

Singh, M., Rawat, A.K.S. and Govindrajan, R. 2007.Antimicrobial activity of some Indian mosses.*Fitoterapia,* 78: 156-158.

Singh, M., Singh, S., Nath, V., Sahu, V. And Rawat, A.K. 2011. Antibacterial activity of some bryophytes used traditionally for the treatment of burn infections. *Pharmaceut. Biol.,* 49(5): 526-530.

Ulka, J.S. and Karadge, B.A. 2010. Antimicrobial activity of some bryophytes (liverworts and a hornwort) from Kolhapur district. *Phcog. J.,* 2(16): 29-32.

Vats, S. and Alam, A. 2013.Antibacterial Activity of *Atrichum undulatum* (Hedw.) P. Beauv. against some pathogenic bacteria. *J. Biol. Sci.,*13(5): 427-431.

Wallach, P., Lofez, L., Oberpaur, C., Vacarezza, F. and Maier, L. 2010. Estudio preliminary de effectors antimicrobianos "*in vitro*" delmusgo *Sphagnum magellanicum* Brid. *Agro. Sur.,* 38: 80-86.

Wu, P.C. 1982. Some uses of mosses in China. Bryol. Times, 13: 5.

Zinsmeister, H.D. and Mues, R. 1980. The flavonoid chemistry of Bryophytes. *Rev. Latinoam. Quim.,* 11: 23-29.

Chapter 10

Biochemical Traits for Cold Tolerance of Chickpea (*Cicer Arietinum* L.) Genotypes

Mahesh Kumar*, Neeraj Kumar, Parveen Kumar, Ajeev Kumar

Department of Botany and Plant Physiology, CCS Haryana Agricultural University, Hisar-125004, Haryana, India

**Corresponding author E-mail: maheshk.100290@gmail.com*

Abstract

The present investigation was carried out with six chickpea genotypes viz. ICCV 88506 (Tolerant, National check), HC-1, H03-56, H07-120, H08-71 and H09- 96 differing in their cold sensitivity, to study the low temperature (<5°C) induced changes in biochemical traits under field conditions. Sampling was done at 4 day exposure to below (stress) and above (control) 5 °C temperature. The proline content of leaves increased significantly from 0.781 to 0.913 and 0.605 to 0.798 mg g-1 dry weight, respectively in HC-1 and H03-56 genotypes in stress (<5°C) conditions. Likewise, the total soluble carbohydrate (TSC) content of leaves also increased from 18.35 to 22.74 and 16.83 to 19.69 mg g-1 dry weight in HC-1 and H03-56, respectively. A marked increase in hydrogen peroxide H2O2, lipid peroxidation (MDA content) and relative stress injury (RSI %) was noticed in leaves which was much higher in H03-56 than all other six chickpea genotypes in stress (<5 °C) condition. Better accumulation of osmotically active solutes, i.e. proline, TSC of HC-1 and H07-120, helped in maintaining the higher relative water content percentage of these organs than noticed in other genotypes in stress conditions. The specific activities of ROS scavenging enzymes such as SOD, CAT and POX also increased in leaves of all the six chickpea genotypes, in stress (<5 °C) condition. The increase was more in HC-1 and H07-120 as compared to other chickpea genotypes. Higher activities of antioxidant enzymes, lower accumulation H2O2 and MDA content in HC-1 and H07-120 than all others indicated that these enzymes play a key role in removal of ROS better in these genotypes than all other six chickpea genotypes, thus minimizing the cellular damage caused by ROS under chilling stress. Hence, the mechanism *of cold tolerance was better in HC-1 and H07-120 than other six chickpea genotypes as found biochemical traits studied and these genotypes can be used for cold stress management.*

Keywords: *Chickpea, low temperature stress, specific activity*

1. Introduction

Chickpea (*Cicer arietinum* L.) is mainly grown for human consumption and also used as feed for animals. It is the main source of protein for vegetarian people and the most important pulse crop in the Indian sub-continent grown during cool season. Chickpea straw, the main by-product produced after chickpea grain threshing, is used for animal feeding by the farmers (Kumar *et al.*, 2016).

Chickpea production is constrained by low temperatures across much of its geographical range. The parts of the north Indian sub-continent are the most affected regions due to chilling as the temperature is below 15 ^{0}C at flowering (Berger 2007). Cold related stress is among the least understood of the adversities that affects the plants and can be defined in terms of either chilling between 0 ^{0}C and 10 ^{0}C or freezing below 0 ^{0}C without snow cover (Wery *et al.*, 1993). Low temperature (0-10 ^{0}C) prevails in the month of December, January and early February is a major factor leading to unstable yield in many current production areas. Being a cool season food legume, chickpea faces low temperature to the tune of 0-5 ^{0}C for about 15-20 days in the northern states as crop is highly sensitive to mean temperature below 15 °C at flowering leading to flower dropper pod abortion that results in yield loss.

The factor responsible is cold stress for the greatest agricultural loss by affecting the biochemical activities in plants and thus limiting the yield and productivity. Temperature below 10 ^{0}C is known to alter a variety of biochemical processes ranging from photosynthesis to reactive oxygen species (ROS) and solute accumulation and has adverse effect on chickpea yield and results in losses from 15-20 % (Chaturvedi *et al.*, 2009).

A number of biochemical changes occur during cold-acclimation like alteration in membrane lipid composition, the appearance of new isozymes, increased level of sugars and soluble proteins contents and increased levels of proline and other organic acid etc. Wery *et al.* (1994) observed the altered membrane functions during chilling stress by lipid peroxidation and leakage of electrolytes from chickpea cells. Survival at low temperature depends to a large degree on the ability of plants to maintain the integrity of membranes as they constitute the main damage site. Some of these changes, such as alteration in lipid composition, appear to have a role in bringing about the increased cold tolerance of acclimated plants. The most advantageous temperature range for normal flowering, fertilization and seed set is 10 to 14 0C as minimum temperature and 25 to 31 0C as maximum temperature.

The adverse effects of cold stress in chickpea can be mitigated by developing new improved cultivars having low temperature tolerance. Information on tolerance to low temperature stress in chickpea genotypes is still limited. Keeping this in view, the present study was undertaken to evaluate different chickpea genotypes for biochemical traits under chilling stress i.e. below 5^{0}C for low temperature tolerance.

2. Materials and Methods

2.1 Experimental Details

The present investigation was done in six chickpea (*Cicer arietinum* L.) genotypes namely ICCV 88506 (Tolerant, National check), HC 1, H 03-56, H 07-120, H 08-71 and H 09- 96 (differing in their sensitivity to low temperature) and evaluation for low temperature tolerance for biochemical traits. The experiment was conducted under natural agro-climatic conditions during *Rabi* 2015-16 in the research area of Pulses Section, Department of Genetics and Plant Breeding, CCS Haryana Agricultural University, Hisar in the Randomized Block Design with three replications. Chickpea genotypes were sown in the plot size of 1.2 m x 4 m on 15th October, 2015 (early sown condition in Haryana) to expose the plants to chilling temperature at flowering stage.

2.2 Climate and Weather Conditions of Experimental Location

Hisar (Haryana) is located in Indo-Gangetic plains of North-West India at 215.2 meters above mean sea level with latitude of 29° 10' North and longitude of 75° 46' East. The climate of Hisar can be classified as tropical, semi-arid and hot which is mainly dry with very hot summer and cold winter except during monsoon. The winter season starts late in November and remains upto first week of March. The average rainfall varies from 300-500 mm and the total rain as well as its distribution is subjected to great variations. About 80-90 per cent of the total rain is received from South-West monsoon during the month of July to September. The minimum temperature in this area reaches upto 0.5 ºC normally after mid-December and up to January. Weekly weather parameters recorded at observatory of the CCS Haryana Agricultural University, during *Rabi* 2015-16 are depicted in (Fig. 10.1). Weekly mean maximum temperature during crop season ranged from 15.3°C to 38.6 °C, whereas, the weekly mean minimum temperature varied from 2.4°C to 21.2 °C (40th to 15th Standard Meteorological Week).

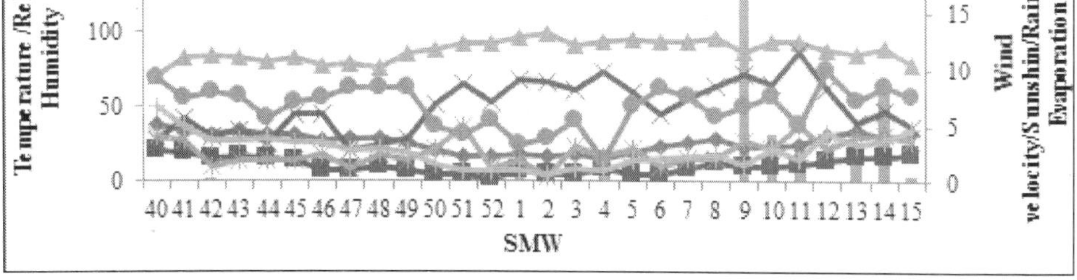

Fig. 10.1: Mean weekly meteorological data (October 2015- April 2016).

Source- Department of Agricultural Meteorology, CCS HAU, Hisar, Haryana, India.

2.3 Observations

In this experiment, the various observations with regard to biochemical parameters were determined. Lipid peroxidation [In terms of malondialdehyde MDA content] MDA is a product of lipid peroxidation and was measured by thiobarbituric acid (TBA) reaction with minor modifications of the method of Heath and Packer (1968). Hydrogen peroxide (H_2O_2) content of the leaves was determined by a modified Patterson *et al.* (1984) method. Proline content was estimated by using the method of Bates *et al.* (1973). Total soluble carbohydrates (TSC) were determined with the method of Yemm and Willis. (1954) using anthrone reagent. Antioxidant enzymes, Superoxide dismutase (EC1.15.1.1) SOD activity was estimated by the method of Giannopolitis and Reis (1997) with little modifications. Catalase (EC 1.11.1.6) (CAT) activity was estimated by the UV method of Aebi (1984). Peroxidase (EC 1.11.1.7) (POX) activity was estimated by the method of Shannon *et al.* (1966).

2.4 Statistical Analysis

Data were subjected to Analysis of Variance (ANOVA) using Online Statistical Analysis Package (OPSTAT) Computer Section, CCS Haryana Agricultural University, Hisar.

3. Results and Discussion

3.1 Lipid Peroxidation

Lipid peroxidation was measured in terms of malondialdehyde (μ mol MDA g^{-1} DW) content. The mean values of MDA content in chilling stressed (<5°C) plants showed marked increase (20.26) over the control (>5°C) plants (14.23) in all the chickpea genotypes (Fig. 10.2). Maximum MDA content was measured in H 03-56 (24.18) and minimum in HC-1 (17.34) in chilling stressed plants. There were significant differences among genotypes, temperature and their interaction.

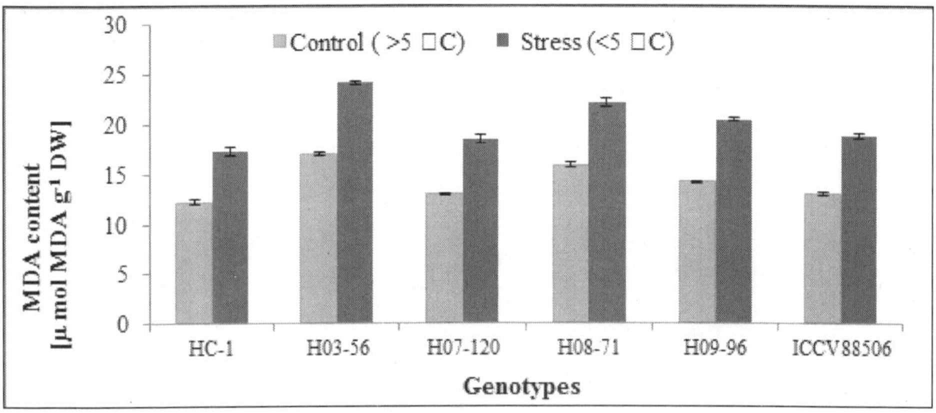

Fig. 10.2: Changes in malondialdehyde (MDA) content in leaves of chickpea genotypes under low temperature (<5 °C) condition.

[Vertical bars indicate ± SE for each treatment]

Lipid peroxidation (LP) is the symptom most easily ascribed to oxidative damage. It is an attack upon polyunsaturated fatty acids of the membrane (Heath and Packer, 1968) which lead to the breakdown of lipid and impairment of membrane function. It most frequently measured in terms of malondialdehyde (MDA) content and is most easily ascribed as an indicator of oxidative damage (Mandhania *et al.*, 2006). Wery *et al.* (1994) observed that altered membrane functions during chilling stress by lipid peroxidation and leakage of electrolytes in the chickpea. In present results membrane stability was affected after 4 days exposure to low temperature below 5 °C. The damage of membrane was reflected in the lipid peroxidation as it was increased at 4 DAE to low temperature. Among six genotypes studied the maximum MDA content was observed in H 03-56 and minimum was noticed in HC 1 in stress (<5 °C) (Fig. 2). Chilling temperature causes the loss of membrane integrity due to lipid peroxidation and MDA is a common product of lipid peroxidation that reflects the extent of oxidative injury.

3.2 Hydrogen Peroxide (H_2O_2) Content

It is obvious from the data presented in (Fig. 10.3) that the hydrogen peroxide content (μ moles g^{-1} DW) increased after 4 days of exposure to low temperature (<5 °C) in all the chickpea genotypes over the control (>5 °C) plants. The H_2O_2 content observed maximum in H 03-56 (0.636) and minimum in HC 1 (0.556) in stress (<5 °C) condition. The overall interaction between genotypes and temperature was statistically significant.

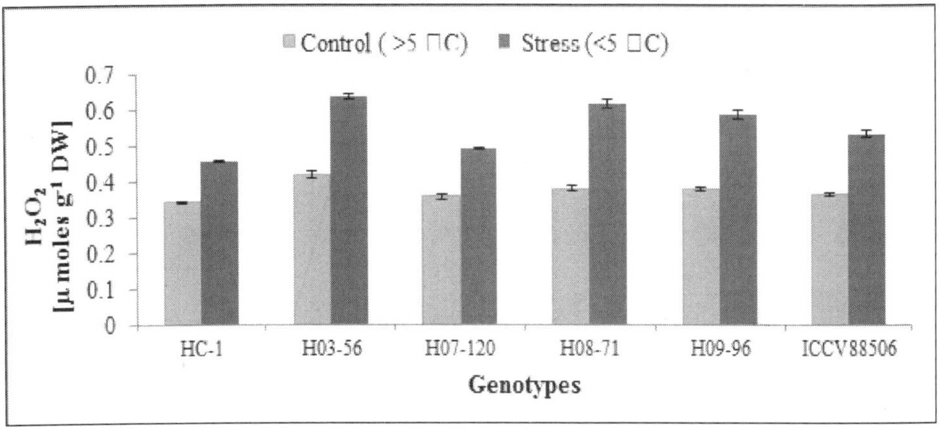

Fig. 10.3: Changes in hydrogen peroxide (H_2O_2) content in leaves of chickpea genotypes under low temperature (<5 °C) condition.

[Vertical bars indicate ± SE for each treatment]

Hydrogen peroxides is a toxic compound produced as a result of the dismutation of the superoxide radical, and a higher concentration is injurious to the cell and the plant, resulting in lipid peroxidation and membrane injury. Our results also showed an increased H_2O_2 levels in leaves after 4 days exposure to low temperature (stress) condition in all the chickpea genotypes. However, the H_2O_2 accumulation increased was more in leaves of H 03-56 than those in HC-1 and H 07-120 (Fig. 3). Increase in H_2O_2 production in chickpea under low temperature has also been reported by Manoharlal (2013).

3.4 Proline Content

The changes in proline content (mg g^{-1} DW) after 4 days exposure to low temperature (<5 °C) in leaves of chickpea genotypes are presented in (Fig. 10.4). There was significant differences in proline content of leaves and it increased from 0.691 to 0.850 with increasing 4 days after exposure to low temperature (<5 °C) as compared to control (>5 °C). The highest proline content was observed in HC1 (0.847) and lowest in H 03-56 (0.702) in stress (<5 °C) as compared to control (>5 °C) plants. There were significant differences among genotypes, temperature and their interaction was also non significant.

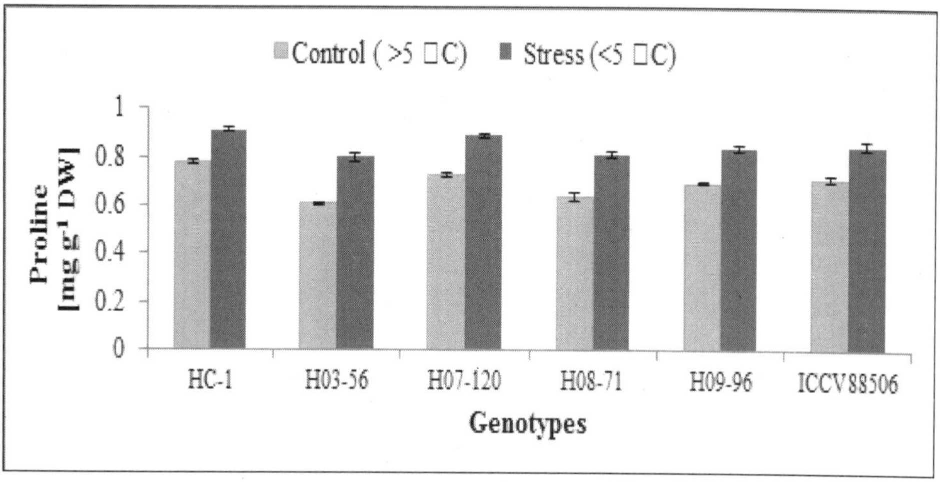

Fig. 10.4: Changes in proline content in leaves of chickpea genotypes under low temperature.

(<5 °C) condition [Vertical bars indicate ± SE for each treatment]

Proline is an amino acid with potent role in osmotic regulation under abiotic stresses. It also acts as an antioxidant under oxidative stress. Proline is considered to be a compatible molecule that has the ability to stabilize subcellular structures and scavenge radicals. This investigation revealed that proline accumulation in leaves of six chickpea genotypes enhanced under low temperature (<5 °C) (Fig. 4). Among six genotypes, proline was more accumulated in HC 1 and H 07-120 as compared to others with increasing exposure to low temperature (<5 °C). Varietal evaluation for proline content was conducted at temperature below 5°C and above 5°C. A relationship was found between levels of proline and chilling resistance. It has been believed that the accumulation of organic salts together with proline is implicated in protection mechanisms such as restoring cell volume and turgor, reduction in cell damage by free radicals etc which is in accordance with our results.

3.5 Total Soluble Carbohydrates (TSC)

The change in the level of TSC content after 4 days exposure to low temperature (<5°C) in leaves of chickpea genotypes is shown in (Fig. 10.5). The TSC increased significantly 4 days after exposure to low temperature (<5°C) as compared to control (>5°C). The genotype HC 1 maintained higher (20.55) TSC and lowest in H03-56 (18.26) in stress condition (<5 °C).The interaction of genotypes and temperature was found to be significant.

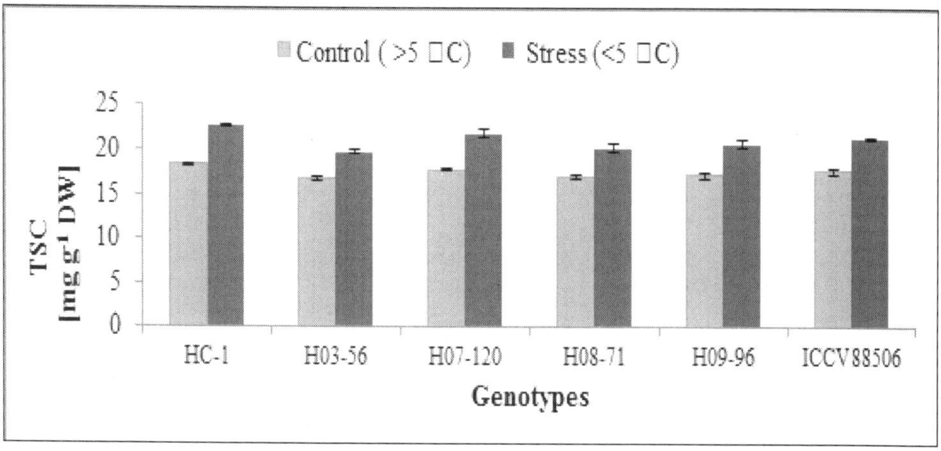

Fig. 10.5: Changes in total soluble carbohydrates (TSC) content in leaves of chickpea genotypes under low temperature (<5 °C) condition.

[Vertical bars indicate ± SE for each treatment]

At low temperature stress, total soluble carbohydrates content increased in early maturing chickpea genotypes (Nayyar, 2005). In the present investigation total soluble carbohydrates also increased in stress plants compared to their respective controls in all the six chickpea genotypes (Fig. 5). The accumulation of total soluble carbohydrates was more in HC-1 than other genotypes in the present study. Taffouo *et al.* (2009) also reported increase in TSC accumulation in stressed plants compared to control in *Vigna unguiculata* L.

3.6 Antioxidant Enzymes:

3.6.1. Superoxide Dismutase (SOD)

After 4 days exposure to low temperature (< 5°C) the specific activity of SOD (mg^{-1} protein min^{-1}) in leaves was found to be increased as compared to control (> 5°C) temperature condition in all the six genotypes. The specific activity of SOD was observed maximum in HC1 (16.50 units mg^{-1} protein min^{-1}) and minimum in H 03-56 (11.87 units mg^{-1} protein min^{-1}) in low temperature stress (<5 °C) as shown in (Fig. 10.6).

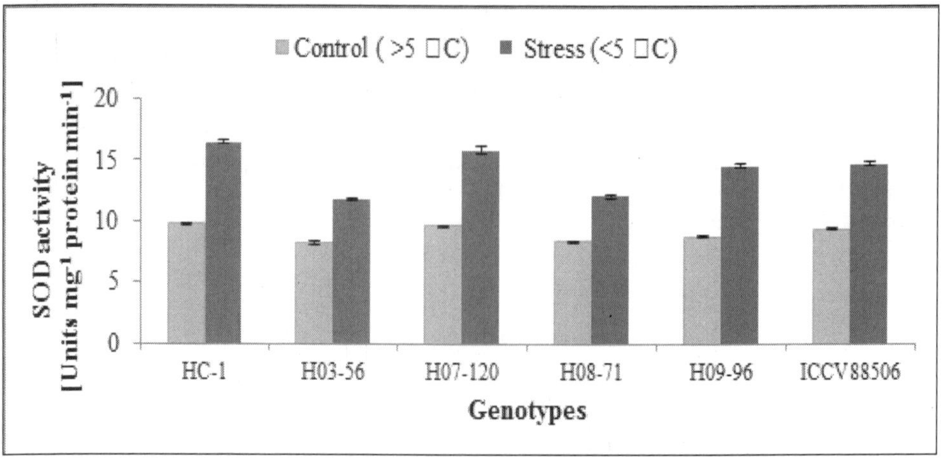

Fig. 10.6: Changes in specific activity of superoxide dismutase (SOD) in leaves of chickpea genotypes under low temperature (<5 °C) condition.

[Vertical bars indicate ± SE for each treatment]

3.6.2. Catalase (CAT)

The specific activity of catalase [mg^{-1} (protein) min^{-1}] in leaves of six genotypes after 4 days exposure to low temperature (<5°C) is represented in (Fig. 10.7). The specific activity of CAT increased significantly after 4 days after exposure to low temperature (<5 °C) in all six genotypes as compared to control (>5°C). The activity of CAT was highest in HC 1(7.65) and lowest in H 03-56(5.82) genotype. The statistical data revealed that the significant differences among genotypes, temperature levels and their interaction were found to be significant.

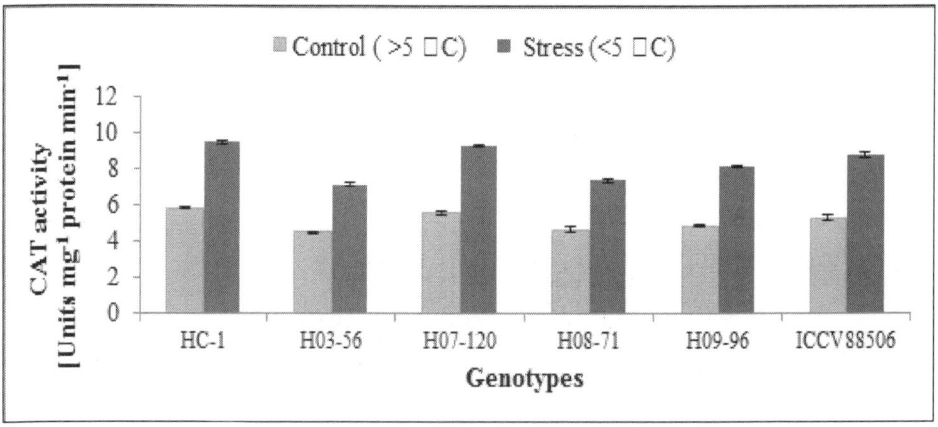

Fig. 10.7: Changes in specific activity of catalase (CAT) in leaves of chickpea genotypes under low temperature (<5 °C) condition.

[Vertical bars indicate ± SE for each treatment]

3.6.3. Peroxidase (POX)

The specific activity of POX measured as units [mg^{-1} (protein) min^{-1}] in leaves significantly varied among various chickpea genotypes (Fig. 10.8) showed that low temperature stress resulted in an increase in the specific activity POX in leaves of all six genotypes. It was highest in HC-1 (0.130) and lowest in H03-56 (0.083) under low temperature (<5°C) stress conditions. The interaction between genotype and temperature was found significant.

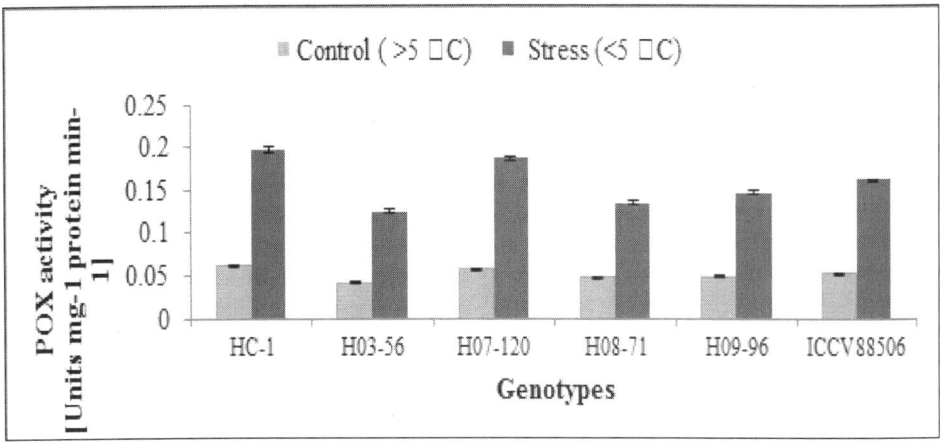

Fig. 10.8: Changes in specific activity of peroxidase (POX) in leaves of chickpea genotypes under low temperature (<5 °C) condition.

[Vertical bars indicate ± SE for each treatment]

3.7 Antioxidants Defense System

Chilling temperatures lead to oxidative stress which occurs due to the overproduction of reactive oxygen species (ROSs), [such as superoxide (O^-_2), singlet oxygen ($^1O_2$), hydrogen peroxide (H_2O_2) and hydroxyl redicals ($^\cdot OH$)] that cause peroxidations of membrane lipids, denaturation and aggregation of proteins, DNA fractures and inactivation of enzymes (Gill and Tuteja, 2010). It is suggested that oxidative stress may be a significant factor in relation to chilling injury. Higher plants have defense system to scavenge reactive oxygen species (ROS).

The activity of SOD, which is responsible for scavenging O^{-2} to produce H_2O_2, increased under cold stress. This increased activity may reflect both the enhanced amount of O^{-2} production and the scavenging ability of SOD in response to cold stress and could be an important protective mechanism for the photosynthetic apparatus. The specific activity of SOD was also increased in leaves of all the six chickpea genotypes after 4 days exposure to low temperature (>5 °C), which was found to be higher in genotype HC 1 followed by H 07-120 and minimum in H 03-56 in response to cold stress (>5 °C) (Fig. 10.6) during present investigation.

CAT is the principal H_2O_2-scavenging enzyme located in peroxisomes and scavenges photorespiratory H_2O_2 without consuming reducing power. Our data showed that the specific activity of CAT in leaves was increased with the cold stress (<5 °C) as compared to control (>5 °C) condition. The specific activity of CAT was higher in HC 1 followed by H 07-120 and minimum in H 03-56 in cold stress (<5 °C). The cold stress enhanced higher specific activity of CAT in HC-1 and H 07-120 may suggest its effective scavenging mechanism to remove H_2O_2 (Fig. 10.7) and imparting tolerance against cold stress. The capacity of plants to withstand cold stress is not constant but increases noticeably upon exposure to progressive cold stress (Gill and Tuteja, 2010).

An important enzyme regulating intracellular H_2O_2 levels is peroxidase (POX). Peroxidases, besides their main function in H_2O_2 elimination, can also catalyse O_2 and H_2O_2 formation by a complex reaction in which NADH is oxidized using trace amounts of H_2O_2 (Blokhina *et al.*, 2003). In the present investigation, the specific activity of POX in leaves increased after 4 days exposure to low temperature (<5 °C), which was found to be higher in HC-1 followed by H07-120 and minimum in H03-56 genotypes (Fig. 10.8). In creating cold tolerance, plant cells induce cascades of alterations in metabolic pathways including the activity of antioxidant enzymes (Maali-Amiri *et al.*, 2007). In acclimated plants, antioxidants and antioxidant enzymes such as SOD, CAT, POX had an important role in creating greater cold tolerance. In our study we hypnotized that an increased activity of these antioxidant enzymes (SOD, CAT and POX) contributes to the protection of chickpea genotypes.

4. Conclusion

The chickpea genotype HC 1 and H 07-120 showed significantly higher accumulation of proline and TSC contents than other genotypes. The H_2O_2 accumulation increase was more in leaves of H 03-56 than those in HC 1 and H 07-120 in stress (<5 °C) which confirms that chilling stress lead to excessive ROS production in H 03-56. This indicated that H 03-56 is chilling sensitive as compared to HC 1 and H 07-120. The genotypes HC 1 and H 07-120 also showed higher activity of antioxidant enzymes as compared to all other genotypes. Higher specific activity of SOD, CAT, POX, lower accumulation H_2O_2 and MDA content in HC 1 and H 07-120 than all other genotypes indicated that these enzymes play a better key role in removal of ROS in HC 1 and H 07-120 and thus minimizing the cellular damage caused by ROS under low temperature (<5°C). Therefore, the chickpea genotypes HC 1 and H 07-120 were found to be relative tolerant to low temperature stress than other genotypes based upon various biochemical attributes studied. These chickpea genotypes can be used in cold tolerance for crop improvement programme of chickpea.

References

Aebi, H. 1984. Catalase *in vitro*. *Methods* in *Enzymology*, 105: 121-126.

Bates, L., Waldren, R.P. and Teare, I.D. 1973. Rapid determination of free proline for water-stress studies. *Plant Soil*, 39: 205-207.

Berger, J.D. 2007. Ecogeographic and evolutionary approaches to improving adaptation of autumn-sown chickpea (*Cicer arietinum* L.) to terminal drought: the search for reproductive chilling tolerance. *Field Crops Res.*, 104: 112-122.

Blokhina, O., Eija, V. and Kurt, V.F. 2003. Antioxidants, oxidative damage and oxygen deprivation stress: a review. *Annals of Botany*, 91: 179-194.

Chaturvedi, S.K., Mishra, D.K., Vyas, P. and Mishra, N. 2009. Breeding for cold tolerance in chickpea. *Trends in Biosciences,* 2: 1-6.

Giannopolitis, C.N. and Reis, S.K. 1997. Superoxide dismutase I: occurrence in higher plants. *Plant Physiol.,* 59: 309-314.

Gill, S.S. and Tuteja, N. 2010. Reactive oxygen species and antioxidant machinery in abiotic stress tolerance in crop plants. *Plant Physiol. Biochem.,* 48: 909–930.

Heath, R.L. and Packer, L. 1968. Photoperoxidation in isolated chloroplasts. I. Kinetics and stoichiometry of fatty acid peroxidation. *Archives Biochem. Biophy.,* 125: 189-198.

Kumar, K., Lal, R. and Jat, B.L. 2016. Evaluation of chickpea (*Cicer arietinum* L.) genotypes for yield and their susceptibility reaction against *Helicoverpa armigera* (Hub.). *Forage Research,* 41: 237-241.

Maali-Amiri, R., Goldenkova-Pavlova, I.V., Yur'eva, N.O., Pchelkin, V.P., Tsydendambaev, V.D., Vereshchagin, A.G., Deryabin, A.N., Trunova, T.I., Los, D.A. and Nosov, A.M. 2007. Lipid fatty acid composition of potato plants transformed with the 12-desaturase gene from Cyanobacterium. *Russian J.Plant Physiol.,* 54: 600–606.

Mandhania, S., Madan, S. and Sawhney, V. 2006. Antioxidant defense mechanism under salt stress in wheat seedling. *Biologia Plantarum,* 50: 227-231.

Manohar, L. 2013. Physiological and biochemical responses to low temperature in chickpea (*Cicer arietinum* L.). M.Sc. thesis, Chaudhary Charan Singh Haryana Agricultural University, Hisar, India.

Nayyar, H. 2005. Putrescene increase floral retention, pod set and seed yield in cold stressed chickpea. *Journal of Agronomy and Crop Science,* 191: 340-45.

Patterson, B.D., Macrae, E.A. and Ferguson, I.B. 1984. Estimation of hydrogen peroxide in plant extracts using titanium. *Analytical Biochemistry,* 139: 487–492.

Shannon, L.M., Kay, E. and Lew, J.Y. 1966. Peroxidase isozymes from horseradish roots: isolation and physical properties. *J. Biol. Chem.,* 241: 2166–2172.

Taffouo, V.D., Kemdem-Kouamou, J., Tchiengue-Ngalangue, L.M., Nandjou-Ndjeudji, B.A. and Amougou, A. 2009. Effects of salinity stress on growth, ions partitioning and yield of some cowpea (*Vigna unguiculata* L. Walp.) cultivars. *Int. J. Bot.,* 5: 135-145.

Wery, J., Silim, S.N., Knight, E.J., Malhotra, R.S. and Cousin, R. 1994. Screening techniques and sources of tolerance to extremes of moisture and air temperature in cool season food legumes. *Euphytica,* 73: 73-83.

Wery, J., Turc, O. and Lecoeur, J. 1993. Mechanisms of resistance to cold, heat and drought in cool-season legumes, with special reference to chickpea and pea. In:. Singh, K. B. and Saxena, M.C. (Eds). *Breeding for Stress Tolerance in Cool- Season Food Legumes.* Wiley, Chichester, pp. 271-291.

Yemm, E.W. and Willis, A.J. 1954. The estimation of carbohydrates in plant extract by anthrone. *Biochem. J.,* 57: 508-514.

Chapter 11

Screening and Isolation of Potent Dye Decolorizing Bacteria from Textile Effluent

Rashmi and Bindu Battan*

Department of Biotechnology, Kurukshetra University, Kurukshetra-136119, India
**Corresponding Author E-mail: bin_battan@yahoomail.com*

Abstract

A total of 22 bacterial isolates were isolated from different samples (Dry sludge, Wet sludge, Soil & Textile effluent) collected from Nahar Textile Mill, Lalru, and Punjab and screened for dye decolorizing ability. All 22 isolates appeared positive in response to decolorization of 16 different dyes. Three bacterial strains were identified to be most potent decolorizers among all, that showed complete decolorization of almost all the dyes. The identified strains were Fictibacillis gelatini, Bacillus subtilis inaquosorum & Bacillus subtilis subtilis. Furthermore, isolated strains showed appreciable ability to decolorize different types of textile dyes under optimized conditions and were found to be 100 % decolorizers till 500 mg/l within minimum time of 5-9 hours to maximum 1 week in static conditions of growth at pH 7 and at temperature 37 °C. This study thus reveals that some bacteria inhabit in textile effluent whereby utilize the dyes as their source of energy and nutrition and imply their importance in the treatment of industrial effluents.

Keywords: *Decolorization, Potent bacteria, Textile effluents, Bacillus* sp., *Isolation & screening.*

1. Introduction

One of the main sources with severe pollution problems worldwide is the textile industry and its dye-containing wastewaters. Worldwide 10,000 types of textile dyes with an estimation of 7.105 metric tonns are annually produced (Baban *et al.*, 2010, Robinson *et al.*, 2002, Soloman *et al.*, 2009). Major classes of synthetic dyes used include azo, anthraquinone and triarylmethane dyes and azo dyes represent 70 % of all organic dyes used by textile industries (Stolz, 2001). Color elimination in waste water is today the principal problem and has to be removed before discharging into receiving water body. About 2-20 % of dyes are directly discharged as aqueous effluents in different environmental components (Verma *et al.*, 2015). Improper disposal of these dyes in aquatic ecosystem leads to decreased photosynthetic activity, water quality, and dissolved oxygen. These dyes can also affect the human health by causing nausea, ulceration of the skin, hemorrhagic and severe damage to reproductive system, kidney, liver, brain and central nervous system (Sandhya *et al.*, 2005, Vandevivere *et al.*, 1998).

A conventional waste water treatment system is not efficient to remove recalcitrant, because of high cost, low efficiency and in-applicability to wide variety of dyes (Stephonson *et al.*, 1996). The reuse of industrial effluents can be an attractive option for meeting the increasing demand of water. For this purpose microbial decolorization is efficient, eco-friendly and more cost-effective (Phugare *et al.*, 2011). Due to ubiquitous nature of bacteria, they can be used as invaluable tools for the biological treatment of textile effluent. As a preliminary step for biological treatment, there is a need to isolate bacterial strains with a potential to decolorize and degrade textile dyes and remove other pollution parameters (Olukanni *et al.*, 2006). A number of bacterial species belonging to genera *Bacillus, Micrococcus, Proteus, Pseudomonas, Sphingomonas* and *Staphylococcus* are reported to have been isolated for bio-treatment of textile dyes (Ali *et al.*, 2009, Zhang *et al.*, 2010, Ayed *et al.*, 2011, Chen *et al.*, 2011). Microorganism isolated from soils and sludge sample belonging to *Bacillus sp* (Horitsu *et al.*, 1977), *Alcaligenes sp.* and *Aeromonas sp.* were found to have good dye decolorization activity (Sharma *et al.*,2004).

The aim of the present study was to isolate and characterize efficient bacterial strains with the remarkable ability to decolorize/degrade various classes of textile dyes. The use of isolated bacteria either individually or as consortium was envisaged to develop efficient biological process for the treatment of effluents containing different dyes.

2. Materials and Methods

2.1 Chemical Collection

All the chemicals of analytical grade were purchased from Hi-Media Laboratories, India.

2.2 Sample Collection

Different samples of soil, wet sludge, dry sludge and effluent of reactive dyes were obtained from Nahar Textile Industry, Lalru (Punjab). All the samples were collected in sterile bottles and stored at room temperature and samples were tested within 24 h of collection.

2.3 Dye Collection

Textile dyes, Congo red was purchased from Hi- Media laboratories, India. Rest of the 15 dyes namely; Trypan blue, Disperse orange 1, Disperse yellow198, Direct brown 95, Disperse red 167, Reactive yellow 15, Reactive black 5, Disperse orange 30, Disperse blue 183, Reactive blue 222, Reactive yellow 145, Reactive red 195,

Disperse blue 79, Reactive red 120, Disperse red 1 were procured from Bhiwani Textile Mill, Haryana, India.

2.4 Isolation of Bacteria and Maintenance of Pure Colonies

The bacterial colonies were isolated from the reactive dye effluent, sludge and soil sample by serial dilution and spreading method. Plates were incubated at 37 C for 24 h. The pure cultures of bacterial isolates were maintained on nutrient agar slants at 4 C and in glycerol stock solution. Decolorization experiments were carried out in nutrient broth.

2.5 Screening of Dye Decolorizing Bacteria

Prepared the 50 ml nutrient broth in 250 ml borosilicate glass flask for each pure isolate, then inoculated the media with these isolates and incubated at 37 C for 24 h. The tests tubes of 25 ml containing 10 ml of nutrient broth and amended with 500 mg/l of dye solution and were inoculated with 24 h old cultures. The percent of decolorization was determined by decrease in absorbance of nutrient broth.

2.6 Identification of Dye Decolorizing Bacteria

The cultures were identified by performing biochemical tests in the laboratory. Further identification was studied by 16 S r-DNA sequencing; and was done by MTCC IMTECH, Chandigarh.

2.7 Decolorization of Textile Dyes

Decolorization activity was expressed in terms of percentage decolorization and was determined by monitoring the decrease in absorbance at absorption maxima (λ max) of respective dyes. The Uninoculated medium supplemented with respective dye was used as reference. The culture suspension was centrifuged at 10,000 rpm for 10 min for removal of the biomass as they interfere with the measurement. The degree of decolorization of dyes was measured by reading the absorbance of supernatants at λ max of specific dyes by UV-visible spectrophotometer. The decolorization was calculated according to the following formula:

Decolorization (%) = Initial absorbance- observed absorbance/ Initial absorbance 100

Average decolorization rate = C % D 100/100 t

Where C = initial concentration of dye mg/l,

% D = dye decolorization % after time t (Pant *et al.*, 2009).

3. Results and Discussion

3.1 Isolation of Dye Decolorizing Bacteria

After spreading the effluent on the nutrient agar plate, many bacterial colonies were developed after 24 h of incubation period, out of which 22 isolates showed dye decolorizing activity. They were designated as RS6, WS6, DS6, DS4, RS6I, RS6II, RD6, DS4I, RS6III, RS6IV, RS6IV, DS6I, RS6VI, RS6VII, RS6VIII, DS6II, WS6I, WS6II,

WS6III, DS6III, RD6I, and RD6II and were assessed for the determination of dye decolorization activity.

3.2 Selection and Screening of More Potent Dye Decolorizing Bacterial Isolates

Screening was carried out using selected dyes in nutrient broth medium. All 22 colonies were tested individually for their ability to decolorize 16 different dyes by using 5 mg/10 ml of dyes. All isolates from four samples decolorized the dyes with different capacities. Most of the isolates showed complete decolorization but only 3 were selected for further studies on the basis of maximum decolorization of dyes in minimum time of 4 to 9 hrs as given in (Table 11.1-4).

There were total of 9 cultures isolated from the soil sample (Table 11.1) designated as RS6, RS6I, RS6II, RS6III, RS6IV, RS6V, RS6VI, RS6VII, RS6VIII but out of these only 2 (RS6 & RS6 IV) were selected for the bacterium consortium. Isolate **RS6** decolorized almost all the dyes completely like DO1, DY198, DR167, RY15, DO30, DB183, RB222, RY145, RR195, DB79, RR120 and DR1. Also it showed complete decolorization of dye RY15, RY145 and RR195 in minimum time of 9 hr, 19 hr and 12 hr respectively. Among the incomplete decolorized dyes were; CR which was decolorized upto 45% in 4 days, DB95 upto 35 % in 3 days, RB5 upto 54 % in 4 days and no decolorization of TB was recorded. Isolate RS6 decolorized Direct brown 95 dye upto 35% which was not decolorized by any other culture (Fig.11.1, a). Pokharia *et al.*, 2013, isolated 45 bacterial cultures from soil and effluent samples collected from Nahar Oswal Denim, Mohali, 3 isolates from effluent sample collected from Abhishek Industries Ltd., Barnala, and 9 isolates obtained from soil samples near local dyers.

Isolate **RS6IV** also showed tremendous capability for complete decolorization of dyes where TB which was not decolorized by RS6 was completely decolorized in 5 days and also showed complete decolorization of other dyes like DO1, RY15, RB5, DO30, RY145, RR195, RR120, DR1, and among partially decolorized dyes were, CR 63% in 4 days, DY198 37 % in 6 days, DR167 41 % in 4 days and RB222 73 % in 5 days whereas dyes DB95,DB183 and DB79 were not decolorized by this isolate and the isolate S6IV also showed 100 % decolorization of RY15, RY145 and RR195 dyes in 12 hr, 5 hr and 19 hr respectively (Fig. 11.1, b). Also RB5 was 100 %

decolorized by only S6IV.

Table 11.1: Decolorizing bacteria isolated from soil sample.

Dyes → Strains ↓	CR	TyB	Dis.01	Dis.Y 198	Dir.B 95	Dis.R 167	RY15	RB5	Dis. 030	Dis.B 183	RB222	RY145	RR195	Dis. B79	RR120	Dis.R1
RS6	45/4d	-	*8d	*6d	35/3d	*3d	*9h	54/4d	*3d	*5d	*6d	*9h	*12h	*5d	*2d	*2d
RS6I	12/1d	*7d	82/4d	80/3d	-	*7d	*8d	-	*3d	27/7d	51/7d	*1d	95/4d	*7d	28/9d	56/9d
RS6 II	11/1d	-	53/2d	87/3d	-	*6d	*8d	81/7d	92/6d	-	85/9d	*2d	*2d	*7d	*3d	54/7d
RS6 III	86/7d	*7d	*3d	32/6d	-	26/5d	*4d	10/4d	24/1d	-	49/5d	*1d	*3d	*6d	*3d	*3d
RS6 IV	63/4d	*5d	*3d	37/6d	-	41/4d	*18h	100/4d	*3d	-	73/5d	*4h	*12h	-	*2d	*4d
RS6V	80/7d	*7d	90/5d	*6d	-	55/5d	50/7d	-	*3d	-	-	*1d	*7d	21/6d	58/7d	*4d
RS6 VI	22/2d	-	10/2d	59/6d	-	-	95/4d	86/2d	23/4d	-	-	*2d	*6d	-	-	14/4d
RS6 VII	-	-	16/2d	50/6d	-	-	*5d	14/2d	-	-	-	*2d	*5d	-	-	53/6d
RS6VIII	42/2d	-	10/2d	*7d	-	74/5d	*7d	81/2d	22/4d	-	51/5d	*2d	*6d	-	33/3d	*4d

♦ Data in numbers represents decolorization in percentage.

*Represents complete decolorization (100%), rest of them were incompletely decolorized; d-days.

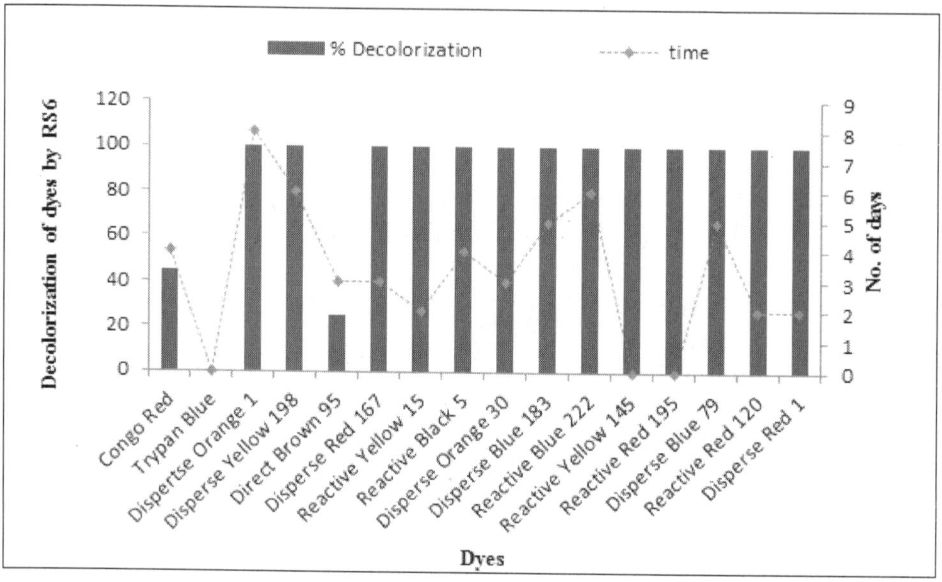

Fig 11.1: a. Percent decolorization of textile dye by bacterial isolate RS6.

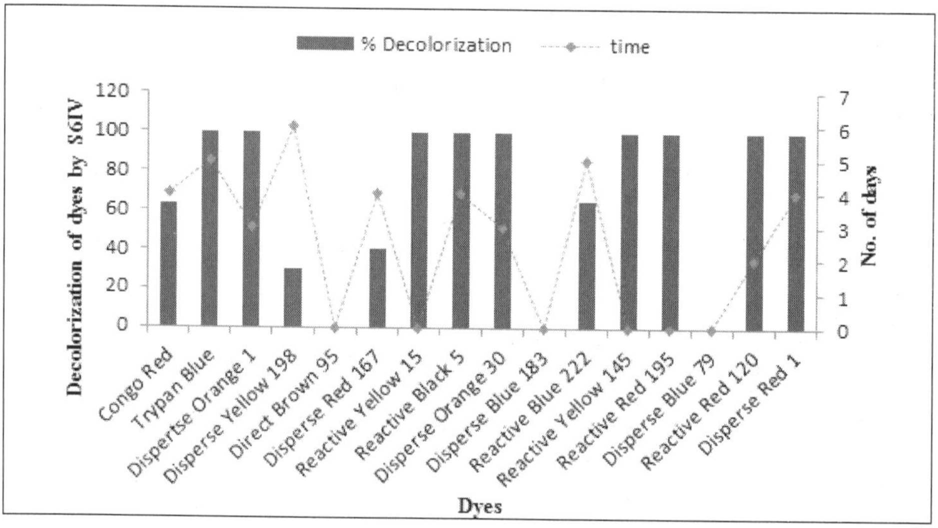

Fig.11.1: b. Percent decolorization of textile dye by bacterial isolate S6IV.

Table 11.2: Decolorizing bacteria isolated from wet sludge sample.

Dyes → / Strains ↓	CR	Ty B	Dis. 01	Dis. Y 198	Dir. B 95	Dis. R 167	RY 15	RB 5	Dis.O 30	Dis.B 183	R B 222	RY 145	RR 195	Dis. B 79	RR 120	Dis.R1
WS6	*5d	-	*8d	*6d	31/3d	*3d	*8d	-	*2d	44/7d	63/7d	*9h	*2d	*5d	92/8d	*3d
WS6I	26/2d	-	21/2d	*7d	-	-	34/4d	91/2d	-	-	-	*2d	10/5d	-	15/2d	88/6d
WS6II	-	-	-	11/3d	-	-	-	-	-	-	-	-	-	16/4d	-	51/4d
WS6III	-	-	-	16/3d	-	-	65/4d	-	-	*3d	-	83/2d	*3d	12/3d	-	66/4d

♦Data in numbers represents decolorization in percentage.

*Represents complete decolorization (100%), rest of them were incompletely decolorized; d-days.

Table 11.3: Decolorizing bacteria isolated from dry sludge sample.

Dyes → / Strains ↓	CR	Ty B	Dis. 01	Dis. Y 198	Dir. B 95	Dis. R 167	RY 15	RB 5	Dis.O 30	Dis.B 183	RB 222	RY 145	RR 195	Dis. B 79	RR 120	Dis. R1
DS6	54/4d	*5d	*8d	*6d	-	*3d	*18h	80/4d	*3d	62/8d	*2d	*9h	*18h	*5d	*2d	*2d
DS4	60/4d	*6d	*8d	*6d	-	*7d	*8d	-	-	*5d	-	*2d	*2d	53/3d	*5d	*6d
DS4I	17/1d	-	52/4d	90/6d	-	*6d	*2d	80/3d	87/6d	-	67/7d	*4d	*1d	*9d	95/7d	-
DS6I	11/8d	*5d	58/5d	34/6d	-	77/3d	*7d	*3d	39/2d	-	-	*3d	*5d	-	56/7d	*3d
DS6II	46/3d	-	19/2d	31/6d	-	71/5d	81/6d	80/2d	-	-	-	*2d	*6d	-	*2d	58/7d
DS6III	-	-	*5d	57/3d	-	-	-	-	-	-	-	-	-	-	-	-

♦Data in numbers represents decolorization in percentage.

*Represents complete decolorization (100%), rest of them were incompletely decolorized; d-days.

Six cultures were isolated from the sample dry sludge; these are DS6, DS4, DS4I, DS6I, DS6II, and DS6III (Table 11.3). Among these isolates **DS6** showed the maximum potential of decolorizing the dyes it showed complete decolorization of dyes TB, DO1, DY198, DR167, RY 15, DO30, RB222, RY145, RR195, DB79, RR120 and DR1. Whereas dye CR was decolorized upto 54 % in 4 days, RB5 upto 80 % in 4 days, DB183 upto 62 % in 8 days and DB95 showed no decolorization at all. Isolate DS6 was selected for the consortium as it also showed complete decolorization of the three dyes: RY15, RY145, and RR 195 in minimum time of 19hr, 9hr, and 19 hr respectively (Fig. 11.1, c). Verma *et al.*, 2015, isolated total of six different isolates (DDB 1-DDB-6) from sludge, soil and liquid samples which were screened for azo dye degradation. Acid red (22, 60, 97, 68, 23 and 68 %) & Direct black (26, 63, 100, 52, 31 and 62 %) were decolorized in mineral salt medium containing 100 mg/l of each dye at 35 C within 5 to 6 days.

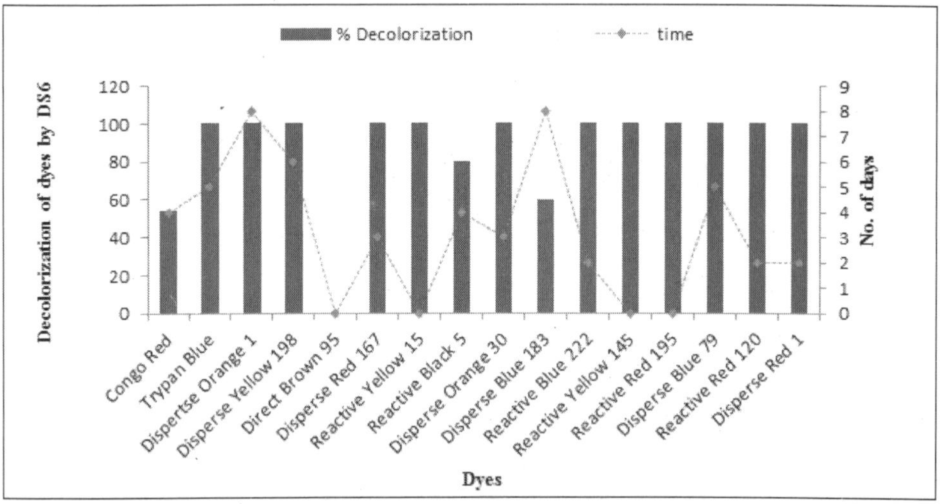

Fig 11.1: c. Percent Decolorization of textile dye by bacterial isolate DS_6.

**Represents complete decolorization (100%), rest of them were incompletely decolorized; d-days.*

Three cultures were isolated from Reactive dye effluent sample, were named as: RD6, RD6I, and RD6II (Table 11.4). Out of all 22 isolates only 3 were selected on the basis of maximum decolorization in minimum time of hours for further research study. Dwivedi *et al.*, 2017 isolated 4, 6 and 5 colonies from Acid red, Remazol black and Reactive blue dyes respectively. The identified bacterial cultures decolorized all three dyes completely in 96 hrs, which were evaluated in nutrient broth.

The isolation of different microorganisms from the sample indicates the natural adaptations of microorganisms to survive in presence of toxic dyes and difference in their decolorization rate is due to the loss of ecological interaction. Similar results also obtained by Saikia N. and Gopal M. (2004).

Table 11.4: Decolorizing bacteria isolated from reactive dye effluent sample.

Dyes / Strains	CR	Ty B	Dis. O 1	Dis. Y 198	DB 95	RY 15	RB 5	Dis.O 30	Dis.B 183	RB 222	RY 145	RR 195	Dis. B 79	RR 120	RR 120	Dis. R1
RD6	12/1d	-	33/2d	84/4d	-	*6d	*8d	96/5d	80/6d	28/7d	85/4d	*3d	94/7d	*7d	88/3d	56/7d
RD6I	-	-	-	56/3d	-	-	-	-	*1d	90/4d	-	*1d	92/4d	13/4d	*3d	24/4d
RD6II	-	-	-	55/3d	-	-	-	-	-	*1d	-	-	*1d	*3d	*3d	*3d

◆ *Data in numbers represents decolorization in percentage.*

The difference in decolorization rate is also due to the difference in their specificity, structure, complexity and nature and the position of substituent in the aromatic rings and interaction with azo bond with different dyes (Carliell, C.M.,Barclay *et al*; 1995). Mohamed (2016), isolated 37 bacterial colonies, out of which, 5 identified bacterial strains *Pseudomonas* sp., *Salmonella* sp., *Aeromonas* sp., *Klebisiella* sp. and *Bacillus* sp., isolate *Pseudomonas* sp. could decolorize Reactive black 5 and Reactive blue 222 upto 83% and 70% respectively. According to Arulazhagan (2016), out of 8 isolates, 1 isolate recognized as *Bacillus subtilis* showed maximum degradation of Reactive red M8B (0.1 %) upto 97.13 % after 5th day.

3.3 Identification of Microorganisms

A preliminary identification of the culture was based upon biochemical tests and final identification was done by 16 S rDNA sequences. The homology found in the GenBank database indicated that the strain RS6, DS6, and S6IV were in the Phylogenetic branch of the genus *Fictibacillus*, *Bacillus* and *Bacillus* respectively. These isolates were identified as *Fictibacillus gelatini*, *Bacillus subtilis* subsp. *inaquosorum* and *Bacillus subtilis* subsp. *subtilis*.

4. Conclusions

Textile effluent and sludge produced by effluent treatment plant is rich source of dye decolorizing bacterial population. Samples were dominated with Gram +ve & Endospore forming species. Initially, all 22 isolates were tested for their ability to decolorize 16 dyes. Afterwards three potential isolates were selected on the basis of their maximum dye decolorizing ability in minimum time for industrial purpose.

Acknowledgements

Authors would like to acknowledge Department of Biotechnology, Kurukshetra University Kurukshetra, Haryana, India, for providing all help and required facilities.

References

Ali, N., Hameed, A. and Ahmed, S. 2009. Physiological characterization and bioremediation perspective of textile effluent, dyes and metals by indigenous bacteria. J. Hazard Mater. 164: 322-328.

Arulazhagan, P. 2016. A study on microbial decolorization of reactive red M8B by Bacillus subtilis isolated from dye contaminated soil samples. Int. J. Curr. Res. Biol. Med. 1: 1-13.

Ayed, L., Mahdhi, A., Cheref, A. and Bakhrouf, A. 2011. Decolorization and degradation of azo dye Methyl Red by an isolated Sphingomonas paucimobilis: Biotoxicity and metabolites characterization. Desalination. 274: 272-277.

Baban, A., Yediler, A., Avaz, G. and Hostede, S.S. 2010. Biological and oxidative treatment of cotton textile dye-bath effluents by fixed and fluidized bed reactors. Biores. Technol. 101: 1147-52.

Carliell, C.M., Barclay, S. J., Naidoo, N., Burckley, C.A., Mulholland, D.A. and Senior, E. 1995. Microbial decolorization of reactive azo dye under anaerobic conditions. Water Res. Commission. 22: 61-69.

Chen, B., Huueh, C., Chen, W. and Li, W. 2011. Exploring decolorization and halo tolerance characteristics by indigenous acclimatized bacteria: Chemical structure of azo dyes and dose- response assessment. J. Taiwan Inst. Chem. E. 42: 816-825.

Dwivedi, P. and Tomar, R.S. 2017. Microbial degradation and decolorization of azo and anthraquinone textile dyes. Int. J. Pharma. Bio. Sci. 8: 989-998.

Horitsu, H., Takada, M., Idaka, E., Tomoyeda, M. and Ogawa, T. 1977. Degradation of amino azo-benze near by Bacillus subtilis. Eur. J. Appl. Microbiol., 4: 217-224.

Mohamed, W. S. 2016. Isolation and screening of reactive dye decolorizing bacterial isolates from textile industry effluent. Int. J. Micro Res. 7: 1-8.

Olukanni, O.D., Osuntoki, A.A. and Gbenle, G.O. 2006. Textile effluent biodegradation potentials. Afr. J. Biotechnol. 5: 1980-1984.

Pant, D. and Adholeya, A. 2009. Nitrogen removal from bio-methanated spent wash using hydroponics treatment followed by fungal decolorization. Environ. Eng. Sci. 26: 559.

Phugare, S.S., Kalyani, D.C., Surwase, S.N. and Jadhav, J.P. 2011. Ecofriendly degradation, decolorization and detoxification of textile effluent by a developed bacterial consortium. Ecotox. Environ. Safe. 74: 1288-96.

Pokharia, A. and Ahluwalia, S.S. 2013. Isolation and screening of dye decolorizing bacterial isolated from contaminated sites. Textile light Industrial Sci. Technol. 2: 54-61.

Robinson, T., Chandran, B. and Nigam, P. 2002. Removal of dyes from an artificial textile dye effluent by two agricultural residue, corncob and barley husk. Environ. Int. 28: 29-33.

Saika, N. and Gopal, M. 2004. Biodegradation of β-cyfluthrin by fungi. J. Agri. Food Chem. 52: 1220-1223.

Sandhya. S., Padmavathy, S., Swaminathan, K., Subrahmanyam, Y.V. and Kaul, S.N. 2005. Microaerophilic–aerobic sequential batch reactor for treatment of azo dyes containing simulated wastewater. Process. Biochem. 40: 885-90.

Sharma, D.K., Saini, H.S., Singh, M., Chimni, S.S. and Chadha, B.S. 2004. Biotreatment of Simulated Textile Dye Effluent Containing malachite green by an Up-Flow Immobilized Cell Bioreactor. W. J. Microbiol. Biotechnol. 20: 431-434.

Soloman, P.A., Basha, C.A., Velan, M., Ramamurthi, V., Koteeswaran, K. and Balasubramanian, N. 2009. Electrochemical degradation of Remazol Black B dye effluent. CLEAN–Soil, Air, Water. 37: 889-900.

Stephenson, R.J. and Sheldon, J.B. 1996. Coagulation and precipitation of a mechanical pulping effluent. I Removal of carbon and turbidity. Wat. Res. 30: 781-792

Stolz., A. 2001. Basic and applied aspects in the microbial degradation of azo dyes. Applied Microbiol. Biotechnol. 56: 69-80.

Vandevivere, P.C., Bianchi, R. and Verstraete, W. 1998. Review: Treatment and reuse of wastewater from the textile wet-processing industry: Review of emerging technologies. J of Chemical Technol. and Biotechnol. 72: 289-302.

Verma, S.K., Yadav, S. and Singh, P. 2015. Isolation, Screening and Characterization of Azo Dye Decolorizing Bacteria from Contaminated Soil. Intl. J. Engg. Sci. Adv. Research. 1: 215-222.

Zhang, M., Chen, W., Chen, B., Chang, C., Hsuesh, C., Ding, Y., Lin, K. and Xu, H. 2010. Comparative study on characteristics of azo dye decolorization by indigenous decolorizers. Biores. Technol. 101: 2651-2656.

Chapter 12

Celiac Disease: Pathogenesis, Diagnosis, Treatment and Novel Therapeutics

Ritu Saini[1,*], Harnek Singh Saini[2], Rajesh Kumar[2] and Anjali Dahiya[1]

[1]Department of Chemistry and Biochemistry, CCS Haryana Agricultural University, Hisar

[2]Department of Biotechnology, University Institute of Engineering and Technology, Kurukshetra University, Kurukshetra

**Corresponding Author E-mail: ritusaini214@gmail.com*

Abstract

*Celiac disease (CD) is characterized by an immune response to ingested wheat gluten and related proteins of rye and barley that leads to inflammation, villous atrophy, and crypt hyperplasia in the intestine. This immune response is controlled by CD4+ T cells in the lamina propria that recognize gluten peptides in the context of disease predisposing HLA-DQ2 and HLA-DQ8 molecules. Most patients with CD carry the human leukocyte antigen HLA-DRB1*03 allele (usually associated with HLA-DQ2) or HLA-DRB1*04 (associated with HLA-DQ8). Transglutaminase 2 (TG2) appears to be an important component of the disease, both as a deamidating enzyme that can enhance the immunostimulatory effect of gluten and as a target autoantigen in the immune response. While gluten peptides bind poorly to HLA-DQ2/8, modification by tissue tranglutaminase converts the neutral amino acid glutamine into glutamic acid, introducing a negative charge that allows high affinity binding. The pathogenesis of disease involves interactions between environmental, genetic, and immunologic factors. The classic form of celiac disease typically presents in early childhood with abdominal pain and diarrhea, malabsorption, and nutrient deficiencies. Its diagnosis includes a serological screening and an upper gastrointestinal endoscopy with multiple biopsies. CD is associated with increased rates of anemia, osteoporosis, cancer, neurologic deficits, and additional autoimmune disorders. Gluten-free diet is the only effective treatment.*

Keywords: *Gluten, Celiac disease, pathogenesis, gluten free diet and transglutaminase 2.*

1. Introduction

Gluten, a protein found in wheat, barley, and rye, forms when water is added and the protein molecules glutanin and gliadin combine (Biesiekierski, 2017). Gluten gives elasticity to dough, helping it rise and keep its shape and often gives the final product a chewy texture. Gluten causes health problems for those with gluten-related disorders, including celiac disease (CD), non-celiac gluten sensitivity (NCGS), gluten ataxia, dermatitis herpetiformis (DH), and wheat allergy (Ludvigsson *et al.*, 2013). Gluten causes inflammation in the small intestines of people with CD. In these people, the gluten-free diet is demonstrated as an effective treatment (Volta *et al.*, 2015). Gluten-free foods can either be naturally gluten-free or commercially gluten-free. Naturally gluten-free would include foods such as meat, fish, eggs, milk and dairy products, legumes, nuts, fresh fruits, vegetables, potatoes, rice, and maize and other foods that by nature contain no gluten (Saturni *et al.*, 2010). Commercially gluten-free products are those that would normally contain gluten, but have been modified to meet the needs of the gluten-free consumer. An example of this would include rice flour bread, where rice flour is used as a substitute for wheat flour or gluten-free cookies (Table 12.1).

Table 12.1: Gluten containing products and gluten free products.

Gluten containing grains & grain products	Gluten free grains & grain products	Other gluten-free foods
Barley, bulgar, durum, einkorn, emmer, filler, farro, graham flour, kamut, malt, roux, rye, semolina, triticale and wheat	Amaranth, arrowroot, cassava, buckwheat, corn (maize), dahl, flax, legumes, millet, nuts, poi, rice, polenta, potatoes, quinoa, sago, sorghum, soy, tapioca, teff and wild rice	Fresh meat, poultry, fish and seafood, fresh vegetables and fruits, eggs cheese and other dairy and all products made with gluten-free ingredients

2. Celiac Disease

Celiac disease (CD) is a syndrome characterized by damage of the micro-villi of small intestinal mucosa caused by the gliadin fraction of wheat gluten and similar alcohol-soluble proteins (prolamines) (Fasano and Catassi, 2001). Micro-villi are very small finger-like projections in the small intestine that increase surface area in order to absorb as many nutrients as possible. The damaged villi will no longer be able to perform its role and will result in a lack of vitamins and other nutrients being absorbed into the body (Fig. 12.1). This can cause serious health problems such as osteoporosis, iron deficiency anemia, infertility, cancer and other auto-immune diseases. The typical changes of the duodenal mucosa include: raised intra-epithelial lymphocyte, crypt hyperplasia and various degree of villous atrophy (Oberhuber *et al.*, 1999). The damaged micro-villi will begin repairing themselves once gluten has been removed from the diet and in most cases, full recovery is expected within 1-2 years, depending on how long it has been un-diagnosed.

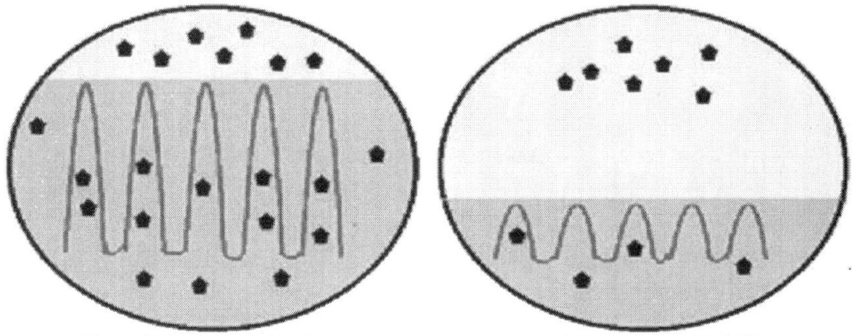

Fig. 12.1: Normal villi create a larger surface area allowing for the absorption for the absorption of nutrients while with CD the villi become inflamed and flattened, this greatly reduce the surface area available for absorption.

One challenge of diagnosing CD is that there are no classic signs or symptoms. For each person, it may be different. Most people with CD have complaints of bloating, abdominal pain and diarrhea. Some people develop anemia, joint pain, stomach pain, irritability or depression, dental and/or bone disorders like osteoporosis, foul smelling or oily stools, skin rash, weight loss and stunted growth in kids (Table 12.2). These symptoms can also imitate other conditions. Further testing of these symptoms is needed to determine the root problem. People with CD can also have silent CD, meaning they have no outward symptoms of the disease.

Table 12.2: Symptoms of CD.

Classic Adult Onset Symptoms	Atypical Adult Onset Symptoms	Symptoms in Children
Diarrhea	Mouth ulcers	Delayed puberty
Constipation	Anemia	Dental enamel defects
Gas and bloating	Bone and joint pain	Diarrhea/constipation
Abdominal pain	Osteoporosis/osteopenia	Difficulty in learning
Nausea and vomiting	Chronic hepatitis	Failure to thrive
	Infertility/recurrent miscarriages	Short Stature

3. Pathogenesis of CD

Environmental, genetic and immunologic factors are important in the pathogenesis of celiac disease (Kagnoff, 2002). A component of gluten, gliaden of the dietary cereal grains wheat, rye and barley, interacts with a specific genetic form of HLA receptor on an antigen presenting cell (Molberg *et al.*, 2003). The disease is associated with HLA alleles DQA1*0501/DQB1*0201 encoding DQ2 molecules (Sollid and Thorsby, 1993). An increased risk of celiac disease has been observed among persons who carry two DQB1*02 alleles (Abadie *et al.*, 2011; Liu *et al.*, 2014). Dieterich *et al.* (1997) recently demonstrated that one of the targets of the autoimmune response in CD is the tissue transglutaminase (TTG). Transglutaminase 2 (TG2) is a 78 kDa, ubiquitious, calcium dependent enzyme in mammals. The transglutaminase activity of TG2 must also be tightly controlled at the post-translational level. TG2, in the small intestinal mucosa is predominantly inactive under normal physiological conditions. Ca^{2+} ions, guanine nucleotides (GTP & GDP) and the redox environment of the protein are the post-translational regulators of the TG2 (Klöck *et al.*, 2012). Transglutaminase activity of full length TG2 requires Ca^{2+} ion binding. Up to six Ca^{2+} ions can bind to each molecule of TG2 (Bergamini, 1988). In contrast, TG2 activity is inhibited upon binding to one molecule of GTP or GDP. Activation is accompanied by a large conformational change from a compact to an extended structure (Begg *et al.*, 2005). The deamidating activity of this enzyme seems to generate gliadin peptides that bind to DQ2 to be recognized by disease-specific intestinal T cells (Molberg *et al.*, 2000). TTG deamidates glutamine residues to glutamic acid residues making an even more potent antigen. TTG is released by fibroblasts during injury. Deamidation results in higher binding affinity to HLA DQ2/8 molecule on the surface of antigen presenting cells. T helper cells are activated and, in turn, activate B and killer T cells. Plasma cell antibodies bind to gliadin bound to enterocytes, TTG and reticular fibers surrounding gut smooth muscle (endomysial ab's). T cells release (inappropriate) inflammatory cytokines as well as inflict tissue damage (Fig. 12.2).

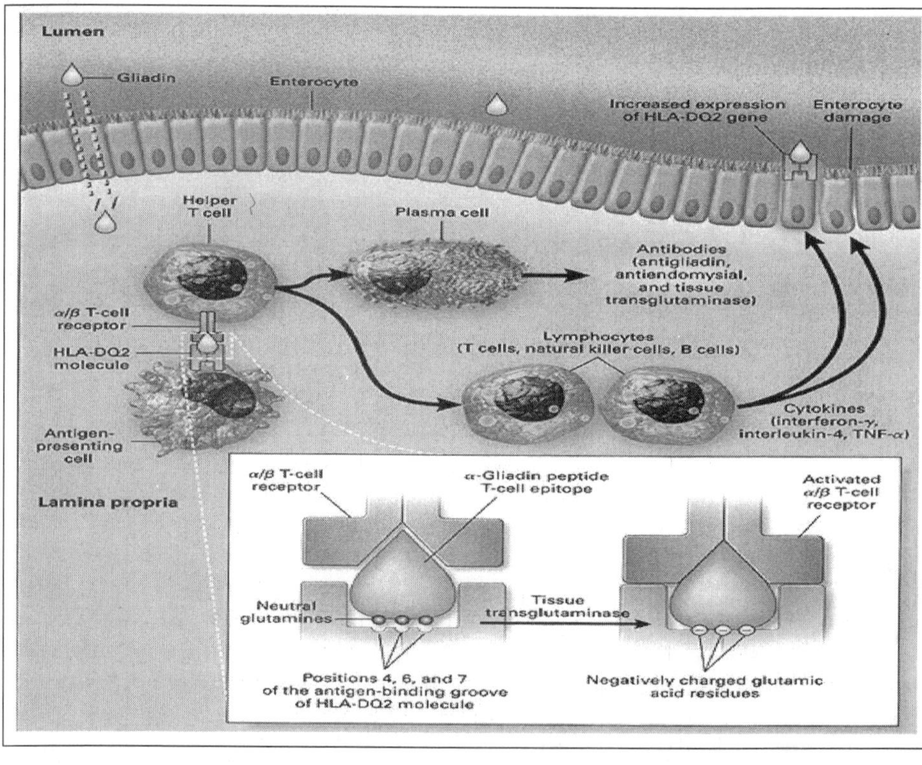

Fig. 12.2: Gliadin is absorbed into the lamina propria and presented in conjunction with HLA-DQ2 or DQ8 cell-surface antigens by antigen presenting cells, probably dendritic cells, to sensitized T cells expressing the a/b T-cell receptor. Tissue transglutaminase deamidates gliadin peptides, generating acidic, negatively charged residues of glutamic

acid from neutral glutamines (inset). Because negatively charged residues are preferred in positions 4, 6, and 7 of the antigen-binding groove of HLA-DQ2, deamidated gliadin elicits a stronger T-cell response. These lymphocytes then activate other lymphocytes to generate cytokines, such as interferon-g, interleukin-4, and tumor necrosis factor a (TNF-a), which damage the villi, resulting in enteritis. Induction of aberrant HLA class II cell-surface antigens on the enterocytes may permit these cells to present additional antigens to the sensitized lymphocytes (Farrell and Kelly, 2002).

4. Diagnosis of CD

Celiac disease is a frequent disorder (1% prevalence) in populations of mainly Caucasoid descent. The disease develops at all ages, and most patients are diagnosed in adolescence. According to the Celiac Sprue Association, there are three major steps that are taken in order to diagnose a person with CD (Fig. 12.3).

The first step includes a thorough physical examination to look for any signs of CD (fatigue, weight loss, etc). A blood panel is also conducted, sometimes referred to as a Celiac Blood Panel, to look at other indictors of CD, such as antibodies. Untreated celiac patients typically have increased immunoglobulin A (IgA) and immunoglobulin G (IgG) antibodies to wheat gluten and to the autoantigen transglutaminase 2 (TG2) (Dieterich *et al.*, 1997). Notably these antibodies disappear when the patients are commencing the gluten-free diet.

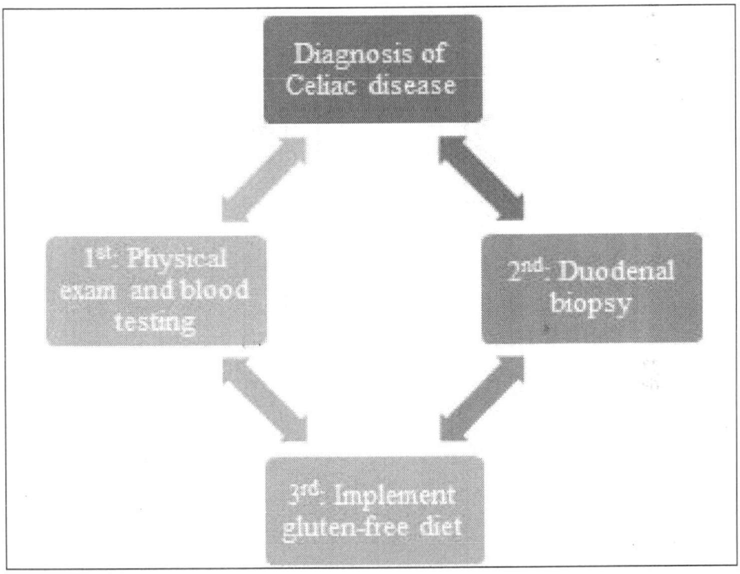

Fig. 12.3: Steps involved in diagnosis of celiac disease.

Second, a multi-sample biopsy of the duodenal portion of the intestine is done in order to look at possible damage from gluten. The established celiac lesion shows a complex interplay of inflammatory changes, and the typical morphological features including increased numbers of intraepithelial lymphocytes (IELs), infiltration of the lamina propria with increased density of T cells and plasma cells, decreased enterocyte height, villous atrophy, and crypt hyperplasia (Rostom *et al.*, 2006).

The third step includes implementing the gluten-free diet. When the person's health begins to improve, symptoms start to diminish and the intestinal villi start to repair themselves, CD is diagnosed.

5. Treatment of CD

Gluten-free diet (GFD) is the only current treatment for this condition that allows the recovery of the intestine and the gradual decrease of digestive symptoms, signs and systemic manifestations (Bascuñán *et al.*, 2017; Vici *et al.*, 2016). The gluten-free diet implicates a diet without wheat, rye barley, and all closely related cereals including spelt (a wheat variant). Worldwide, there is debate regarding the accepted definition for what constitutes "gluten-free." It appears that a gluten amount in the range of 50 - 100 mg per day is safe for most celiac disease patients (Stern *et al.*, 2001). Collin *et al.* (2004) have estimated a safe and rational threshold for daily gluten at 100 ppm gluten (100 mg gluten/1 kg), providing that the total daily GF flour intake does not exceed 300 grams. This level was determined by taking into consideration the amount of residual gluten found in GF products and the total intake of these products. At this level of intake, studies indicate good mucosal recovery.

The GF diet should never be started without an accurate diagnosis of celiac disease. Avoidance of food products that contain gluten proteins. It is essential that the diagnosis be confirmed before submitting patients to this therapy. Key elements to successful treatment include the motivation of the patient. Formal consultation with a trained dietitian is necessary. The dietitian plays a vital role in helping the patient successfully adapt to the necessary behavioral changes and may provide much of the required follow-up. National celiac disease support organizations can provide patients invaluable resources for information and support.

6. Disorders Associated with CD

CD shares important features with other autoimmune diseases like type 1 diabetes mellitus and rheumatoid arthritis: it is chronic, multifactorial, with a female to male ratio of roughly 2 to 1. These diseases all have a strong HLA-association, indicative of the involvement of the adaptive immune system and the presence of autoantibodies is characteristic (Koning, 2005). Endocrine disorders most commonly, Type 1 diabetes and thyroid disease conditions affects 5% to 10% of patients with celiac disease (Collin *et al.*, 2002). Some patients with coeliac disease are affected by a loss of bone mass grossly proportional to the degree of malabsorption (Kemppainen *et al.*, 1999). In children, osteoporosis can be reversed by a gluten-free diet with restoration of normal peak densitometric values (Mora *et al.*, 2001) however, in adults, bone-mineral density improves but rarely normalizes. Accordingly, a higher risk of fractures has been reported in adults with coeliac disease (Ludvigsson *et al.*, 2007).

A wide range of neurological and psychiatric disorders such as ataxia, peripheral neuropathy, epilepsy (mainly with occipital calcifi cations), headache, dementia, depression, autism, and schizophrenia has been reported in association with CD (Bushara, 2005) (Fig. 12.4).

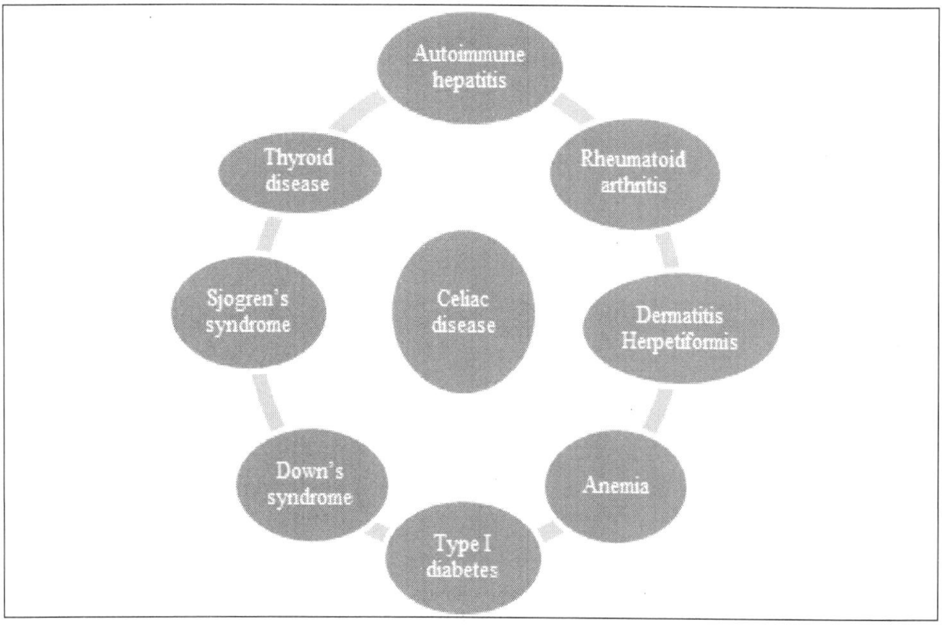

Fig. 12.4: Autoimmune Disorders Associated with Celiac Disease.

Furthermore, women with coeliac disease more frequently experience recurrent spontaneous miscarriage, delayed menarche, early menopause, and amenorrhoea. Fertility problems and loss of libido can also be present in male patients (Rostami *et al.*, 2001). Gluten sensitivity is sometimes expressed in the form of dermatitis herpetiformis, a pruritic, chronic skin disease characterized by symmetrical papulovesicular lesions and presence of granular deposits of IgA in the dermal papillae. This condition affects about 10% to 20% of patients with celiac disease (Fry, 2002).

7. Emerging therapy for CD

The new insight in the pathogenesis has given a number of new targets for intervention. There are many targets in the inflammatory cascade of events, which can be modulated in order to halt the process of immunological injury (Table 12.3).

Table 12.3: Novel therapeutics for CD.

1. Reduction in the exposure of Gluten	Decrease exposure to immunogenic gluten in genetically susceptible person is to grow genetically modified grains, which are devoid of immunogenic epitopes, *eg.* Genetic deletion of α-gliadin in Chinese spring wheat, led to the decrease in the number of T cell- activating epitopes without affecting the baking quality of the wheat flour dough.	Sollid and Khosla, (2011); Crespo *et al.* (2012); Schuppan *et al.* (2009); Stoven *et al.* (2012); Van den *et al.* (2009)

2.Inhibition of TG2	acivicin or alternatively 6-diazo-5-oxo-norleucine (DON), novel trans-cinammoyl derivatives, cinnamoyl benzotriazolyl amides and the3-(substituted cinnamoyl) pyridines, they display Reversible inhibition of TG2. L682777, 2-[(2-oxopropyl) thioimidazolium inhibitor, was able to prevent the in situ cross linking of gluten peptides to endogenous proteins and alsoprevented T-cell activation induced by the non-deamidated form of an immuno dominant gluten peptide.	Hausch *et al.* (2003); Pardin *et al.* (2008); Maiuri *et al.* (2005)
3. Assisted digestion of Gluten	To detoxify the gluten either in the Gastrointestinal lumen using endopeptidases (Intra-luminal enzymatic digestion). Example- Prolyl endopeptidases (PEPs) can break bonds between proline and glutamine, ALV003 is a combination of acysteine-endoprotease derived from germinated barley seeds and a PEP from S. capsulate targeting gluten.	Shan *et al.* (2004); Pyle *et al.* (2005); Tye-Din *et al.* (2010)
	Using predigested gluten where gluten is digested partially before ingestion. Wheat flour-fermented with selected lactobacilli and fungal proteases (routinely used in bakeries) led to decrease in Gluten content to <10ppm (gluten-free). VSL#3,a highly concentrated mixture of lactic acid and bifidobacteria, was able to hydrolyze completely the α2-gliadin derived epitopes 62–75 and 33-mer peptide.	Di Cagno *et al.* (2010); De Angelis *et al.* (2006)
4.Prevention of passage of immunogenic peptides through the tight junctions	The zonulin inhibitor, AT1001 is reducing gluten-induced barrier dysfunction. Zonulin, is a regulator of epithelial permeability and is highly expressed in the mucosa and blood of patients with celiac disease.	Fasano *et al.* (2000); Paterson *et al.* (2007)
5.Blocking of HLA-DQ2 to prevent binding of immunogenic peptides	Amino- acid substitutions within gliadin T cell stimulatory sequences can convert the immunogenic epitope to either an agonist or antagonist.	Ellis *et al.* (2003); Silano *et al.* (2007)
	Substitution of alanine amino-acid at key positions (3, 8 and 10) in the immune dominant peptide of the wheat α2-gliadin led to abolition of the immunogenicity of the peptide when tested against T cell clones.	
	A decapeptide (QQPQDAVQPF) obtained from an alcohol-soluble protein fraction of durum wheat was shown to be antagonistic to gluten toxicity.	
6.Immune-modulation and induction of tolerance to gluten	NexVax2 is a therapeutic vaccine uses three gluten Peptides with the goal of inducing a “tolerogenic” response in patients with celiac disease. NexVax2 has showed efficacy in HLA-DQ2 transgenic mice having gluten-sensitive T cells.	De Kauwe *et al.* (2009); Brown *et al.* (2011)
	Vaccinated subjects showed presence of IFN-γ producing anti-gluten T cells	

7.Gluten-Sequestering Polymer	Polymeric binders like poly (hydroxyethyl methacrylate-co-styrene sulfonate) [P(HEMA-co-SS)], by binding with α-gliadin counteracts the toxic effects of gliadin on intestinal epithelial cells in vitro	Liang *et al.* (2009); Liang *et al.* (2010)
8. Anti-inflammatory drugs	Glucocorticoids like prednisone, fluticasone propionate & budesonide showed inhibitory effect on both B and T cell proliferation and protects from harmful effects of gluten.	Wall *et al.* (1970); Mitchison *et al.* (1991); Ciacci *et al.* (2009)
	Anti-interferon-γ, antiTNF-α and Anti-interleukin-15 antibodies were capable of efficiently reversing the intestinal damage in celiac patients.Interleukin-10 (IL-10) is an immunoregulatory cytokine in the intestinal tissue and was shown to suppress gliadin- Induced T cell activation.	Bethune *et al.* (2009); Gillett *et al.* (2002); Yokoyama *et al.* (2009); Salvati *et al.* (2005); Rashtak *et al.* (2012)

8. Conclusion

Celiac disease is not a food allergy, it is an autoimmune disorder caused by the abnormal functioning of the immune system that produces antibodies to your own tissue. Both the adaptive and innate arms of the immune system are involved in the response to gluten and the subsequent action of lamina propria and intraepithelial lymphocytes in driving the autoimmune response that eventually leads to mucosal damage. Expression of HLA-DQ2 and HLA-DQ8 molecules is an essential genetic component of the disease, being necessary for the immune reaction against gluten. A better understanding of the underlying mechanism of pathogenesis of celiac disease and associated disorders will help in devising new strategies for diagnosis and treatment of the disease, including prevention of its long-term complications, and serve as a model for investigation of other autoimmune disorders.

References

Abadie, V., Sollid, L.M., Barreiro, L.B. and Jabri, B. 2011. Integration of genetic and immunological insights into a model of celiac disease pathogenesis. *Annu. Rev. Immunol.* 29: 493-525.

Bascuñán, K.A., Vespa, M.C. and Araya, M. 2017. Celiac disease: understanding the gluten-free diet. *European Journal of Nutrition.* 56(2): 1-11.

Begg, G.E., Holman, S.R., Stokes, P.H., Matthews, J.M., Graham, R.M. and Iismaa, S.E. 2006. Mutation of a critical arginine in the GTP-binding site of transglutaminase 2 disinhibits intracellular cross-linking activity. *J. Biol. Chem.* 281(18):12603-12609.

Bergamini, C.M. 1988. GTP modulates calcium binding and cation-induced conformational changes in erythrocyte transglutaminase. *FEBS Lett.* 239(2): 255-258.

Bethune, M.T., Siegel, M., Howles-Banerji, S. and Khosla, C. 2009. Interferon-gamma released by gluten-stimulated celiac disease-specific intestinal T cells enhances the transepithelial flux of gluten peptides. *J. Pharmacol. Exp. Ther.* 329: 657-668.

Biesiekierski, J.R. 2017. "What is gluten?". *J. Gastroenterol. Hepatol.* (Review). 32(1): 78-81.

Brown, G.J., Daveson, J., Marjason, J., Ffrench, R.A., Smith, D., Sullivan, M., Tye-Din, J.A. and Anderson, R.P. 2011. A phase I study to determine safety, tolerability and bioactivity of Nexvax2® in HLADQ^{2+} volunteer swith celiac disease following along-term, strict gluten- free diet. *Gastroenterology.* 140: 437-8.

Bushara, K.O. 2005. Neurologic presentation of celiac disease. *Gastroenterology.* 128: 92-97.

Ciacci, C., Maiuri, L., Russo, I., Tortora, R., Bucci, C., Cappello, C., Santonicola, A., Luciani, A., Passananti, V. Iovino, P. 2009. Efficacy of budesonide therapy in the early phase of treatment of adult celiac disease patients with malabsorption: an in vivo/ in vitro pilot study. *Clin. Exp. Pharmacol. Physiol.* 36: 1170-1176.

Collin, P., Kaukinen, K., Valimaki, M. and Salmi, J. 2002. Endocrinological disorders and celiac disease. *Endocr. Rev.* 23: 464-483.

Collin, P., Thorell, L., Kaukinen, K. and Mäki, M. 2004. The safe threshold for gluten contamination in gluten-free products. Can trace amounts be accepted in the treatment of coeliac disease? *Aliment. Pharmacol. Ther.* 19: 1277-1283.

Crespo Pérez, L., Castillejode Villasante, G., Cano Ruiz, A. and León, F. 2012. Non-dietary therapeutic clinical trials in celiac disease. *Eur. J. Intern. Med.* 23: 9-14.

De Angelis, M., Rizzello, C.G., Fasano, A., Clemente, M.G., De Simone, C., Silano, M., De Vincenzi, M., Losito, I. and Gobbetti, M. 2006. VSL#3 probiotic preparation has the capacity to hydrolyze gliadin polypeptides responsible for Celiac Sprue. *Biochim. Biophys. Acta.* 1762(1): 80-93.

De Kauwe, A.L., Chen, Z., Anderson, R.P., Keech, C.L., Price, J.D., Wijburg, O., Jackson, D.C., Ladhams, J., Allison, J. and Mc Cluskey, J. 2009. Resistance to celiac disease in humanized HLA-DR3-DQ2-transgenic mice expressing specific anti-gliadin CD^{4+} T cells. *J. Immunol.* 182: 7440-7450.

Di Cagno, R., Barbato, M., Di Camillo, C., Rizzello, C.G., De Angelis, M., Giuliani, G., De Vincenzi, M., Gobbetti, M. and Cucchiara, S. 2010. Gluten-free sour dough wheat baked goods appear safe for young celiac patients: apilotstudy. *J. Pediatr. Gastroenterol. Nutr.* 51(6): 777-783.

Dieterich, W., Ehnis, T., Bauer, M., Donner, P., Volta, U., Riecken, E. and Schuppan, D.1997. Identification of tissue transglutaminase as the autoantigen of celiac disease. *Nat. Med.* 3: 797-801.

Ellis, H.J., Pollock, E.L., Engel, W., Fraser, J.S., Rosen-Bronson, S., Wieser, H. and Ciclitira, P.J. 2003. Investigation of the putative immune dominant T cell epitopes in celiac disease. *Gut.* 52: 212-217.

Farrell, R.J. and Kelly, C.P. 2002. Celiac Sprue. *N. Engl. J. Med.* 346(3): 180-188.

Fasano, A. and Catassi, C. 2001. Current approaches to diagnosis and treatment of celiac disease: An evolving spectrum. *Gastroenterology,* 120: 636-651.

Fasano, A., Not, T., Wang, W., Uzzau, S., Berti, I., Tommasini, A. and Goldblum, S.E. 2000. Zonulin, a newly discovered modulator of intestinal permeability, and its expression in coeliac disease. *Lancet.* 355: 1518-1519.

Fry, L. 2002. Dermatitis herpetiformis: problems, progress and prospects. *Eur. J. Dermatol.* 12: 523-31.

Gillett, H.R., Arnott, I.D., Mc Intyre M., Campbell, S., Dahele, A., Priest, M., et al. 2002. Successful infliximab treatment for steroid-refractory celiac disease: a case report. *Gastroenterology.* 122: 800-805.

Hausch, F., Halttunen, T., Maki, M. and Khosla, C. 2003. Design, synthesis, and evaluation of gluten peptide analogs as selective inhibitors of human tissue transglutaminase . *Chem. Biol.* 10: 225-231.

Kagnoff, M.F. 2002. Celiac disease pathogenesis: the plot thickens. *Gastroenterology.* 123:939-943.

Kemppainen, T., Kroger, H., Janatuinen, E., Arnala, I., Kosma, V.M., Pikkarainen, P., Julkunen, R., Jurvelin, J., Alhava, E. and Uusitupa, M.1999. Osteoporosis in adult patients with celiac disease. *Bone.* 24: 249-55.

Klöck, C., Diraimondo, T.R. and Khosla, C. 2012. Role of transglutaminase 2 in celiac disease pathogenesis. *Semin Immunopathol.* 34(4): 513-522.

Koning, F.1., Schuppan, D., Cerf-Bensussan, N. and Sollid, L.M. 2005. Pathomechanisms in celiac disease. *Best Pract Res Clin Gastroenterol.* 19(3): 373-387.

Liang, L., Pinier, M., Leroux, J.C. and Subirade, M. 2009. Interaction of alpha-gliadin with poly (HEMA-co-SS): structural characterization and biological implication. *Biopolymers.* 91: 169-178.

Liang, L., Pinier, M., Leroux, J.C. and Subirade, M. 2010. Interaction of alpha-gliadin with polyanions: design considerations for sequestrants used in supportive treatment of celiac disease. *Biopolymers.* 93: 418-428.

Liu, E., Lee, H.S. and Aronsson, C.A. 2014. Risk of pediatric celiac disease according to HLA haplotype and country. *N. Engl. J. Med.* 371: 42-49.

Ludvigsson, J.F., Leffler, D.A., Bai, J.C., Biagi, F., Fasano, A., Green, P.H., Hadjivassiliou, M., Kaukinen, K., Kelly, C.P., Leonard, J.N., Lundin, K.E., Murray, J.A., Sanders, D.S., Walker, M.M., Zingone, F. and Ciacci, C. 2013. "The Oslo definitions for coeliac disease and related terms". *Gut.* 62(1): 43-52.

Ludvigsson, J.F., Michaelsson, K., Ekbom, A. and Montgomery, S.M. 2007. Coeliac disease and the risk of fractures-a general population-based cohort study. *Aliment. Pharmacol. Ther.* 25: 273-285.

Maiuri, L., Ciacci, C., Ricciardelli, I., Vacca, L., Raia, V., Rispo, A., Griffth, M., Issekutz, T., Quaratino, S. Londei, M. 2005. Unexpected role of surface transglutaminase type II in celiac disease. *Gastroenterology.* 129: 1400-1413.

Mitchison, H.C., Al Mardini, H., Gillespie, S., Laker, M., Zaitoun, A. Record, C.O. 1991. A pilot study of fluticasone propionate in untreated celiac disease. *Gut.* 32: 260-265.

Molberg, O., McAdam, S.N. and Sollid L.M. 2000. Role of tissue transglutaminasein celiac disease. *J. Pediatr. Gastroenterol. Nutr.* 30: 232-240.

Molberg, O., Solheim-Flaete, N., Jensen, T., Lundin, K.E., Arentz-Hansen, H., Anderson, O.D., Kjersti-Uhlen, A. and Sollid, L.M. 2003. Intestinal T-cellresponses to high-molecular-weight glutenins in celiac disease. *Gastroenterology*. 125: 337-344.

Mora, S., Barera, G., Beccio, S., Menni, L., Proverbio, M.C., Bianchi, C. and Chiumello, G. 2001. A prospective, longitudinal study of the long-term effect of treatment on bone density in children with celiac disease. *J. Pediatr.* 139: 516-521.

Oberhuber, G., Granditsch, G. and Vogelsang, H. 1999. The histopathology of coeliac disease: time for a standardized reportscheme for pathologists. *Eur. J. Gastroenterol. Hepatol.* 11: 1185-1194.

Pardin, C., Pelletier, J.N., Lubell, W.D. and Keillor, J.W. 2008. Cinnamoyl inhibitors of tissue transglutaminase . *J. Org. Chem.* 73: 5766- 5775.

Paterson, B.M., Lammers, K.M., Arrieta, M.C., Fasano, A. and Meddings, J.B. 2007. The safety, tolerance, pharmacokinetic and pharmacodynamic effects of single doses of AT-1001 in coeliac disease subjects: a proof of concept study. *Aliment. Pharmacol. Ther.* 26: 757-766.

Pinier, M., Verdu, E.F., Nasser-Eddine, M., David, C.S., Vézina, A., Rivard, N. and Leroux, J.C. 2009. Polymeric binders suppress gliadin-induced toxicity in the intestinal epithelium. *Gastroenterology*. 136(1): 288-298.

Pyle, G.G., Paaso, B., Anderson, B.E., Allen, D.D., Marti, T., Li, Q., Siegel, M., Khosla, C. and Gray, G.M. 2005. Effect of pretreatment of food gluten with prolyl endopeptidase on gluten-induced malabsorption in celiac sprue. *Clin. Gastroenterol. Hepatol.* 3: 687-694.

Rashtak, S. and Murray, J.A. 2012. Review article: celiac disease, new approaches to therapy. *Aliment. Pharmacol. Ther.* 35: 768-781.

Richard, J. F. and Ciaran P. K. 2002. Celiac Sprue. *The New England Journal of Medicine*. 346(3): 180-188.

Rostami, K., Steegers, E.A., Wong, W.Y., Braat, D.D. and Steegers-Theunissen, R.P. 2001. Coeliac disease and reproductive disorders: a neglected association. *Eur. J. Obstet. Gynecol. Reprod. Biol.* 96: 146-149.

Rostom, A., Murray, J.A. and Kagnoff, M.F. 2006. American Gastroenterological Association (AGA) institute technical review on the diagnosis and management of celiac disease. *Gastroenterology*. 131: 1981-2002.

Salvati, V.M., Mazzarella, G., Gianfrani, C., Levings, M.K., Stefanile, R., De Giulio, B., Iaquinto, G., Giardullo, N., Auricchio, S., Roncarolo, M.G. and Troncone, R. 2005. Recombinant human interleukin 10 suppresses gliadin dependent T cell activation in ex vivo cultured celiac intestinal mucosa. *Gut*. 54: 46-53.

Saturni, L., Ferretti. G. and Bacchetti, T. 2010. "The gluten-free diet: safety and nutritional quality". *Nutrients* (Review). 2(1): 16-34.

Schuppan, D., Junker, Y. and Barisani, D. 2009. Celiac disease: from pathogenesis to novel therapies. *Gastroenterology*. 137: 1912-1933.

Shan, L., Marti, T., Sollid, L.M., Gray, G.M. and Khosla, C. 2004. Comparative biochemical analysis of three bacterial prolyl endopeptidases: implications for celiac sprue. *Biochem. J.* 383 (2): 311-318.

Silano, M., Di Benedetto, R., Trecca, A., Arrabito, G., Leonardi, F. and De Vincenzi, M. 2007. A decapeptide from durum wheat prevents celiac peripheral blood lymphocytes from activation by gliadin peptides. *Pediatr. Res.* 61: 67-71.

Sollid, L. and Thorsby, E. 1993. HLA susceptibility genes in celiac disease:genetic mapping and role in pathogenesis. *Gastroenterology*. 105: 910-922.

Sollid, L.M and Khosla, C. 2011. Novel therapies for celiac disease. *J. Intern. Med.* 269: 604-613.

Sollid, L.M. 2002. Coeliac disease: dissecting a complex inflammatory disorder. *Nat. Rev. Immunol.* 2(9): 647-655.

Sollid, L.M. and Khosla, C. 2005. Future therapeutic options for celiac disease . *Nat. Clin. Pract. Gastroenterol. Hepatol.* 2: 140-147.

Stern, M., Ciclitira, P.J., van-Eckert, R., Feighery, C., Janssen, F.W., Méndez, E., Mothes, T., Troncone, R. and Wieser, H. 2001. Analysis and clinical effects of gluten in coeliac disease . *Eur. J. Gastroenterol. Hepatol.* 13: 741-747.

Stoven, S., Murra,y J.A. and Marietta, E. 2012. Celiac disease: advances in treatment via gluten modification. *Clin. Gastroenterol. Hepatol.* 10: 859-862.

Tye-Din, J.A., Anderson, R.P., Ffrench, R.A., Brown, G.J., Hodsman, P., Siegel, M., Botwick, W. and Shreeniwas, R. 2010. The effects of ALV003 pre-digestion of gluten on immune and symptoms in celiac disease in vivo. *Clin. Immunol.* 134: 289-295.

Van den Broeck, H.C., van Herpen, T.W., Schuit, C., Salentijn, E.M., Dekking, L., Bosch, D., Hammer, R.J., Smulders, M.J.M., Gilissen, L.J.W.J. and Van der Meer, I.M. 2009. Removing celiac disease- related gluten proteins from bread wheat while retaining technological properties: a study with Chinese Spring deletion lines. *BMC Plant Biol.* 9: 41.

Vici, G., Belli, L., Biondi, M. and Polzonetti, V. 2016. Gluten free diet and nutrient deficiencies: A review. *Clinical Nutrition*. 35 (6): 1236-1241.

Volta, U., Caio, G., De Giorgio, R., Henriksen, C., Skodje, G. and Lundin, K.E. 2015. "Non-celiac gluten sensitivity: a work-in-progress entity in the spectrum of wheat-related disorders". *Best Pract. Res. Clin. Gastroenterol.* 29(3): 477-491.

Wall, A.J., Douglas, A.P., Booth, C.C. and Pearse, A.G. 1970. Response of the jejuna mucosa in adult celiac disease to oral prednisolone. *Gut.* 11: 7-14.

Yokoyama, S., Watanabe, N., Sato, N., Perera, P.Y., Filkoski, L., Tanaka, T., Miyasaka, M., Waldmann, T.A., Takachika, H. and Peera, L.P. 2009. Antibody-mediated blockade of IL-15 reverses the autoimmune intestinal damage in transgenic mice that overexpress IL-15 in enterocytes. *Proc. Natl. Acad. Sci. USA.* 106: 15849-15854.

Chapter 13

Synergistic Effect of Arbuscular Mycorrhizal Fungi with *Trichoderma Viride* and *Pseudomonas Fluorescens* in Promoting Growth and Yield of Sunflower Plant for Sustainable Agriculture

Alpa Yadav[1,*], Esha Jangra[2] and Ashok Aggarwal[2]

[1]*Department of Botany, Indira Gandhi University, Meerpur, Rewari (Haryana) - 122502 India*

[2]*Department of Botany, Kurukshetra University, Kurukshetra-136119*

**Corresponding Author E-mail: alpa.yadav21@gmail.com*

Abstract

Microorganisms that sustain the fertility of soils, resulting in improved plant nutrition; have continued to magnetize attention because of the increasing cost of agricultural inputs and some of their negative impacts on environmental sustainability. A pot experiment was performed to see the interactive potential of arbuscular mycorrhizal fungi (Glomus mosseae and Acaulospora laevis) alone or in combination with Trichoderma viride and Pseudomonas fluorescens on Sunflower under glass house conditions. Various morphological and biochemical parameters were measured after 120 days. Although, all co-inoculation treatments showed beneficial effects but inoculation with a combination of G. mosseae + A. laevis + T. viride + P. fluorescens showed maximum increment in all the growth, physiological and yield parameters. The overall results demonstrate that the co-inoculation of bioinoculants like P. fluorescens and T. viride with AM fungi promotes higher AM colonization, spore numbers enhancing nutrient acquisition especially phosphorus (P), producing plant growth hormones resulting in improvement of rhizospheric condition of soil, altering the physiological and yield properties of sunflower plant.

Keywords: *AM fungi, Helianthus annuus, Trichoderma viride, Pseudomonas fluorescens, interaction.*

1. Introduction

Helianthus annuus L. (family: Compositae) is a major oil seed crop in the world, ranked second after soybean (Weiss, 2000). Sunflower is valued for its monounsaturated fats and contains soluble vitamins A, D, E and K and therefore considered as a useful source of dietary fat in preventing heart disease. The seeds contain hulls with high fibres and wax contents and low protein content, considered as a big obstacle for obtaining a better yield of oil and high-quality protein meal. The oil content of sunflower seed ranges between 26-36 %, whereas the protein content is 28-32%.Sunflower contains monounsaturated fats, used in manufacturing of margarine and sunflower cake is used as cattle feed (Jackoben *et al.*, 1992). The indiscriminate use of chemical fertilizer/ pesticides leads to disastrous consequences of environmental quality, promoting the study of natural sources of fertilizer, bio-stimulants and soil amendments (Tanwar *et al.*, 2013 and Yadav *et al.*, 2013). The absence of effective symbiotic growth promoting microbes in the rhizosphere of the host plant is a limiting factor for growth and yield. Organic wastes and biofertilizers like arbuscular mycorrhizal fungi (AMF), *Trichoderma viride* and *Pseudomonas fluorescens* have received considerable attention in the recent years owing to their beneficial response in improving crop productivity (Meena and Marimuthu, 2012). AMF contains useful microorganisms living in the rhizosphere of plant, promoting its growth and yield by increasing the supply or availability of essential nutrients to the plants mainly phosphorus. AMF colonization have great potential in nutrient uptake, water absorption, bicontrol of pathogens, synthesis of growth hormones and establishment of plantlets in adverse soil and climatic conditions (Yadav *et al.*, 2013). *Pseudomonas* and other plant growth promoting rhizobacteria (PGPR) influence plant growth by increasing nutrient availability, reducing pathogenic infection and abiotic stress to plants. *Trichoderma* species also known as Mycorrhiza Helper Organisms (MHO) act as endophytic plant symbionts by altering the plant physiology and provide resistance to pathogens (Hoyos-Carvajal, 2009). In the light of the above, the present study was carried out to assess the consequent effect of dominant AM fungi alone or in various combinations with other bioinoculants (*Trichoderma viride* and *Pseudomonas fluorescens*) to find out the best combination having the maximum capability of increasing plant growth and better yield of sunflower.

2. Materials and Methods

2.1 Collection of Soil Sample

Composite soil sample from rhizospheric soil of sunflower was collected from Botanical Garden, Kurukshetra University, Kurukshetra and kept in sterilized polythene bags at 10°C till further processing. The soil characteristics are as follows: sand-64.2 %, silt-21.81 %, clay-3.90 %, pH-6.890, EC- 1.00 dS/m, organic carbon-0.40 %, total N0.042 %, P-0.0018 Kg/m2, K-0.022 Kg/ m2, and S-14.80 ppm.

2.2 Isolation & Mass Multiplication of Dominant AM Fungi from Soil Samples

Isolation of dominant AM spores were done by using 'Wet Sieving and Decanting Technique' of Gerdemann and Nicolson (Gerdemann and Nicolson, 1963). Spores were then picked by hypodermic needle under stereobinocular microscope.

Two dominant AM spores i.e. *Glomus mosseae* Nicol. and Gerd. and *Acaulospora laevis* Gerd. & Trappe were mass multiplied by the funnel technique of Menge and Timmer, (1982) using maize as host plant for a period of 3 months.

2.3 Mass Culture of *Trichoderma Viride*

T. viride was mass produced by using wheat bran and saw dust medium which was prepared by using wheat bran, saw dust and distilled water in the ratio of 3:1:4.

2.4 Mass Culture of *Pseudomonas Fluorescens*

Pseudomonas fluorescens (MTCC No. 103) was procured from IMTECH (Institute of Microbial Technology, Chandigarh, India) and multiplied in nutrient broth medium (1.25 g peptone, 0.75 g beef extract, 1.25g NaCl, 250 mL distilled water) for 24 h for suitable bacteria growth.

2.5 Preparation of Pot Mixture

The experiment was laid out in a randomized complete block design, with five replicates of each treatment. Soil from experimental site was collected and mixed with sand in a proportion of 3:1 (soil: sand). This mixture was then sieved through 2-mm sieve and autoclaved at 121°C for two hours for two consecutive days. Surface sterilized seeds of sunflower were grown in earthen pots (25 25 cm) under polyhouse conditions. To each pot 10 percent of inoculum of each AM fungi (*G. mosseae* or *A. laevis*) was added alone and in combination with *T. viride*, *P. fluorescens* and then the effect of these bioinoculants were recorded on different growth parameters of sunflower after 120 days of inoculation. Different treatments used during the present investigation were as follows:

1. Control (autoclaved sterile sand: soil without any bioinoculant)
2. *Glomus mosseae* (G)
3. *Acaulospora laevis* (A)
4. *Trichoderma viride* (T)
5. *Pseudomonas fluorescens* (P)
6. *G. mosseae* + *A. laevis* (G + A)
7. *G. mosseae* + *T. viride* (G + T)
8. *G. mosseae* + *P. fluorescens* (G + P)
9. *A. laevis* + *T. viride* (A + T)
10. *A. laevis* + *P. fluorescens* (A + P)
11. T. viride + P. fluorescens (T + P)
12. *G. mosseae* + *A. laevis* + *T. viride* (G + A + T)
13. *G. mosseae* + *T. viride*+ *P. fluorescens* (G + T + P)
14. *G. mosseae* + *A. laevis* + *P. fluorescens* (G + A + P)
15. *A. laevis* + *T. viride* + *P. fluorescens* (A + T + P)

16. *G. mosseae* + *A. laevis* + *T. viride* + *P. fluorescens* (G + A + T + P)

2.6 Quantification of AM Spores

It was done by 'Grid Line Intersect Method' of Adholeya and Gaur (1994) Spores were counted under stereo binocular microscope by using a counter.

2.7 Identification of AM Fungi

For identification of AM spores, the keys of Walker (1983), Schenik and Perez (1990), Morton and Benny (1990) and Mukerji (1996) were followed.

2.8 Growth Parameters and Mycorrhizal Root Colonization

Plants were harvested after 120 days by uprooting them from the soil and various morphological and physiological parameters were measured. After harvest, roots and shoots were weighed separately to determine fresh weight, and then placed in an oven to dry at 70 C until a constant dry weight was obtained. Mycorrhizal root colonization was studied by 'Rapid Clearing and Staining Method' of Phillips and Hayman (1970). Percentage AM colonization of roots was calculated using: (Number of root segments colonized / number of root segments studied) 100.

2.9 Estimation of Total Chlorophyll

The chlorophyll content was estimated by using Arnon's method (Arnon, 1949) by using 80 % acetone as solvent.

2.10 Stomatal Conductance Estimation

The stomatal conductance of all experimental plants was measured by using Porometer (AP4Delta T devices, Cambridge, UK) after 120 days of inoculation in morning and evening.

2.11 Phosphorus Estimation

The phosphorus content of roots and shoots of all experimentally plants was estimated by 'Phospho-vanadomolybdate yellow colour method (Jackson, 1973).

2.12 Phosphatase Estimation

Phosphatase activity was assayed by using pnitrophenyl phosphate (PNPP) as substrate, hydrolyzed by the enzyme to p-nitrophenol. For this, ice cold sodium acetate buffer (0.05 M with pH 4.8) for acid phosphatase and sodium carbonate-bicarbonate buffer (0.05 M with pH 10) for alkaline phosphatase activity was used (Tabatabai and Bremner, 1969)

Protein was estimated by the method of Bradford (Bradford, 1976) using coomassive brilliant blue G-250 dye.

2.12 Oil Extraction

Oil was extracted by petroleum ether of boiling range between 40-60°C using the Soxhlet's procedure (AOCS., 1997). Five replicates of each treatment were taken.

2.13 Statistical Analysis

All results were analysed using Analysis of Variance (ANOVA), followed by post hoc test through computer software SPSS 11.5 version. Means were ranked at $P < 0.05$ level of significance using Duncan's Multiple Range Test for comparison.

3. Results and Discussion

3.1 Plant Growth Response of Sunflower

It is evident from Table. 1 that all inoculated plants showed significant increment in growth parameters in comparison to the control. Analysis of the research showed that soil inoculated with *G. mosseae* + *A. laevis* + *T. viride* + *P. fluorescens* increased maximum plant height with fresh and dry biomass followed by triple combination of *G. mosseae* + *A. laevis* + *T. viride*. So, *G. mosseae* and *T. viride* were found to be common most effective bioinoculants causing an increase in root surface area and efficiency of roots to absorb more water. The main absorption apparatus of mycorrhiza is the extension of hyphae that can penetrate soil pore inaccessible to root hairs and absorb water that is not available to non - mycorrhizal plants (Gong *et al.*, 2000). It can be said that as a result of AMF treatment, root elongation was observed which ultimately absorb more of the nutrients especially away from the P depletion zone, resulting in better growth of the plant over uninoculated ones. Vinale *et al.* (2008) demonstrated that stimulation of plant growth by *Trichoderma* species treatment might be due to production of secondary metabolites which may act as an auxin like compound, leading to the development of plant growth. Our results are in accordance with the work of Cagras *et al.* (2000); Ortas (2010) and Yaseen *et al.* (2012).

3.2 Plant Leaf Area, Stomatal Conductance and Plant Chlorophyll Content

The analysis of variance (Table. 2) showed the significant effect of bioinoculants and their interaction effects on leaf area, stomatal conductance and chlorophyll content. Among all of treatments, highest stomatal conductance (morning and evening) was detected in triple combination of *G. mosseae* + *A. laevis* + *T. viride* followed by *G. mosseae* + *A. laevis* + *T. viride* + *P. fluorescens*. One of the most important indicators of physiological activity is the rate of photosynthesis, which is related to the chlorophyll content of plants. The results presented in Table. 2 revealed that chlorophyll content in the inoculated plants were higher than those in non-inoculated plants.

Table-1: Influence of AM Fungi, *T. viride* and *P. fluorescens* on different growth parameters of Sunflower after 120 days

Sr. No.	Parameters Treatments	Plant height (cm)	Shoot biomass (gm)		Root biomass (gm)		Root length (cm)	Root colonization (%)	AM spore number/ 10gm soil
			Fresh	Dry	Fresh	Dry			
1.	**Control**	079.98 ± 4.41^{f}	10.8 ± 1.54^{f}	2.92 ± 0.66^{f}	0.63 ± 0.08^{f}	0.40 ± 0.01^{f}	7.6 ± 0.84^{f}	0^{f}	0^{f}
2.	**G**	126.36 ± 4.31^{d}	20.64 ± 2.13^{d}	6.00 ± 1.58^{de}	4.21 ± 0.13^{cd}	1.13 ± 0.10^{de}	18.4 ± 0.53^{cd}	74.62 ± 2.02^{d}	70.82 ± 2.91^{cd}
3.	**A**	109.40 ± 5.34^{e}	16.42 ± 5.15^{e}	5.76 ± 1.23^{e}	3.62 ± 0.56^{e}	1.02 ± 0.07^{ef}	12.26 ± 1.59^{e}	49.16 ± 4.40^{ef}	45.56 ± 2.22^{ef}
4.	**T**	111.30 ± 0.81^{e}	18.38 ± 2.27^{dc}	5.08 ± 1.61^{e}	3.70 ± 0.13^{de}	1.05 ± 0.03^{ef}	13.52 ± 0.35^{d}	0^{f}	0^{e}
5.	**P**	105.54 ± 5.01^{ef}	14.26 ± 2.94^{e}	4.86 ± 1.28^{ef}	3.55 ± 0.24^{ef}	1.00 ± 0.31^{e}	11.24 ± 0.23^{e}	0^{f}	0^{ef}
6.	**G+A**	131.48 ± 2.66^{cd}	24.52 ± 3.73^{c}	7.74 ± 0.77^{d}	4.29 ± 0.13^{cd}	1.11 ± 0.08^{de}	16.72 ± 1.02^{cd}	79.28 ± 5.64^{c}	75.34 ± 5.01^{cd}
7.	**G+T**	119.36 ± 5.06^{de}	23.32 ± 3.55^{cd}	7.14 ± 0.11^{d}	3.91 ± 0.05^{d}	1.09 ± 0.07^{ef}	19.56 ± 1.48^{c}	64.77 ± 2.96^{de}	63.30 ± 1.59^{d}
8.	**G+P**	135.46 ± 5.07^{cd}	25.66 ± 3.40^{bc}	8.46 ± 0.51^{cd}	4.44±0.03c	1.16 ± 0.02^{cd}	20.44 ± 1.47^{bc}	83.64 ± 2.16^{bc}	83.46 ± 1.65^{c}
9.	**A+T**	122.34 ± 3.61^{d}	24.14 ± 2.89^{c}	8.02 ± 1.24^{cd}	4.01 ± 0.21^{d}	1.15 ± 0.03^{d}	17.42 ± 0.25^{cd}	69.72 ± 5.01^{de}	67.30 ± 3.04^{cd}
10.	**A+P**	115.56 ± 3.43^{de}	21.70 ± 1.77^{c}	6.64 ± 0.47^{de}	3.90 ± 0.37^{d}	1.12 ± 0.01^{de}	15.34 ± 0.57^{cd}	60.88 ± 2.27^{e}	57.42 ± 3.36^{de}
11.	**T+P**	112.56 ± 3.00^{de}	20.16 ± 3.56^{d}	5.72 ± 0.41^{e}	3.77 ± 0.07^{de}	1.11 ± 0.35^{de}	14.34 ± 2.96^{d}	0^{f}	0^{de}
12.	**G+A+T**	150.62 ± 6.77^{ab}	34.44 ± 1.35^{ab}	10.8 ± 0.44^{ab}	4.82 ± 0.01^{b}	1.20 ± 0.02^{bc}	24.46 ± 2.45^{ab}	90.08 ± 5.68^{b}	97.40 ± 2.58^{bc}
13.	**G+T+P**	137.50 ± 4.09^{cd}	28.44 ± 3.34^{b}	9.0 ± 0.79^{b}	4.50 ± 0.15^{c}	1.17 ± 0.06^{c}	20.68 ± 0.22^{bc}	87.43 ± 1.62^{bc}	91.34 ± 3.64^{bc}
14.	**G+P+A**	145.70 ± 2.93^{b}	32.22 ± 2.58^{ab}	10.2 ± 0.70^{ab}	4.68 ± 0.06^{bc}	1.25 ± 0.04^{b}	22.86 ± 0.83^{b}	94.45 ± 2.72^{ab}	100.3 ± 1.66^{ab}
15.	**A+T+P**	141.40 ± 3.81^{b}	29.30 ± 2.43^{b}	9.64 ± 0.67^{b}	4.58 ± 0.05^{bc}	1.20 ± 0.02^{bc}	21.38 ± 0.27^{b}	89.78 ± 0.89^{bc}	95.24 ± 2.28^{bc}
16.	**G+A+T+P**	153.72 ± 4.69^{a}	37.38 ± 4.00^{a}	11.9 ± 0.80^{a}	4.97 ± 0.05^{a}	1.38 ± 0.05^{a}	24.58 ± 1.50^{a}	96.37 ± 2.80^{a}	103.96 ± 2.51^{a}
L.S.D($P\leq0.05$)		10.812	71.158	2.467	0.992	0.476	3.309	9.6872	7.506
ANOVA($F_{15,32}$)		102.86	1.549	30.962	35.936	6.748	70.321	131.635	288.818

‡G: *Glomus mosseae*, A: *Acaulospora laevis*, T: *Trichoderma viride*, P: *Pseudomonas fluorescens*, ±: Standard deviation,
ɳMean value followed by different alphabet/s within a column do not differ significantly over one other at($P\leq0.05$) Duncan's Multiple Range Test)

Table-2: Influence of AM Fungi, *T. viride* and *P. fluorescence* on different physiological parameters of Sunflower after 120 days

Sr. No.	Parameters Treatments	Leaf area (Sq. cm)	Chlorophyll content (mg gm^{-1} Fresh weight)			Stomatal conductance (mmol^{-2}s^{-2})			
			Chlorophyll a	Chlorophyll b	Total Chlorophyll	Morning		Evening	
						Lower	Upper	Lower	Upper
1.	**Control**	57.39±5.529^{f}	0.828±0.183^{f}	0.607±0.069^{f}	1.435±0.185^{f}	121.26±3.02^{f}	9.84±1.00^{f}	115.38±3.17^{f}	6.46±0.84^{f}
2.	**G**	97.61±5.459^{d}	1.337±0.030^{d}	1.093±0.003^{d}	2.430±0.027^{c}	433.08±0.81^{a}	39.32±5.10^{ab}	341.36±1.55^{bc}	32.24±1.45^{bc}
3.	**A**	65.19±2.160^{ef}	1.154±0.068^{e}	0.810±0.017^{ef}	2.039±0.18^{-ef}	217.16±1.73^{ef}	21.16±1.83^{d}	202.36±2.55^{ef}	8.24±2.46^{ef}
4.	**T**	70.56±4.649^{e}	1.156±0.017^{e}	0.923±0.043^{e}	2.080±0.059^{ef}	234.34±4.15^{de}	24.40±2.96^{c}	216.34±3.16^{ef}	10.46±3.32^{e}
5.	**P**	68.24±2.594^{ef}	1.136±0.061^{ef}	0.804±0.086^{ef}	1.940±0.029^{ef}	212.38±3.21^{ef}	23.36±2.11^{c}	208.48±3.92^{ef}	9.36±1.55^{ef}
6.	**G+A**	101.38±4.819^{cd}	1.405±0.069^{cd}	1.126±0.046^{cd}	2.533±0.102^{cd}	376.4±3.66^{c}	36.32±2.79^{ab}	331.22±2.40^{b}	29.38±5.75^{bc}
7.	**G+T**	85.63±8.674^{de}	1.245±0.034^{de}	1.059±0.009^{d}	2.304±0.042^{de}	230.30±1.84^{de}	25.22±3.29^{c}	212.34±4.45^{e}	12.08±1.21^{de}
8.	**G+P**	107.38±3.564^{cd}	1.424±0.088^{cd}	1.135±0.063^{cd}	2.559±0.041^{cd}	338.28±1.60^{cd}	32.24±1.58^{b}	274.30±3.60^{c}	18.30±1.60^{d}
9.	**A+T**	95.18±7.329^{d}	1.305±0.068^{d}	1.086±0.005^{d}	2.392±0.069^{de}	247.06±1.56^{d}	28.42±3.90^{bc}	241.34±1.50^{de}	16.70±4.09^{de}
10.	**A+P**	81.51±2.581^{de}	1.224±0.037^{de}	1.009±0.367^{de}	2.432±0.475^{d}	331.12±1.90^{cd}	30.24±3.08^{b}	269.24±3.02^{cd}	15.46±1.45^{de}
11.	**T+P**	74.71±4.047^{e}	1.116±0.003^{ef}	1.003±0.316^{de}	2.119±0.317^{e}	222.04±2.00^{e}	27.24±1.60^{bc}	259.28±1.58^{d}	14.46±5.08^{de}
12.	**G+A+T**	123.02±5.630^{b}	1.935±0.052^{b}	1.234±0.046^{b}	3.177±0.100^{b}	439.4±3.27^{ab}	49.04±1.31^{ab}	347.36±4.83^{a}	38.24±2.23^{a}
13.	**G+T+P**	111.75±2.606^{c}	1.474±0.032^{c}	1.138±0.015^{c}	2.612±0.048^{c}	382.12±1.65^{b}	38.36±1.42^{ab}	323.22±6.22^{bc}	30.64±5.05^{bc}
14.	**G+P+A**	119.31±6.146^{bc}	1.665±0.047^{bc}	1.207±0.085^{bc}	2.874±0.052^{bc}	431.02±1.58^{ab}	37.30±1.53^{ab}	330.12±3.10^{b}	27.30±1.39^{bc}
15.	**A+T+P**	115.27±2.570^{bc}	1.504±0.004^{bc}	1.205±0.054^{bc}	2.710±0.052^{bc}	369.26±5.58^{c}	35.3±3.03^{ab}	328.20±2.38^{bc}	21.54±5.25^{c}
16.	**G+A+T+P**	131.50±4.917^{a}	2.006±0.005^{a}	1.660±0.019^{a}	3.656±0.018^{a}	436.24±1.57^{ab}	41.1±2.57^{a}	344.30±3.63^{ab}	34.10±1.13^{b}
L.S.D. ($P \leq 0.05$)		12.450	0.166	0.383	0.417	6.907	6.756	7.834	8.133
ANOVA ($F_{15, 32s}$)		108.280	107.096	11.880	49.156	6876.034	62.467	2405.772	51.165

‡G: *Glomus mosseae*, A: *Acaulospora laevis*, T: *Trichoderma viride*, P: *Pseudomonas fluorescens*, ±:Standard deviation
ηMean value followed by different alphabet/s within a column do not differ significantly over one other at($P \leq 0.05$) Duncan's Multiple Range Test)

This is in conformity with the observations made earlier in cowpea and citrus (Auge, 2001; Duan, 1996 and Selvaraj, 2009). The increase in total chlorophyll content may be due to increased uptake of phosphorus which increases the photosynthetic activity of the plants and ultimately increase the chlorophyll content in plants. Similar findings were reported already by Arumugam *et al.* (2010), Bhattacharjee and Sharma (2012) and Rajasekaran *et al.* (2006). Akladious and Abbas (2012) revealed that the *Trichoderma* treatments significantly increased photosynthetic pigments of maize plants as compared to control.

3.3 AM Spore Number and Root Colonization of AM

A variety of studies suggest that mineral uptake by plant root can be enhanced, when infected by AM fungi. Data given in Table. 2 indicate that maximum AM spore number and percent AM root colonization was detected in all the AM treated plants. After 120 days of inoculation, percent mycorrhizal root colonization and AM spore number were highest in plants treated with *G. mosseae* + *A. laevis* + *T. viride* + *P. fluorescens* followed by the combination *G. mosseae* + *A. laevis* + *T. viride*. It was found that root colonization and AM spore number were greatly enhanced synergistically by interaction of AM fungi, *P. fluorescens* and *T. viride*. *Trichoderma viride* is highly effective in root colonization by producing secondary metabolites and these metabolites enhance AMF growth and thus mycorrhizal spore number and colonization (Jalaluddin *et al.*, 2008; Soleimanzadeh, 2012). *Pseudomonas* species are also effective in root colonization as they produce secondary metabolites (Akladious, A.S. and Abbas, 2012). The mycorrhizal inoculation (*G. fasciculatum* and *A. laevis*) with *Pseudomonas striata* increased root colonization in sesame plant (Rosa, 2009). The phosphate solubilizing bacteria behaved as mycorrhiza helper bacteria because they promote root colonization when associated with mycorrhizal fungi and phosphate solubilizing bacteria combination (Frey-Klett *et al.*, 1997; Garbaye, 1994; Azco´n-Aguilar and Barea, 1992).

3.4 Plant Phosphorus Content

Data pertaining to phosphorus content are presented in Table 3. which indicates highest shoot and root P content in plants treated with *G. mosseae* + *A. laevis* + T. *viride* + *P. fluorescens* followed by *G. mosseae* + A. *laevis* + *T. viride*. The results of the present study coincide with the findings of Jackson (1973), who reported that the fungal hyphae growing beyond the rhizospheric soil increase the absorptive surface area of the root resulting in a greater efficiency of nutrient absorption, especially slowly diffusing minerals ions like phosphorus (Kothari *et al.*, 1991). Many workers have suggested that phosphorus uptake by plant root can be enhanced when they are infected by AM fungi. (Doley and Jite, 2012; Kaushish *et al.*, 2012; Prasad *et al.*, 2012).

Table-3: Influence of AM Fungi, *T. viride* and *P. fluorescens* on phosphorus content, phosphatase activity and plant yield of Sunflower after 120 days

Sr. No.	Parameters Treatments	Phosphorus content (%)		Phosphatase activity (IU/gm Fresh weight)		Yield		
		Shoot	Root	Acidic	Alkaline	Head diameter	Protein content (%)	Oil content (%)
1.	**Control**	1.042 ± 0.007^{f}	1.124 ± 0.184^{f}	0.096 ± 0.004^{f}	0.232 ± 0.015^{f}	10.02 ± 1.12^{f}	16.65 ± 0.58^{f}	32.39 ± 3.23^{f}
2.	**G**	1.825 ± 0.193^{cd}	1.892 ± 0.056^{cd}	0.455 ± 0.015^{d}	1.067 ± 0.008^{cd}	16.30 ± 1.18^{cd}	22.11 ± 1.59^{cd}	45.72 ± 4.16^{cd}
3.	**A**	1.421 ± 0.044^{e}	1.503 ± 0.035^{de}	0.221 ± 0.040^{ef}	0.799 ± 0.036^{de}	12.73 ± 1.16^{ef}	18.63 ± 2.31^{ef}	40.48 ± 4.82^{ef}
4.	**T**	1.524 ± 0.035^{e}	1.533 ± 0.023^{de}	0.227 ± 0.021^{ef}	0.864 ± 0.017^{de}	13.06 ± 1.25^{e}	19.86 ± 1.94^{e}	41.15 ± 4.89^{e}
5.	**P**	1.774 ± 0.084^{d}	1.403 ± 0.036^{e}	0.195 ± 0.015^{ef}	0.633 ± 0.037^{e}	11.70 ± 1.10^{ef}	18.26 ± 0.11^{ef}	39.94 ± 4.76^{ef}
6.	**G+A**	1.792 ± 0.120^{d}	1.981 ± 0.008^{cd}	0.499 ± 0.029^{d}	1.086 ± 0.009^{cd}	16.72 ± 1.13^{cd}	22.51 ± 2.34^{cd}	46.66 ± 4.49^{cd}
7.	**G+T**	1.661 ± 0.046^{de}	1.692 ± 0.043^{d}	0.394 ± 0.065^{ae}	1.043 ± 0.022^{cd}	14.86 ± 1.25^{de}	20.94 ± 1.90^{d}	44.56 ± 4.33^{d}
8.	**G+P**	1.892 ± 0.062^{cd}	2.094 ± 0.033^{c}	0.503 ± 0.061^{cd}	1.106 ± 0.016^{d}	17.68 ± 1.11^{c}	22.84 ± 2.22^{cd}	47.51 ± 4.81^{c}
9.	**A+T**	1.701 ± 0.054^{d}	1.961 ± 0.042^{cd}	0.373 ± 0.007^{de}	1.099 ± 0.018^{d}	15.79 ± 1.20^{d}	22.55 ± 1.97^{cd}	45.08 ± 4.83^{cd}
10.	**A+P**	1.617 ± 0.049^{de}	1.791 ± 0.072^{d}	0.303 ± 0.014^{de}	$0.984\pm\pm0.007^{d}$	14.06 ± 1.11^{de}	20.71 ± 2.48^{d}	43.80 ± 2.80^{de}
11.	**T+P**	1.573 ± 0.031^{de}	1.753 ± 0.036^{d}	0.293 ± 0.030^{e}	0.962 ± 0.023^{d}	13.38 ± 1.17^{e}	19.78 ± 2.09^{e}	42.68 ± 2.71^{de}
12.	**G+A+T**	2.078 ± 0.056^{ab}	2.984 ± 0.007^{b}	0.894 ± 0.044^{b}	1.163 ± 0.002^{b}	19.51 ± 1.11^{a}	24.80 ± 2.12^{ab}	50.02 ± 1.57^{ab}
13.	**G+T+P**	1.933 ± 0.057^{c}	2.481 ± 0.058^{bc}	0.534 ± 0.032^{c}	1.137 ± 0.005^{bc}	18.04 ± 1.09^{bc}	23.09 ± 3.61^{c}	48.72 ± 3.21^{bc}
14.	**G+P+A**	2.039 ± 0.367^{ab}	2.581 ± 0.037^{bc}	0.763 ± 0.031^{bc}	1.145 ± 0.007^{bc}	18.87 ± 1.15^{b}	24.77 ± 0.04^{ab}	49.79 ± 2.76^{b}
15.	**A+T+P**	2.006 ± 0.005^{b}	2.131 ± 0.015^{c}	0.653 ± 0.040^{bc}	1.127 ± 0.005^{c}	18.35 ± 0.72^{bc}	24.05 ± 4.06^{b}	49.00 ± 2.55^{b}
16.	**G+A+T+P**	2.104 ± 0.022^{a}	3.863 ± 0.045^{a}	0.989 ± 0.009^{a}	1.173 ± 0.006^{a}	19.14 ± 1.02^{ab}	25.09 ± 0.06^{a}	50.96 ± 1.55^{a}
L.S.D($P\leq0.05$)		0.294	0.100	0.086	0.046	39.922	5.460	9.539
ANOVA($F_{15,32}$)		28.583	1429.927	282.851	915.107	1.059	6.864	8.190

‡G: *Glomus mosseae*, A: *Acaulospora laevis*, T: *Trichoderma viride*, P: *Pseudomonas fluorescens*, ±: Standard deviation,
ŋMean value followed by different alphabet/s within a column do not differ significantly over one other at($P\leq0.05$)lead by Duncan's Multiple Range Test

3.5 Root Phosphatase Activity

As illustrated in Table 3, the acid phosphatase activity was appreciably higher in inoculated plants roots in comparison to non-inoculated control sunflower plants. Highest value of both acidic and alkaline phosphatase activity was observed in the plants inoculated with *G. mosseae + A. laevis + T. viride + P. fluorescens* followed by the combination *G. mosseae + A. laevis + T. viride*. Alkaline phosphatase activity was found higher as compared to acidic phosphatase activity. It was found that plants with higher mycorrhizal root colonization had maximum phosphatase activity. This conclusion could be derived from the reports that mycorrhizal treated roots have recorded greater acidic and alkaline phosphatse activities than non-inoculated roots (Kapoor, 1988). The increase of phosphatase activity in mycorrhizal plants which has been reported by several other researchers may be due to higher phosphatase activity of the internal hyphae produced by mycorrhizal fungi (Devi and Reddy, 2001; Dodd *et al.*, 1987; Saito, 1995). Plant yield results depicted in Table. 3 showed that inoculated plants showed significant increase in head diameter, protein and oil content (%) in comparison to the control. After 120 days, higher value of head diameter, protein and oil content was in combination, *G. mosseae + A. laevis + T. viride + P. fluorescens* followed by triple inoculation of *G. mosseae + A. laevis + T. viride*. According to the results on the yield G. mosseae was found to be common most effective bioinoculant. Accordingly the positive effect of mycorrhiza and other bioinoculants on yield, it can concluded that AMF helps to absorb more of the nutrients especially away from the P depletion zone and resulted in the better increment in yield of sunflower plant in comparison to the nonmycorrhizal plants. Our results confirm the recent findings of Zachee *et al.* (2008) and Zolfaghari *et al.* (2013). The beneficial effects of VAM- fungi on inoculation has been reported to increase oil and protein content in *Linum usitatissimum*, which is in conformity with our results of Neetu *et al.* (2011; 2012).

4. Conclusion

In conclusion, the results of this study clearly reveled that the inoculation with AM fungi and other bioinoculants is more effective in increasing of growth parameters and yield of sunflower plant. This increment could be attributed to the enhanced uptake of nutrients, better water absorption, increased surface area of roots and also secretion of some enzymes by inoculated microorganisms. Inoculation of *G. mosseae + A. laevis + T. viride + P. fluorescens* proved to be the best combination for inoculating sunflower seeds under polyhouse condition to get maximum yield and growth. This combination can be tested further in the field conditions at seed sowing stage and can be recommended to farmers after proper confirmation.

Acknowledgements

The authors are thankful to Kurukshetra University, Kurukshetra, India for providing laboratory facilities.

References

Adholeya, A. and Gaur, A. 1994. Estimation of VAM fungal spores in soil. *Mycorrhiza News*. 6: 10-11.

Akladious, A.S. and Abbas, M.S. 2012. Application of *Trichoderma harziunum* T22 as a biofertilizer supporting maize growth. *Afr. J. Biotechnol.* 11: 8672-8683.

AOCS. 1997. Official methods and Recommended Practice of AOCS. The American Oil Chemist's Society. Washington DC 5th Edn.

Arnon, D.T. 1949. Copper enzyme in isolated chloroplasts polyphenol oxidase in *Beta vulgaris. Plant Physiol.* 24: 1-5.

Arumugam, R., Rajasekaran, S. and Nagarajan, S.M. 2010. Response of arbuscular mycorrhizal fungi and *Rhizobium* inoculation on growth and chlorophyll content of *Vigna unguiculata* (L.) Walp Var. Pusa 151. *J. Appl. Sci. Environ.* 14: 113-115.

Auge, R.M. 2001. Water relations, drought and vesicular-arbuscular mycorrhizal symbiosis. *Mycorrhiza.* 11: 3-42.

Azco´n-Aguilar, C. and Barea, J.M. 1992. Interactions between mycorrhizal fungi and other rhizosphere microorganisms. In: M.J. Allen (ed.), Mycorrhizal functioning: an integrative plant-fungal process. Chapman and Hall, New York, pp. 163-198.

Bhattacharjee, S. and Sharma, G.D. 2012. Effect of dual inoculation of arbuscular mycorrhiza and Rhizobium on the chlorophyll, nitrogen and phosphorus contents of Pigeon Pea (*Cajanus cajan* L.). *Adv. Microbiol.* 2: 561-564.

Bradford, M.M. 1976. To estimate protein utilizing the principles of protein-dye binding. *Anal. Biochem.* 72: 248-254.

Cagras, S., Sari, N. and Ortas, I. 2000. The effects of vesicular arbuscular mycorrhizae on the plant growth and nutrient uptake of cucumber. *Turk. J. Agric. Forest.* 24: 571-578.

Dell'Amico, J., Torrecillas, A., Rodriguez, P., Morte, A. and Sanchez-Blanco, M.J. 2002. Responses of tomato plants associated with the arbuscular mycorrhizal fungus Glomus clarum during drought and recovery. J. Agric. Sci. 138: 387-393.

Devi, C. and Reddy. 2001. Growth response of groundnut to VAM fungus and Rhizobium inoculation. *Plant Pathology Bulletin.* 10: 71-78.

Dodd, J.C., Burton C.C., Burns, R.G. and Jeffries, P. 1987. Phosphatase activity associated with the rhizosphere of plants infected with vesicular-arbuscular mycorrhizal fungi. *New Phytol.* 7: 163-172.

Doley, K. and Jite, P.K. 2012. Response of Groundnut cultivar to Mycorrhiza inoculation and phosphorus application. *Not. Sci. Biol.* 4: 118-125.

Duan, X., Neuman, D.S., Reiber, J.M., Green, C.D., Saxton, A.M. and Auge, R.M. 1996. Mycorrhizal influence on hydraulic and hormonal factors implicated in the control of stomatal conductance during drought. *J. Exp. Bot.* 47: 1541-1550.

Frey-Klett, P., Pierrat, J.C. and Garbaye, J. 1997. Location and survival of mycorrhiza helper *Pseudomonas fluorescens* during establishment of ectomycorrhizal symbiosis between Laccaria biocolor and Douglas fir. *Appl. Environ. Microbiol.* 63: 139-144.

Garbaye, J. 1994. Helper bacteria: a new dimension to the mycorrhizal symbiosis. *New Phytol.* 128: 197-210.

Gerdemann, J.W. and Nicolson, Y.H. 1963. Spores of mycorrhizae Endogone species extracted from soil by wet sieving and decanting. *Trans. Br. Mycol. Soc.* 46: 235-244.

Gong, Q., Xu, D., and Zhong, C. 2000. Study on biodiversity of mycorrhizae and its application. Beijing: Chinese forest Press. pp. 51-61.

Hoyos-Carvajal, L., Ordua, S. and Bissett, J. 2009. Growth stimulation in bean (Phaseolus vulgaris L.) by *Trichoderma*. *Biol. Control*. 51: 409-416.

Hussain, M.K., Rasul, E. and Ali, S.K. 2000. Growth analysis of sunflower under drought conditions. *Int. J. Agri. Biol.* 2: 136-140.

Jackoben, B., Abbott, L.K. and Robson, A.D. 1992. External hyphae of vesicular-arbuscular mycorrhizal fungi associated with *Trifolium subterrancum* L. I. Spread hyphae and phosphorus inflow into roots. *New Phytol.* 120: 371-380.

Jackson, M.L. (1973). "Soil Chemical Analysis Prentice Hall, New Delhi, pp. 485.

Jalaluddin, M., Hajra, N.B., Firoza, K. and Shahina, F. 2008. Effect of *Glomus callosum,* Meloidogyne incognita and soil moisture on growth and yield of sunflower. *Pak. J. Bot.* 40: 391396.

Kapoor, A., Singh, V.P. and Mukherji, K.G. 1988. Studies on the physiology of mycorrhizal and non- mycorrhizal *Trigonella* roots (Abstract). In: First Asian Conference on mycorrhiza. Jan 29th -31st, University of Madras, India.

Kaushish, S., Kumar, A. and Aggarwal, A. 2012. Influence of inoculation with the endomycorrhizal fungi and *Trichoderma viride* on morphological and physiological growth parameters of *Rauwolfia serpentine* Benth. Ex. Kurtz. *Ind. J. Microbiol.* 52: 295-299.

Kothari, S.K., Marschner, H., Romheld, V. 1991. Contribution of VAM hyphae in acquisition of phosphorus and zinc by maize grown in a calcareous soil. *Plant Soil.* 131: 177-185.

Meena, B. and Marimuthu, T. 2012. Effect of application methods of *Pseudomonas fluorescens* for the late leaf spot of groundnut management. *J. Biopest*. 5: 14-17.

Menge, J.A. and Timmer, L.M. 1982. Procedure for inoculation of plants with VAM in the laboratory, greenhouse and field, In: N.C. Schenck (Ed.), Methods and Principles of Mycorrhizal Research, American Phytopathological Society, St. Paul, pp. 59-68.

Morton, J.B. and Benny, G.L. 1990. Revised classification of arbuscular mycorrhizal fungi (Zygomycetes): A new order, Glomales, two new suborder, Glomineae and Gigasporineae, with an emendation of Glomaceae. *Mycotaxon*. 37: 471-491.

Mukerji, K.G. 1996. Advances in Botany, In: Mukerji, K.G., Mathur, B., Chamola, B.P. and Chitralekha, P. (Eds), Taxonomy of endomycorrhizal fungi, APH Pub. Co., pp. 211-221.

Neetu, N., Tanwar, A., Aggarwal, A. and Alpa, A. 2012. Influence of arbuscular mycorrhizal fungi and Pseudomonas fluorescens with different levels of superphosphate on the growth response of linseed (*Linum usitatissimum* L.). *Chil. J. Agr. Res.* 72: 237-243.

Neetu, Tanwar, A. and Aggarwal, A. 2011. Impact of arbuscular mycorrhizal fungi and other bioinoculants on growth promotion in *Linum usitatissimum* L. *J. Ind. Bot. Soc.* 90: 216-223.

Ortas, I. 2010. Effect of mycorrhiza application on plant growth and nutrient uptake in cucumber production under field conditions. *Span. J. Agric. Res.* 8: 116-122.

Phillips, J.M. and Hayman, D.S. 1970. Improved procedures for clearing roots and staining parasitic and VAM fungi for rapid assessment of infection. *Trans. Br Mycol. Soc.* 55: 158-161.

Prasad. K., Aggarwal, A., Yadav, K. and Tanwar, A. 2012. Impact of different levels of superphosphate using arbuscular mycorrhizal fungi and *Pseudomonas fluorescens* on *Chrysanthemum indicum* L. *J. Soil Sci. Plant Nutr.* 12: 451-462.

Rajasekaran, S., Nagarajan, S.M., Arumugam, K., Sravanamuthu, R., Balamurugan, S. 2006. Effect of dual inoculation (AM fungi and *Rhizobium*) on TMV-2. *Plant Archives*, 6: 671672.

Rosa, P.M., Antoniassi, R., Freitas, S.C., Bizzo, H.R., Zanotto, D.L., Oliveira, M.F., Castiglioni, V.B.R. 2009. Chemical composition of Brazillian sunflower varieties. *Helia.* 32: 145-156.

Saito, M. 1995. Enzyme activities of the internal hyphae and germinated spores of an arbuscular mycorrhizal fungus, *Gigaspora margariata* Becker and Hall. *New Phytol.* 129: 425-431.

Schenick, N.C. and Perez, Y. 1990. Manual for the identification of VA- mycorrhizal VAM fungi. 3rd ed., Synergistic Publication, Gainesville, U.S.A.

Selvaraj, T., Nisha M.C. and Rajeshkumar, S. 2009. Effect of indigenous arbuscular mycorrhizal fungi on some growth parameters and phytochemical constituents of *Pogostemon patchouli* Pellet. *Maejo Int. J. Sci. Technol.* 3: 222-234.

Soleimanzadeh, H. 2012. Response of sunflower (*Helianthus annuus* L.) to inoculation with mycorrhiza under different phosphorus levels. *American Eurasian J. Agric. Environ. Sci.* 12: 337341.

Tabatabai, M.A. and Bremner, J.M. 1969. Use of p-nitrophenyl phosphate for assay of soil phosphatase activity, *Soil Biol. Biochem.* 1: 301-307.

Tanwar, A. Yadav, K., Prasad, K. and Aggarwal, A. 2013. Biological Amendments on Growth, Nutritional Quality, and Yield of Celery. *Int. J. Veg. Sci.* 19: 228-239.

Vinale, F., Sivasithamparam, K., Ghisalberti, E.L., Marra, R., Barbetti, M.J., Li, H., Woo, S.L. and Lorito, M. 2008. A novel role for Trichoderma secondary metabolites in the interactions with plants. *Physiol. Mol. Plant Pathol.* 72: 80-86.

Walker, C. 1983. Taxonomic concepts in the Endogonaceae spore wall characteristics in species description. *Mycotaxon.* 18: 443-445.

Weiss, E.A. 2000. Oil Seed Crops, 2nd Ed., World Agriculture Science, Blackwell Sci. Ltd. Paris, Berlin, Victoria, p. 365.

Yadav, A., Yadav, K., Tanwar, A. and Aggarwal, A. 2013. Interaction of VAM fungi with *Bradyrhizobium japonicum* and *Trichoderma viride* on some physiological parameters of soybean. *J. Pure App. Microbiol.* 7: 505-514.

Yaseen, T., Burni, T. and Hussain, F. 2012. Effect of Arbuscular Mycorrhizal inoculation on nutrient uptake, growth and productivity of chickpea (*Cicer arietinum*) varieties. *Int. J. Agron. Plant Production,* 3: 334-335.

Zachee, A., Bekolo, N., Bime, Dooh, N., Yalen, M. and Godswill, N. 2008. Effect of mycorrhizalinoculum and urea fertilizer on diseases development and yield of groundnut crops (*Arachis hypogaea* L.). *Afr. J. Biotechnol.* 7: 2823-2827.

Zolfaghari, M., Nazeri, V., Sefidkon, F. and Rejali, F. 2013. Effects of arbuscular mycorrhizal fungi on plant growth and essential oil content and composition of *Ocimum basilicum* L. *Iranian J. Plant Physiol.* 3: 643-650.

Chapter 14

Tissue culture studies of *Tinospora cordifolia* using different explants

Priti and Sulekha Rani*

Department of Biotechnology, Kurukshetra University, Kurukshetra - 136119, Haryana, India.

**Corresponding author E-mail: sulekha.chahal@yahoo.com*

Abstract

Tinospora cordifolia is a large deciduous climbing shrub found throughout India, especially in tropical part of India and also in certain parts of China. The plant is mainly known for its medicinal properties. The present investigation was carried out to standardize an in vitro regeneration protocol for Tinospora. Different explants such as leaves, nodes and inter-nodes were used for in vitro studies. For in vitro regeneration, explants were cultured on Murashige and Skoog medium supplemented with different concentrations and combinations of cytokinins and auxins for shoots and roots proliferation. Hormonal concentration of media and explants types play important role for in vitro regeneration process. Concentration and combinations of hormones which favour callus induction are different from those which induce organogenesis. The callusing was exhibited by all the three explants but leaf and nodal explants were somewhat better than their inter-nodal counterpart. The direct shoot formation capability of nodal explants was excellent while leaf and inter-nodal explants give inferior results. Hence nodal explants were found to be more potent and used further for direct in vitro regeneration. Different concentration and combinations of hormones for in vitro regeneration studies was worked out for nodal explants. BAP at lower concentration displays callus formation but at higher concentration induce direct shoots while Kinetin at lower concentration promotes formation of shoot buds and at higher concentration stimulates shoots elongation. BAP in combination with Kinetin endorses direct multiple shoots and BAP along with NAA encourages callus production and proliferation. Among different concentration of IAA, IBA and NAA, the highest degree of root induction for nodal explants was found on the MS media containing 1mg/l concentration of IAA. It shows 90% of roots development within 25-28 days.

Keywords: *Tinospora cordifolia, Growth regulators, MS Media, BAP, IAA, Callus.*

1. Introduction

Medicinal plants since times immemorial have been used in virtually all cultures as a source of medicine in the traditional medicinal practices for primary health care needs. As per the traditional medicinal programme of the World Health Organization (WHO) nearly 80% of the world population used phytoproducts (Dubey *et al.*, 2004). There has been an ever-increasing demand especially from developed countries for more and more drugs from plant sources due to the current widespread belief that 'green medicine' is safe and more dependable than the synthetic drugs which are costly and have some side effects.

Tinospora cordifolia (Guduchi) (Willd.) Miers ex Hook. f. & Thoms is a widely used shrub in medicine. *T. cordifolia* belongs to the family Menispermaceae (Anonymous, 2003; Aima *et al.*, 2003; Vaidya *et al.*, 1994) which consists of about 70 genera and 450 species that are found in tropical lowland regions. This family is a rich source of alkaloid and terpenes (Sharma *et al.*, 2010). *Tinospora* is a rich source of variety of natural chemical constituents; like berberine, tinosporin, tinocordifolioside, magnoflorine, palmatine, isocolumbin, tinocordiside, cordifolioside and syringing etc. (Chaudhary *et al.*, 2014; Mittal *et al.*, 2014). Due to these chemical constituents, *Tinospora* is widely used in veterinary folk medicine/ayurvedic system of medicine for its general tonic, anti-periodic, anti-spasmodic, anti-inflammatory, anti-arthritic, anti-allergic, and anti-diabetic (Lueng *et al.*, 2000; Ondrizek *et al.*, 1999; Anonymous, 1976). The plant is used in ayurvedic, "Rasayanas" to improve the immune system and the body resistance against infections. The stems, leaves, and roots are used in medicine (Premila *et al.*, 2006; Pandey *et al.*, 2002; Krishna *et al.*, 2009). The root of this plant is also known for its anti-stress, anti-leprotic and anti-malarial activities (Nadkarni and Nadkarni, 1976; Chopra *et al.*, 1956; Chopra *et al.*, 1982; Zhao *et al.*, 1991; Nayampalli *et al.*, 1982).

The various reports on its multiple uses attracted attention for utilization of the plant for medicinal purposes. Medicinal plants species are getting endangered status at an increasing rate owing to urbanization, deforestation and uprooting of complete plants to procure plant extracts for the production of medicines by pharmaceutical companies (Mohammed and Kumar, 2012). *Tinospora* grows wild in forest and other areas and there is no organized propagation information available so far. Conventional vegetative propagation of this plant has limited potential for large scale propagation of elite plants. Due to high demand of *Tinospora cordifolia*, it has become highly imperative to regenerate, multiply and conserve this medicinal plant with desirable traits. Biotechnology has opened up new vistas in the conservation of medicinal plants by way of '*in vitro* propagation'. *In vitro* propagation ensures not only continuous supply of plant throughout the year but also prevents the destruction of the natural population of medicinal plant (Datta *et al.*, 1993; Hamilton *et al.*, 2004). Among different techniques micropropagation of plants through tissue culture has immersed as an invaluable aid for the production of high quality economically important plant species in agriculture, forestry and medicinal plant industries for the past decades (Rani and Raina, 2000; Sunandakumari *et al.*, 2004; Sujatha *et al.*, 2007). Micropropagation provides an ideal means for large scale propagation and the reintroduction of the plants in its natural habitats (Bhojwani *et al.*, 1989; Lal *et al.*, 1988). Moreover, *T. cordifolia* is a genetically diverse plant and henceforth, the metabolic constituents and other associated characteristics do affected by geographical location, environmental conditions and season. As a result, the study made on the same plant located in different region can vary minutely or heftily.

Thus, *in vitro* regeneration is one of the efficient strategies for conservation of endangered medicinal plants of a particular location. The present study highlights the efficient protocol for *in vitro* rapid propagation of this divine plant located in pious Kurukshetra region.

2. Materials and Methods

2.1 Source of Explant

Plant of *T. cordifolia was* obtained from Botanical Garden of Kurukshetra University, Kurukshetra. The leaf, stems (nodes and internodes) of this plant were selected and used as explants for *in vitro* propagation.

2.2 Selection of Explants

Experiments were conducted employing the leaves, nodal and internodal regions of *T. cordifolia* as explants. The selected explants were cultured on MS Media supplemented with different concentrations and combinations of plant growth regulators (cytokinins and auxins) for checking their response to *in vitro* regeneration conditions i.e. callusing, shoot induction, proliferation, multiple shoot formation and rooting.

2.3 Sterilization of Explants

The fresh plant material (stem and leaves of *T. cordifolia*) brought to the laboratory in sealed plastic bags were thoroughly washed with running tap water to remove soil and other superficial contamination, followed by washing with tween 20 (surfactant) before excision of the explants. Explants excised from the leaves and nodal & inter nodal portion of stems, were then washed with double distilled water and subsequently allowed to stay in a fresh lot of double distilled water for 3–4 h. Pretreatment of explants i.e. soaking in double distilled water for 3-4 hours was essential for good *in vitro* regeneration response because this process removed the oxidized phenolic compounds, leaching out of which in the media during inoculation could be inhibitory for the growth of explants (Thomas and Mehta, 1983). Explants were surface sterilized by rinsing with 70% ethyl alcohol for 2 minutes and consequently with 0.15% $HgCl_2$ for 1 minute. After that, explants were repeatedly washed (9 to 10 times) with sterilezed distilled water and allowed to dry. The edges of the sterilized explants were trimmed before inoculation and subsequently cultured on appropriate media.

2.4 Culture Initiation, Multiplication and Elongation

Explants were cultured on Murashige and Skoog's (MS) medium supplemented with diverse combinations and concentrations of plant growth regulators along with necessary macro-nutrients, micro-nutrients, an iron source, vitamins, 30g/l sucrose (as a carbon source) and 8g/l agar as a gelling agent. The maintained pH range of the media was 5.7-5.8 (adjusted prior to autoclaving). Different cultures were incubated in culture room at 25±2°C with a photoperiod of eighteen hours light and six hours dark. After 4 weeks of culture, callus & axillary shoots developed on primary explants were excised and sub cultured for further proliferation and multiplication.

2.5 *In Vitro* Rooting, Acclimatization and Transfer of Plantlets to Soil

The individual shoots were excised and cultured for rooting in MS medium supplemented with various concentration and combinations of auxins i.e. IAA (Indole-3-acetic acid), IBA (Indole-3-butyric acid) and NAA (Napthalene acetic acid). After differentiation and establishment of plantlets, the rooted healthy plants were removed from culture vessels and transferred to culture tubes containing steriled distilled water. After 10 days, plantlets were transferred to plastic pots containing steriled course sand and garden soil in 1:1 ratio, moistened with autoclaved tap water. The potted plantlets were maintained inside the culture room at 25±2°C, with a photoperiod of 16/8-hour (light/dark). After 30 days the plantlets were transplanted to pots containing normal (non-autoclaved) garden soil and kept under shade in green house for another 2 weeks before transferring to outdoors under full sun.

3. Results and Discussion

The present study was carried out to develop an efficient *in vitro* regeneration protocol for *T. cordifolia*. For *in vitro* propagation, different explants (nodes, inter nodes and leaves) were regenerated via direct and indirect organogenesis (Fig. 14.1, 14.2 & 14.3). But, in the course of *in vitro* culturing of *T. cordifolia*, bacterial and fungal contamination was a big obstacle. Phenolic exudation further limits the establishment of explants *in vitro*. Hence, to control and solvation of these hurdles was the most vital pace of the work. For surface sterilization and to regulate phenol exudation, plant material was washed with tween 20 (surfactant) for 20 minutes before excision of the explants. After excision, explants were washed with double distilled water and subsequently allowed to stay in double distilled water for 3–4 hours. Soaking in double distilled water for 3-4 hours removed the oxidized phenolic compounds. Explants were then surface sterilized by rinsing with 70% ethyl alcohol for 2 minutes and consequently with 0.15% $HgCl_2$ for 1 minute. After that, explants were repeatedly washed (9 to 10 times) with steriled distilled water and allowed to dry before inoculation.

For direct shoot morphogenesis, nodal, inter nodal and leaf explants of this pharmaceutically important plant were tried for reproducible procedure of rapid clonal multiplication. Morphogenetic responses of the different explants of *T. cordifolia* cultured on MS medium containing different plant growth regulator either singly or in combination, were worked out. When explants were grown on MS basal medium without growth regulators, they were minutely responding to bud break even after 30 days however when explants were cultured on MS medium supplemented with different concentration of cytokinins like BAP and Kinetin, shoots were induced. The results were in accordance with Lane *et al.* (1979) & Bhojwani *et al.* (1980) who stated that cytokinins are one of the most important hormones for shoot proliferation and bud breakage & development of shoots from the explants.

Among different explants tested, nodal explants response for shoot regeneration and multiplication was better as compared to internodes and leaves. The nodal cuttings showed visible growth after seven days in culture and most of them grew into shoots within 30 days. At lower concentration of BAP, callus was formed but when the concentration was increased, shoots were formed (Table. 14.1).

Table 14.1: Morphogenic response of nodal explants of *T. cordifolia* to different concentrations of cytokinins (BAP, Kin) and/or GA_3 Data presented as the mean value ± standard error after 30 days of culture from four independent experiments each with 20 replicates.

Growth Regulators (mg/l)	Shoot Development (%)	Shoot no./ Explant	Shoot Length (cm)
BAP			
0.25	Callus	-	-
0.5	Callus	-	-
1.0	Callus	-	-
3.0	95	3.66±1.25	4.0±0.06
5.0	84	3.98±1.40	3.68±0.03
Kinetin (Kin)			
0.25	40	2.80±0.06	1.38±0.06
0.50	60	2.12±0.00	1.82±0.05
1.0	80	2.56±0.12	1.98±0.03
3.0	100	2.76±0.16	2.86±0.07
5.0	90	2.35±0.06	1.55±0.02
GA3			
0.25	25	1.12±0.6	3.8±0.4
0.50	40	1.18±0.7	4.0±0.2
1.0	55	1.80±0.2	4.6±0.8
2.0	70	1.10±0.5	4.8±0.3
5.0	30	1.00±0.3	5.1±0.6
BAP+ Kin			
3+0.50	30	3.0±0.3	3.6±0.7
3+1.00	55	4.0±0.7	4.8±0.6
3+2.00	90	5.5±1.0	5.3±0.9
3+3.00	70	4.5±0.6	5.0±0.1
3+4.00	60	4.0±0.4	4.6±0.4
BAP+ GA_3			
3.0+0.2	85	4.7±0.5	4.8±0.2
3.0+0.3	77	4.4±0.3	5.8±0.6
3.0+0.4	70	4.2±0.7	5.0±0.3
Kin+ GA_3			
3.0+0.2	65	2.66±0.76	6.10±0.78
3.0+0.3	78	2.90±0.54	6.20±0.50
3.0+0.4	72	2.95±0.23	6.00±0.43

Fig.14.1 (a-m): Various stages of regeneration process in nodal explants of *T. cordifolia.*

(a) *Tinospora cordifolia*- medicinal plant (b-d) Nodes, leaf and inter nodal explants grown on MS media (e-j) Shoot bud initiation from nodal explants (e-h) Shoot differentiation and elongation (i) Elongated shoot (j) Roots formation on the excised shoot on MS media. (k) Roots of *Tinospora cordifolia* (l-m) Acclimatized plantlet in pot.

Higher concentrations of kinetin induce shoots in nodal explants of *Tinospora* (Gururaj *et al.*, 2007). In our findings, Kinetin at concentration 3mg/l showed 80% shoots regeneration with two shoots from single explant with 2.86 cm shoot length. When BAP was used with Kinetin then multiple shoot formation took place but at a particular concentration (3:2mg/l). When Kinetin used individually at higher concentration resulted in shoot regeneration and elongation but if kinetin used in combination with BAP, then the less concentration of kinetin help in multiple shoot formation. If the concentration of kinetin in combination with BAP increases then, the multiple shoot formation response decreases as shown in Table 14.1. When BAP was used in combination with NAA, then the callus formation took place (Table. 14.2 and Fig. 14.2). Number and length of shoots were also affected by concentrations of cytokinins (Table. 14.1).

Fig. 14.2 (a-d): Callus formation from nodal explant of *T. cordifolia* (a) Callus induction from nodal explant (b) Callus proliferation from nodal explants (c-d) Callus subcultured and maintained.

Fig. 14.3 (a-d): Multiple shoot formation from nodal explants on MS Media with BAP and Kinetin.

(a) shoot initiation from nodal explants (b) Multiple shoot differentiation from nodal explants (c-d) Multiple shoots formation from nodal explants

Gibberellic acid (GA_3) also helps in shoots regeneration. GA_3 when used individually with a concentration of 2mg/l, showed 70% response for shoots development. But when the combinations of GA_3 with BAP and Kinetin at different concentrations were tried then the small concentration of GA_3 (0.2mg/l and 0.3mg/l) with BAP (3mg/l) and also with Kinetin (3mg/l), showed enhanced results for shoot regeneration in terms of number of shoots and shoot length (Table. 1).

Indirect organogenesis took place via callus formation. Explants *i.e.* leaves, nodes and inter nodes were tried for better response of callus regeneration. Leaves and nodes showed the good response as compared to internodes, when different concentrations of growth regulators were used (Table. 14.2).

Table 14.2: Effect of plant growth regulators on callus induction on nodal explants in *T. cordifolia*.

Plant growth regulators (mg/l)				
BAP	NAA	2,4-D	% Callus formation	Compactness
0.5	2.0	-	42	+
0.5	3.0	-	70	+
0.5	4.0	-	65	++
1.5	2.0	-	48	+
1.5	3.0	-	52	+
1.5	4.0	-	38	+
0.5	-	2.0	45	+
0.5	-	3.0	68	+
0.5	-	4.0	50	+
1.5	-	2.0	78	+
1.5	-	3.0	60	++
1.5	-	4.0	54	+

(+) Fragile callus, (++) compact callus

Growth regulators NAA and 2,4-D alone could initiate callusing from stem, leaf and nodal segments but callus may grow slowly (Nakano *et al.*, 1994). In our study, when auxin like 2, 4-D in combination with BAP (at lower concentration) were used for callus production then the growth of callus enhanced. 2, 4-D at concentration 2.0mg/l with BAP at concentration 1.5 mg/l showed 78% callus regeneration and the concentration of 2,4-D (3.0 mg/l) with BAP (0.5mg/l) showed 68% callus regeneration from nodal explants on MS Media. Callus texture was friable and the colour of the callus was light green. The plant growth regulator BAP when used with NAA, then callus production took place. At a concentration of 1.5mg/l of BAP with 3.0mg/l of NAA showed 52% callus formation and 0.5mg/l of BAP with 3mg/l of NAA showed 70% callus formation. Thus, the combination of BAP and NAA at conc. 0.5: 3.0mg/l was best suited for callus production but, this callus showed less regenerability as compared to the callus formed by 2,4-D and BAP growth regulators (Table.2).

This indirect regeneration via callus mainly utilized for induction of somaclonal variation. To regenerate the true to type plants, direct organogenesis via nodal explants is required. So, in our study the nodal explants were used for the rapid clonal propagation of this medicinally important plant *Tinospora cordifolia*.

About 2-3cm long shoot tips with three to four leaves, harvested from the elongation medium, were used for rooting. Percentage of roots induction and number of roots per shoot were highly influenced by concentration and types of auxins. Among different concentration of IAA, IBA and NAA, the highest degree of root induction was found on the media containing 1mg/l IAA. It showed 90% of roots development within 25-28 days. The 0.5mg/l concentration of IBA showed 80% of roots formation. NAA showed 69% roots formation at 0.5mg/l concentration but this was poor response as compare to IAA and IBA (Table. 3). Although rooting was observed in auxin- free medium also, but frequency was very less. However, the response was increased with subsequent sub culturing. Thus, IAA growth regulator was best suited for the roots development as compare to the IBA and NAA.

Table 14.3: Effect of Auxins on root formation of *T. cordifolia* in *in vitro* formed shoots on MS medium.

Auxins (mg/l) IAA	IBA	NAA	Rooting (%)	Mean no. of roots/ shoot	Mean Root length (cm)
-	-	-	50	2.7±0.5	2.6±0.2
0.5	-	-	70	3.7±0.18	2.9±0.6
1.0	-	-	90	5.0±0.6	3.7±0.5
2.0	-	-	60	2.0±0.3	2.5±0.7
-	0.5	-	80	3.5±0.2	1.8±0.6
-	1.0	-	60	2.7±0.8	2.2±0.8
-	2.0	-	40	1.6±0.21	1.5±0.1
-	-	0.5	69	2.7±0.2	2.2±0.2
-	-	1.0	54	2.4±0.3	2.0±0.1
-	-	2.0	19	1.6±0.2	1.8±0.1

Data represented as mean value ± standard error of roots formed after 30 day of culture from four independent experiments each with 20 replicates.

The rooted plants were transplanted *ex vitro*. After differentiation and establishment of plantlets through successive culturing steps, the rooted healthy plants were removed carefully from the culture vessels, thoroughly washed for removing agar and transferred to culture tubes containing steriled distilled water. After 10 days, plantlets were transferred to plastic pots containing sterile course sand and garden soil in 1:1 ratio, moistened with autoclaved tap water. After 30 days the plantlets were transplanted to pots containing normal (non-autoclaved) garden soil and kept under shade in green house for another 2 weeks before transferring to outdoors under full sun. After two months of growth, 85% plants survived and grew normally in the field.

It is well known that nodal explants were generally stable in culture systems. This regeneration through nodal explant gives true to type plants. By the protocol described here one can get 5-6 rooted plants from single explants in 80-90 days. The rapid multiplication of true to type plants and efficient transplantation of *T. cordifolia* can be useful in propagation and conservation of elite plants for commercial exploitation.

4. Conclusion

The present investigation has resulted in a protocol that could be used for ex-situ conservation and true to type mass propagation of genetically diverse medicinal plant *Tinospora cordifolia,* irrespective of season and unaffected by geographical location & environmental conditions.

Acknowledgement

The authors are thankful to the Department of Biotechnology, Kurukshetra University, Kurukshetra for providing laboratory facilities to carry out this research work.

References

Aima, R.K. 2003. "Pictorial Guide to Plants". Natraj Publishers First Edition Dehradun, 454– 455.

Anonymous, 1976. Wealth of India, Raw materials, X New Delhi, CSIR.

Anonymous, 2003. Wealth of India: A dictionary of Indian Raw Materials and Industrial First Edition. New Delhi: CSIR, 10: 251–2.

Bhojwani, S.S., 1980. *In vitro* propagation of garlic by shoot proliferation. *Scientia Horticulturae* 13: 47–52.

Bhojwani, S.S., Arumugam, N., Arora, R. and Upadhyay, R.P. 1989. *In vitro* conservation of some endangered plant species of India. *Indian J. Plant Genet. Res,* 2: 114-121.

Chaudhary, N., Singh, S., siddique, M.B. and Khatoon, S. 2014. Impact of season and dioecy on therapeutic phytoconstituents of *Tinospora cordifolia,* a Rasayan drug. *Biomed. Res. Int.* 902138.

Chopra, R.N., Chopra, L.C., Handa, K.D. and Kapur, L.D. 1982. Indigenous Drugs of India. 2nd edn. Kolkata, M/S Dhar VN & Sons.

Chopra, R.N., Nayar, S.L. and Chopra, I.C. 1956. Glossary of Indian Medicinal Plants, CSIR, New Delhi.

Datta, P.C. 1993. Biotechnology and Tissue culture of some medicinal plants, In: Govil, J.N., Sing, V.K. and Hshmi (eds.), Gimpses in plant Research, Medicinal plants: New Vista of Research part 2, 11: 337-342.

Dubey, N.K., Kumar, R. and Tripathi, P. 2004. Global promotion of herbal medicine: India's opportunity. *Curr. Sci.,* 86: 37-41.

Gururaj, H.B., Giridhar, P. and Ravishankar, G.A. 2007. Micropropagation of *Tinospora cordifolia* (Willd.) Miers ex Hook. F and Thoms– a multipurpose medicinal plant. *Cur.Sci.,* 92:23-26.

Hamilton, A.C. 2004. Medicinal plants, Conservation and livelihoods. *Biodiversity Conser.,13* :1477-1517.

Krishna, K., Jigar, B. and Jagruti, P. 2009. Guduchi *Tinospora cordifolia*: Biological and Medicinal properties, a review. *The Internet Journal of Alternative Medicine,* 6:1-10.

Lal, N., Ahuja, P.S., Kukreja, A.K. and Pandey, B. 1988. Clonal propagation of Picrorhiza kurroaroyle ex benth. by shoot tip culture. *Plant Cell Rep.*, 7: 202-205.

Lane, W.D. 1979. *In vitro* propagation of *Spirea bumalda* and *Prunus cistena* from shoot apices. *Canadian Journal of Plant Science*, 59: 1025–1029.

Lueng, A. 2000. Scientific studies and reports in the herbal literature: what are we studying and reporting? *Herbal Gram*, 48: 63-64.

Mittal, J., Sharma, M.M. and Batra, A. 2014. *Tinospora cordifolia* a multipurpose medicinal plant-A review. *J. Med. Plants Studs.*, 2: 32-47.

Mohammed, M.S. and Kumar, B.V. 2012. Overexploitation of medicinal plants a setback to gene pool. *Int J Life Sci Pharma Res*, 4: 177-180.

Nadkarni, K.M. and Nadkarni, A.K. 1976. Indian Material Medica, 1. 3rd edn.

Nakano, M., Hoshino, Y. and Mii, M. 1994. Adventitious shoot regeneration from cultured petal explants of carnation. *Plant Cell Tissue and Organ Culture*, 36: 15–19.

Nayampalli, S., Ainapure, S.S. and Nadkami, P.M. 1982. Study of anti-allergic acid broncho dilator effects of *Tinospora cordifolia*. *Ind. J. Pharmacology*, 14: 64-6.

Ondrizek, P.R., Chan, P.J., Patton, W.C. and King, A. 1999. An alternative medicine study of herbal effects on the penetration of zona free hamster oocytes and the integrity of sperm deoxyribonucleic acid. *Fertility and Sterility*, 71: 517-522.

Pandey, G. 2002. Dravyaguna Vijnana .Varuanasi: Krishnadas Academy, Second Edition, 1: 697- 710.

Premila, M.S. 2006. Ayurvedic Herbs: A clinical Guide to the Healing Plants of Traditional Indian Medicine. New York: Haworth Press, 69-76, 175-176.

Rani, V. and Raina, S.N. 2000. Genetic fidelity of organized meristem derived micropropagated plants: a critical reappraisal. *In vitro Cell Dev Biol Plant*, 36: 319-330.

Sharma, A., Gupta, A., Singh, S. and Batra, A. 2010. *Tinospora cordifolia* (Willd.) Hook. F.& Thomson - A plant with immense economic potential. *J. Chem. Pharm. Res.*, 2(5):327-333.

Sujatha, G., Ranjitha, K.B.D. 2007. Effects of phytohormones on micropropagation of *Artemesia vulgaris*. L. *Acta Physiologiae Plantarum*, 29: 189-195.

Sunandakumari, C., Martin, K.P., Chithra, M., Sini, S. and Madhusoodanan, P.V. 2004. Rapid axillary bud proliferation and ex vitro rooting of herbal spice, *Menthapiperita* L. *Ind. J. Biotechnol.*, 3: 108–112.

Thomas, V. and Mehta, A.R. 1983. In: Plant Cell Culture in Crop Improvement (ed.) Sen, S.K. and Giles, K.I., Plenum Press, New York, 451–457.

Vaidya, D.B. 1994. "Materia Medica of Tibetan Medicine". Delhi: Sri Satguru Publications, 163.

Zhao, T.F.,Wang, X., Rimando, A.M. and Che, C. 1991. Folkloric medicinal plant: *Tinospora sagittata* var. cravaniana and *Mohonia bealei*. Planta Med., 57: 505–3.

Chapter 15

Phytochemical Constituents, Urease Inhibition and Antioxidant Effect of a Medicinal Plant: *Cyperus Rotundus*

Neha Sikri and Sunita Dalal*

Department of Biotechnology, Kurukshetra University, Kurukshetra (Haryana) 136119

**Corresponding Author E-mail: sdalal@kuk.ac.in*

Abstract

Cyperus rotundus (Family Cyperaceae) is a traditional medicinal plant appearing among natural drugs used against inflammatory bowel diseases, spasms as well as in stomach disorders. A number of pharmacological and biological activities including anti-inflammatory, antidiabetic, antidiarrhoeal, cytoprotective and antibacterial activities have been reported for this plant. It has already been demonstrated that aerial parts of C. rotundus contain different phenolic compounds but little is known about their antioxidant potentials. In the present study C. rotundus extracts (aqueous and methanol) were evaluated for the presence or absence of the bioactive molecules with medicinal relevance and its substantial antioxidant and urease inhibitory capacities. The present study revealed that C. rotundus contains significant amount of phenols, tannins and flavonoids and also possesses good antioxidant potential. Although the plant exhibited low urease inhibitory potential, but our findings may help to explain the beneficial effects of the plant extracts against various disorders. The crude extracts may be purified for the isolation of compounds with antioxidant and urease inhibitory potential.

Keywords: *C. rotundus, urease inhibition, antioxidant activity, flavoniods*

1. Introduction

Herbal products are the oldest form of medicine known to mankind. These have gained importance in traditional medicine as it contributes to the development of new perspectives in medicine for the effective remedial practices with minimal side effects. According to the WHO (World Health Organisation), in developing countries, millions of people believe in efficacy of plants or plant based products and uses them regularly.

Plant nutrients are exploited for bioactive compounds such as flavonoids, tannins, phenols, saponins which are proven to have remarkable biological and pharmacological activities. *Cyperus rotundus*, a wild grass (commonly named as nut grass) considered in the present study also contains the different bioactive components. The phytochemical surveys revealed that the major constituent of this herb are essential oils, terpenoids, sesquiterpenes and flavonoids. Compounds such as cyperene, cyperol, gurjunene, acyperone are also present (Meena *et al.*, 2010). *C. rotundus* oil mainly consists of β- pinene, -selincne, starch, glycosides and many novel alkaloids (Nima *et al.*, 2008; Bisht *et al.*, 2011). Accordingly, in the present study, aerial parts of the *C. rotundus* were evaluated for the qualitative and the quantitative determination of the different classes of compounds.

Cyperus rotundus is a smooth, erect wild weed of Cyperaceae family, is widely scattered in tropical, warmer temperate regions as well as in Mediterranean basin areas. This widely spread species is known for its rapid growth, high adaptability, herbicide tolerance and tough nature for the ground nutrients. However, fascinating aspects lie in its medicinal properties. *C. rotundus*, one of the world's strong invasive weed is a traditional medicinal herb used among the natural drugs as anti-rheumatic, antiulcer (Gangwar and Ghosh, 2017), anti-inflammatory and anti-cancerous agent (Mannarreddy *et al.*, 2017). Studies revealed that the aerial parts of the *C. rotundus* have strong antioxidant, anti-genotoxic as well as anti-mutagenic activity (Kilani- jaziri soumaya *et al.*, 2014), antidiabetic activity. The rhizome of this plant has been reported to have hepatoprotective, neuroprotective, cardioprotective and antiuropathogenic activity and also used to relieve the diarrhoea, skin and bowel disorders.

Free radicals and reactive oxygen species such as hydroxyl radicals and non-free radical species such as hydrogen peroxide and singlet oxygen are produced in human body as normal body's function like cardiovascular and other metabolic activities. But the uncontrolled production of the excessive amounts of singlet oxygen species and reactive oxygen species leads to the lipid per oxidation. Thus leads to stimulation of oxidation, inactivation of enzymes, DNA damage and other alterations in the cell organelles like cell death or tissue damage. This could also results in the several human disorders like Alzheimer's disease, Parkinson's disease, cancers, diabetes mellitus, ageing, immune suppression, inflammatory diseases, neurodegenerative diseases, cardiovascular diseases, rheumatoid arthritis, neurodegenerative diseases, asthma, cataract, liver damage, ulcerative colitis and age related degenerative processes (Rao *et al.*, 2006). To protect the human body form the deleterious effects of free radicals, use of the antioxidants is highly recommended. In the different industries like food, beverage and cosmetic industries synthetic antioxidants such as BHA, TBHQ (tert-butyl hydroxy quinone) are being used. But because of the toxic effect and not effective at low concentrations, these agents are not much used as alternative antioxidants agents. Thus, there is a need to discover the antioxidants from the natural sources having low side effects. Antioxidants produced form medicinal herbs inhibit polymerization chain reaction initiated by free radicals and also the polyphenols and flavonoids present in them help to get rid of the various disorders. Therefore to explore the natural antioxidants and their mechanism at cellular level is worthwhile.

Ureolytic activity of the enzyme urease has a key role in the pathogenesis of bacteria such as *Clostridium perfringens, Helicobacter pylori, Klebsiella pneumoniae, Proteus mirabilis, Salmonella spp., Staphylococcus saprophyticus, Ureoplasma urealyticum* and *Yersinia enterocolitica* (Burune *et al.*, 2000). Such activity has been reported to be responsible in diseases such as gastritis, peptic ulcers, urinary tract related diseases; mucosa associated lymphoid tissue lymphoma, gastric cancer, urolithiasis, pyelonephritis, ammonia encephalopathy, hepatic coma and gastroduodenal infections (Weeks *et al.*, 2000). It is also responsible for the development of infection-induced reactive arthritis and acute pyelonephritis (Arfan *et al.*, 2010, Mobley *et al.*, 1995). The role of urease in microbial infection has been well established in *Helicobacter pylori*. It is a gram negative micro aerobic bacterium live in the specialized habitat, the human stomach mucosa that have very acidic pH (2) lethal for most microbes but some bacteria are able to survive in these conditions by producing urease enzyme (EC3.5.1.5). Urease hydrolyzes urea, found in the stomach to carbonic acid and ammonia, resulting in the elevated pH and making conditions favourable for different microbes. Treatment of the various pathogens with multiple antibiotics is not much effective as development of the antibiotic resistance, so there is a need to discover inhibitors of the urease enzyme. Already available urease inhibitors have exhibit some unfavorable features like low hydrolytic stability and toxicity, which alter their clinical use.

In agriculture, the high activity of ureolytic bacteria increases the amount of ammonia in the soil via fast urea degradation. Due to which plants are damaged due to lack of necessary nutrients and by the toxicity of the products produced (ammonia and carbon dioxide) by the urease activity (Bremner 1996). During the seed germination process, in the metabolism of nitrogen urease plays a significant role (Mulvaney & Bremner 1981; Khan *et al.*, 2010). In agriculture, NBPT (N-(butyl) thiophosphoric triamide) is only commercially available effective soil urease inhibitor but it is directly related to soil properties so there is a need to discover new, effective and safe urease inhibitors from natural sources.

The aim of the present study was to evaluate quantitative phytochemical composition, *in vitro* antioxidant activities and urease inhibitory activities of the extract of aerial parts of *Cyperus rotundus* in two different solvents *viz.* aqueous and methanol extracts.

2. Materials and Methods

2.1 Chemicals

Urease type 1X (specific activity: 50,000-100000 units/g) from *Canavalia ensiformis* (L.) DC. (Fabaceae) commonly known as Jack-Bean was purchased from Sigma Aldrich (St. Louis, MO). 2, 2'diphenyl1picrylhydrazyl, Folin-ciocalteu's reagent, ascorbic acid, trichloroacetic acid, and glacial acetic acid were purchased from Hi Media Pvt. Ltd. All other chemicals and reagents used in this study were of analytical grade.

2.1 Plant Material and Preparation of Extract

Leaves of the selected flora were collected from surroundings of Kurukshetra University, Kurukshetra, Haryana, India (29° 57′ 31.353″ N, 76°48′52.128″ E). The identifications and authentication of the specimens was done from the Department of Botany, Kurukshetra University, Kurukshetra. The collected leaves were washed under tap water followed by distilled water, shade dried for 7 days and then milled to a coarse powder by a mechanical grinder and stored till the further use.

The powders of dried leaves (10g) were separately soaked in distilled water and methanol (100ml) in a reagent bottle covered with a lid at 37°C for 24h. The powder were packed into soxhlet column for 48 h. Resulting extracts in different solvents were evaporated and concentrated to dryness using the rotatory evaporator at 50°C. The yield of extraction of dried plant material was calculated. The extracts were stored at -4°C.

2.3 Preliminary Phytochemical Screening

Both extracts of the selected plants were evaluated for the screening of the various phytochemicals using standard phytochemical procedures (Khandelwal *et al.*, 2000).

2.4 Estimation of Total Phenolic Content

The content of total phenols was determined according to the Folin-Ciocalteu's method (Singleton *et al.*, 1999). Aliquots of 0.5mL of extracts were mixed with 2.5mL of 10-fold-diluted Folin -Ciocalteu reagent and 2mL of 7.5% sodium carbonate (Na_2CO_3). The mixture was allowed to stand for 90 min at room temperature before the absorbance was taken at 760nm spectrophotometrically. The final results were calibrated to deduct the contribution from ascorbic acid and expressed as Gallic acid equivalent (mg of GA/g of extract).

2.5 Estimation of Total Flavonoid Content

The content of flavonoids in the examined plant extracts was determined using spectrophotometric method by Quettier *et al.*, 2000. A volume of 2.5mL of extract was transferred to a test tube, mixed with 0.15mL of 5% Sodium nitrite for 5min. Then, 0.15mL of 10% aluminium nitrate was added. After 6mins the reaction stopped by adding 1mL of 1M sodium hydroxide. The mixture was further diluted with distilled water up to 5mL. The absorbance of the mixture was immediately measured at 510nm. The flavonoid content was calculated and expressed as rutin equivalent (mg of RU/g of extract).

2.6 Estimation of Total Tannin Content

Total tannin content was measured by Folin-Ciocalteu's method (Singh *et al.*, 2012). 7.5ml of distilled water was added in0.1ml of the extracts followed by addition of 0.5ml Folin -Ciocalteu reagent. 1ml of 35% Na_2Co_3 was added to the above mixture then it was diluted up to 10ml with distilled water. Mixture was kept at room temperature for 30 minutes and absorbance was taken at 725 nm. The final results were calibrated to deduct the contribution from ascorbic acid and expressed as gallic acid equivalent.

2.7 Estimation of Ascorbic Acid

Ascorbic acid content was determined by using method of Roe and Keuther, 1953. Briefly, ascorbic acid was oxidized by activated charcoals to yield dehydroascorbic acid, which further react with 2,4-dinitrophenylhydrazine to form osazone, a light-absorbing substance. The absorbance was read at 490nm spectrophotometrically. Ascorbic acid was taken as standard and the levels of ascorbic acid in the sample was expressed as mg ascorbate/ g sample.

2.8 *In Vitro* Antioxidant Evaluation

2.8.1 2, 2diphenyl1picrylhydrazyl Scavenging Assay

The ability of the plant extract to scavenge DPPH free radicals was assessed by the method described by Mensor *et al.* (2001). The L-ascorbic acid was used as a positive control. The reaction mixture was prepared containing 300 µl of extract of varying concentrations (1-100 µg/ml) and 2 ml of DPPH (0.1 mM in methanol). The reaction mixture was then placed in the cuvette holder of the spectrophotometer and the absorbance was measured at 517 nm against the blank.

The per cent DPPH decolourization of the sample was calculated by the equation:% inhibition = B_0-B_1/B_0 100, Where, B_0 is the absorbance of negative control and B_1 is the absorbance of reaction mixture.

2.8.2 Chelating Effects on Ferrous Ions

The measurement of meal chelating activity of the different extracts was estimated by the method of Dinis *et al.* (1994). The extracts (0.25 ml) were mixed with 1.75 ml of methanol and 0.25 ml of 250 mM $FeCl_2$. This was followed by the addition of 0.25 ml of 2 mM ferrozine, and kept at room temperature for 10 min. before determining the absorbance of the mixture at 562 nm. The chelating effect (%) was calculated from the formula as given for DPPH scavenging assay.

2.8.3 Ferric Reducing Antioxidant Power Assay (FRAP)

FRAP assay of the extracts was determined by a method based on the reduction of ferric-tripyridyltriazine complex to a blue coloured ferrous form described by Benzie and Strain *et al.* (1996). The FRAP reagent containing 2.5mL of 10mM 2,4,6-tripyridy-s-triazine solution in 40mM HCl and 2.5mL of 20mM $Fecl_3$ and 25mL of 0.3M acetate buffer, pH 3.6, was prepared freshly and pre-warmed at 37°C. Aliquots of 40µL of extracts were mixed with 0.2mL of distilled water and 1.8mL of FRAP reagent. The absorbance of reaction mixture at 593nm was measured spectrophotometrically after incubation at 37°C.In this assay, the final result was expressed as the concentration of antioxidants having a ferric reducing ability equivalent to that of 1mM $FeSO_4$.

2.8.4 Coupled Oxidation of β-carotene and Linoleic Acid

β-carotene linoleic acid assay of the listed plants was carried out as described by Miller (1971). BHT was used for comparison. 2 mg of β-carotene dissolved in chloroform (10 ml) was pipette into a small, round-bottom flask. After the chloroform was removed at 40 C under vacuum, 100 µl of linoleic acid, 400 mg of tween 20 and 100 ml of oxygenated water was added to the flask followed by vigorous mechanical stirring. 5 ml of the above prepared emulsion was transferred to a series of test tubes containing 200 µl of extracts of various concentrations. As soon as the emulsion was added to each tube, the zero time absorbance was measured at 470 nm using a spectrophotometer (T60 UV visible). After initial vortexing, the test tubes were incubated in a water bath at 50°C for 120 min. The absorbance of each sample was measured at 470 nm. Blank, devoid of β-carotene was prepared for background subtraction.

Antioxidant activity calculated using the following equation:

$$AA = 100\{1\text{-}A_o - A_t / A_o^{\,o} - A_t^{\,o}\}$$

Where, A_o and $A_o^{\,o}$ are the absorbance at zero time of test and blank, while A_t and $A_t^{\,o}$ are the absorbance at time t of test and blank respectively.

2.8.5 Urease Inhibition Studies

The enzyme activity and inhibition was measured through catalytic effects of urease on urea by measuring change in absorbance in the absence and in the presence of inhibitor at 640nm, using spectrophotometer (T60 UV visible). All phytochemicals were tested for urease inhibition activity at concentration of 1.0 mg/ml. and that exerted significant inhibition, and tested in a concentration range of 100 to 1000µg/ml using Thiourea as standard inhibitor. For urease inhibition assays after addition of 10 ml of phosphate buffer to accurate weight of enzyme, sonication was performed, followed by centrifugation and absorbance of upper solution at 280 nm. By using equation A= bc, where c is concentration of solution (mol/L), b is length of the UV cell and represents molar absorptivity, the concentration of initial urease solution was calculated. After proper dilution, the concentration of enzyme solution was adjusted to 2 mg/ml. Reaction mixture comprising 1.2 ml of phosphate buffer solution (10mM potassium phosphate, 10mM lithium chloride and 1mMethylenediaminetetraacetic acid, pH 8.2 at 37 C), 0.2 ml of urease enzyme solution, and 0.1 ml of test compound were subjected to incubation. Urease activity is determined by measuring the ammonia released during the reaction by modified spectrophotometric method described by Weatherburn (1967). The concentration of compounds that inhibited the hydrolysis of substrate by 50% (IC50) was determined through monitoring the inhibition effect of various concentrations of phytochemicals in the assay followed by kinetics studies. The extent of the enzymatic reaction shall be calculated based on the following equation:

$$I\% = 100 - (T / C * 100)$$

Where I (%) is the inhibition of the enzyme, T (test) is the absorbance of the tested sample in the presence of enzyme C (control) is the absorbance of the solvent in the presence of enzyme.

2.9 Statistical Analysis

All the assays were done in triplicates to test the reproducibility of them. All results are presented as mean ± S.E. SPSS 15.0 (statistical software) was used for statistical analysis of results. The values of $p < 0.05$ were considered statistically significant. Correlations among data obtained were calculated using Pearson's coefficient (r).

3. Results and Discussion

3.1 Extraction Yield

Aqueous and methanol extracts were prepared of the selected medicinal herb to examine phytochemicals qualitatively and quantitatively, antioxidant activity and urease inhibitory activity. The yield of extract obtained from 10g of dry plant material was measured for each extract by formula % yield = weight of extract obtained / weight of powder taken 100 (Table 15.1).

Table 15.1: The yield of extract obtained from 10g of dry plant material

Plant Species	% Yield	
	Aqueous extract	**Methanol Extract**
Cyperus rotundus	9.76%	7.31%

3.2. Preliminary Phytochemical Screening

The selected plant revealed the presence of various phytochemicals *viz.* phenols, flavonoids, saponins, alkaloids and tannins. Out of various phytochemical analyzed, flavonoids and tannins were present in high quantity (Table 15.2).

Table 15.2: Qualitative phytochemical screening of Aqueous and methanol extract.

Plant species	Flavonoids		Phenols		Saponins		Alkaloids		Tannins	
	Aq	Met	Aq	Met	Aq	Met	Aq	Met	Aq	Met
Cyperus rotundus	++	+++	+	+	+	+	+	+	++	+++

+++: highly present, ++: moderately present, +: Low, -: absent. Aq: Aqueous extract, Met: methanol extract.

3.3 Quantitative Phytochemical Screening

Different phytochemicals due to their anti-inflammatory, antimicrobial, anti-diabetic and anticancer properties have been reported to offer protection against various multiple human diseases such as neurodegenerative diseases, ageing, rheumatoid arthritis, metabolic diseases degenerative diseases (Mobley *et al.*, 1995; Khan *et al.*, 2010). To develop a significant relationship between the phenolic compounds and the biological activities of the different extracts, different phytochemicals including total phenolic content, flavonoid, tannins and ascorbic acid content was determined.

Total phenolic content varies from 21.78±1.10mgGAE/gmin aqueous extract to 34.29±1.34mgGAE/gm of extract in methanol extract (Table 15.3). The flavonoid content varies from 46.34±2.37 mgRU/gm in aqueous extract to 173.69±0.68mgRU/g in methanol extract. In present study, total tannin content varies from 59.60±1.14 mgGAE/gm in aqueous extract to 168.70±1.75mgGAE/gm in methanol extract. The ascorbic acid content varies from 43.21±0.42 mgGAE/gm in methanol extract to 56.23±1.42 mgGAE/gm in aqueous extract.

Table 15.3: Quantitative phytochemical screening of aqueous and methanol extract.

Plant Species	Phenolic content(mg GAE/g)		Flavonoid content(mg RU/g)		Tannin content (mgGAE/g)		Ascorbic acid	
	Aq	Met	Aq	Met	Aq	Met	Aq	Met
Cyperus rotundus	21,78 ±0.62	34.29 ±0.06	46.34 ±2.37	173.69 ±1.82	59.60 +1.14	168.70±1.75	56.23±1.42	43.21±0.42

Aq: Aqueous extract, Met: Methanol extract.

3.4 Evaluation of Antioxidant Activity

In the plant extracts, the antioxidant compounds have an important role in scavenging and inhibiting free radicals. Different antioxidant assays *viz.* metal chelating activity, lipid peroxidation inhibition, and reducing power assays were used to determine the antioxidant potential of the wild herb.

In the present study, *Cyperus rotundus* (aq) shows the maximum scavenging activity of 56.71±0.40% in case of DPPH scavenging assay. However the antioxidant evaluation by metal chelating effect, antioxidant activity ranges from 27.47±0.00% to 61.85±0.20%. In case of FRAP assay, antioxidant activity ranges from 71.21±0.40 to 65.43±0.26 %. Aqueous extract shows maximum antioxidant activity by β- carotene bleaching assay (Fig. 15.1, 15.2, 15.3 & 15.4).

DPPH , L-Ascorbic acid as standard:78.26%.

Chelating effect on ferrous ions, EDTA as standard: 72.8%.

FRAP assay, Ferrous sulphate as standard.

β- carotene bleaching method, Butylatedhydroxytoluene (BHT) as standard: 85.24%.

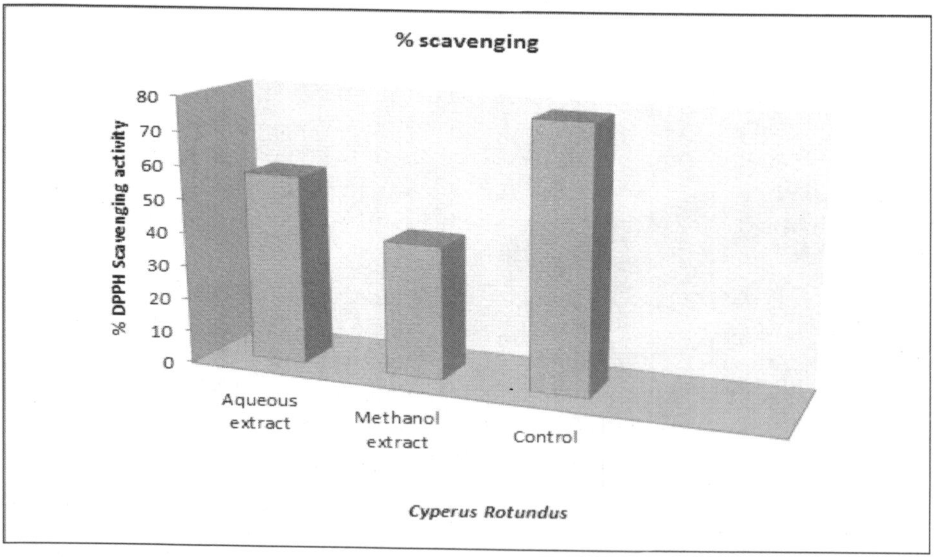

Fig 15.1: Scavenging activity (%) of different plant extracts on DPPH radicals.

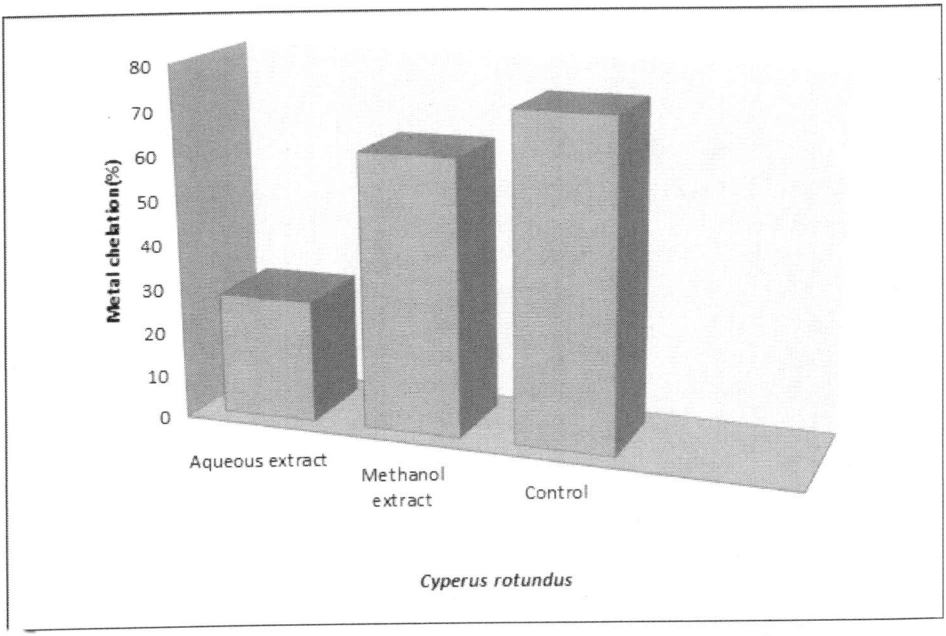

Fig. 15.2: Ferrous ion chelating activity (%) of various extracts.

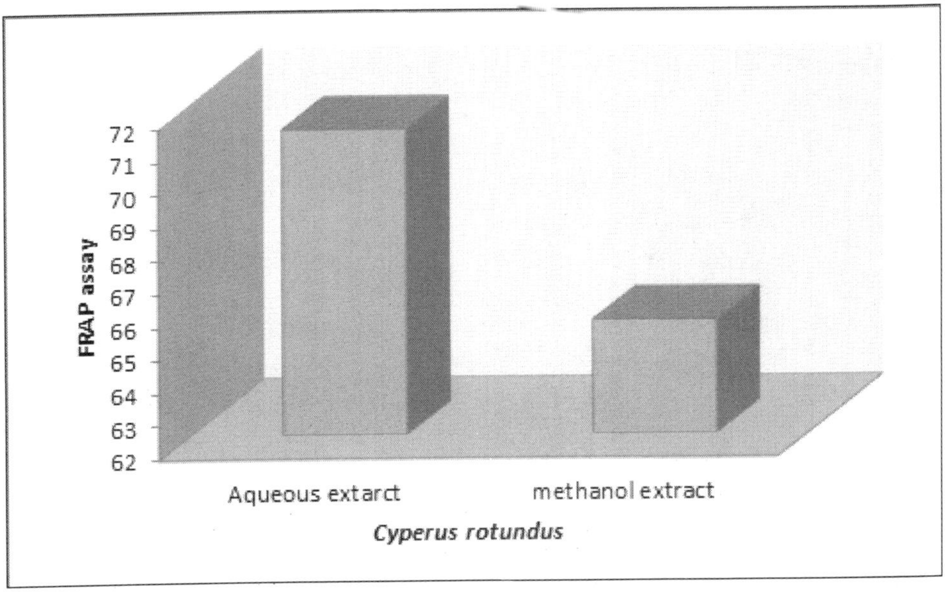

Fig. 15.3: Ferric reducing antioxidant capacity (%) of various plant extracts.

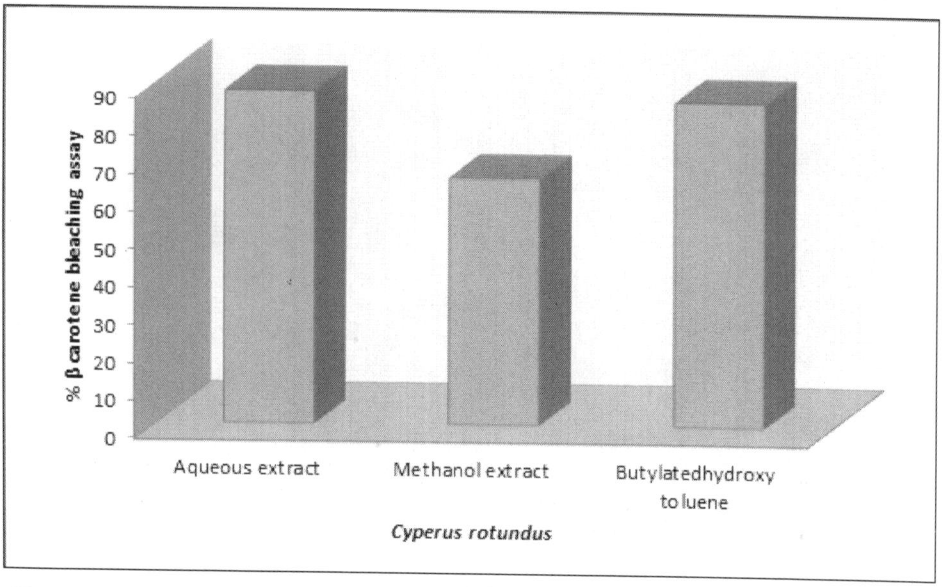

Fig. 15.4: Antioxidant activity (%) of various plant extracts by β- carotene bleaching assay.

3.5 Urease Inhibition Studies

Urease is an enzyme that catalyses the hydrolysis of urea leads to the significant increase in the pH and also has negative effects on human health. Although there are number of synthetic urease inhibitors available in the market but those are not much recommended as they are toxic thus not safe to use. In the present study, urease inhibition activity varies from 36.72±0.67% inhibition to 43.42±0.00% inhibition in different extracts. (Table 15.5).

Table 15.4: Urease inhibitory activity of tested plant extracts at a concentration 1mg/ml.

Plant species	% inhibition*	IC50mg/ml
Cyperus rotundus		
Aq	36.72+0.67	1.32
Met	43.42+0.00	0.76

*Values are expressed as mean±SD of triplicate experiments

Thiourea was taken as standard in case of urease inhibitory profile and % inhibition= 71.47

Aq: aqueous extract; Met: methanol extract.

In the present study, *C. rotundus* plant possesses the significant content of flavonoids and also exhibited good antioxidant potential. Rebaya *et al.* (2015) also shows the significant correlation of flavonoid content with the total antioxidant potential of the plant extracts. Total flavoniod content also significantly related with the urease inhibitory potential of the plant extract (Jalaluddin *et al.*, 2015). Although *C. rotundus* do not have much urease inhibitory potential but the different parts of the plant can be further explored. So, further investigations regarding more biological activities of the plant parts need to be conducted.

Acknowledgements

The authors greatly acknowledge the Haryana State Council for Science & Technology (HSCST) for financial support for the project entitled "Urease Inhibitors from Medicinal plants of Haryana in context to agricultural soil protection and therapy against clinically significant virulence factor" (Endst.No. HSCST/R&D/2015/ 2383).

References

Arfan, M., Ali, M., Ahmad, H., Anis, I., Khan, A., Choudhary, M.I. and Shah, M.R. 2010. Urease inhibitors from *Hypericum oblongifolium* WALL. *J. Enzym. Inhib. Med. Ch.*, 25: 296–299.

Gangwar, A. K. and Ghosh, A.K., 2017. Antiulcer activity of chloroform extracts of *Cyperus rotundus. Int. J. Pharma. Phytochem. Res.*, 9: 780 -782.

Benzie, I.F. and Strain, J.J. 1996. The ferric reducing ability of plasma (FRAP) as a measure of "antioxidant power": The FRAP Assay, *Analytical Biochem.*, 239: 70-76.

Bisht, A., Bisht, G.R., Singh, M., Gupta, R. and Singh, V. 2011. Chemical composition and antimicrobial activity of essential oil of tubers of *Cyperus rotundus* Linn. Collected from Dehradun (Uttarakhand). *Int. J. Res. Pharm. Biomed. Sci.*, 2: 661-665.

Bremner, J.M. 1996. Recent research on problems in the use of urea as a nitrogen fertilizer. *Nitrogen Eco. Trop. Soils.*, 42: 321–329.

Burne, R.A. and Chen, Y.M. 2000. Bacterial ureases in infectious diseases. *Microb. Infect.*, 2: 533–542.

Dinis, T.C.P. and Madeira, V.M.C. 1994. Action of phenolic derivatives (acetaminophen, salicylate and 5-aminosalicylate) as inhibitors of membrane lipid peroxidation and as peroxyl radical sacavangers. *Arch. Biochem. Biophysics.*, 315:161-169.

Jalaluddin, A.J., Awllia,1., Maryam, A.L.G., Etimad H., Sumaira J., Wahab, A.T. Saima, R., and Choudhary, M.I. 2015. Flavonoids as Natural Inhibitors of Jack Bean Urease Enzyme. *Letters in Drug Design & Discovery.*, 13: 1-7.

Khan, I., Ali, S., Hameed, S., Rama, N.H., Hussain, M.T., Wadood, A. and Uddin, R. 2010. Synthesis, antioxidant activities and urease inhibition of some new 1, 2, 4-triazole and 1, 3, 4-thiadiazole derivatives. *Eur. J. Med. Chem.*, 45: 5200–5207.

Khandelwal, K.R. 2000. Practical Pharmacognosy Techniques and Experiments. Niral Prakashan, 2nd edition.

Kilani- Jaziri, S., Ghedira, Z., Nasr., N., Krifa, M., Ghedira, K., Franca, D. M. G. and Ghedira, C. L. 2014. Evaluation of in-vitro antioxidant and apoptotic activities of *Cyperus rotundus. Asian Pacific Journal of Tropical Medicine.*, 105-112.

Mannarreddy, P., Denisb, M., Munireddya, D., Pandurangana , R., Thangavelua, K.P. and Venkatesana, K., 2017. Cytotoxic effect of Cyperus rotundus rhizome extract on human cancer cell lines. *Biomedicine & Pharmacotherapy*, 95: 1375-1387.

Meena, A.K., Yadav, A.K., Niranjan, U.S., Singh, B., Nagariya, A.K. and Verma, M. 2010. Review on *Cyperus rotundus*- A potential herb. *Int. J. Phar. Clin. Res.*, 2: 20-21.

Mensor, L.L., Menzes, F.S., Leitao G.G., Reis A.S., dos Santos, T.C., Coube, C.S. and Leitao, S.G. 2001. Screening of Brazilian plant extracts for antioxidant activity by the use of DPPH free radical method. *Phytotherapy Research.*, 15:127-30.

Miller, H.E. 1971. A Simplified method for the evaluation of antioxidants. *J. Am. Oil Chem. Soc.*, 48: 91-95.

Mobley, H.L., Island, M.D. and Hausinger, R.P. 1995. Molecular biology of microbial ureases. *Microbiol Rev.*, 59: 451–480.

Mulvaney, R.L. and Bremner, J.M. 1981. Control of urea transformations in soils. *New York: Marcel Dekker.*, 5: 153–196.

Nima, Z.A., Jabier, M.S., Wagi, R.I. and Hussain, H.A. 2008. Extraction, identification and antibacterial activity of *Cyperus* oil from Iraqi *C. rotundus. Eng Technol.*, 26: 1156.

Quettier, D.C., Gressier, B., Vasseur, J., Dine, T., Brunet, C., Luyckx, M., Cazin, M., Cazin, J.C., Bailleu, F. and Trotin, F. 2000. Phenolic compounds and antioxidant activities of buckwheat (Fagopyrum esculentum Moench) hulls and flour. *Journal of Ethnopharmacology.*, 72: 35-42.

Rao, A.L., Bharani, M. and Pallavi, V. 2006. Role of antioxidants and free radicals in health and disease. *Adv Pharmacol Toxicol.*, 7: 29-38.

Rebaya, A., Belghith, S.L., Baghdikian, B., Leddet, V.M., Mabrouki, F., Olivier, E., Cherif, J.K. and Ayadi, M.A. 2015. Total Phenolic, Total Flavonoid, Tannin Content, and Antioxidant Capacity of *Halimium halimifolium* (Cistaceae). *J. App. Pharm. Sci.*, 1: 52-57.

Roe, J.H. and Keuther, C.E. 1953. The determination of ascorbic acid in whole blood and wine through 2,4-dinitrophenyl hydrazine derivative of dehydroascorbic acid. *J. Biol. Chem.*, 147:399-405.

Singh, R., Verma, P.K., Singh, G. 2012. Total phenolic, flavonoids and tannin contents in different extracts of *Artemisia absinthium. J. Intercult. Ethnopharmacol.*, 1: 101-104.

Singleton, V.L., Orthofer, R. and Lamuela, R.R.M. 1999. Analysis of total phenols and other oxidation substrates and antioxidants and antioxidants by means of Folin- Ciocalteau reagent. *Methods Enzymology.*, 299: 152-178.

Weatherburn, M.W. 1967. Phenol- hypochlorite reaction for determination of ammonia *Analytical Chemistry.*, 39: 971-974.

Weeks, D.l., Eskandari, S., Scott, D.R. and Sachs, G. 2000. A H+ - gated urea channel: the link between Helicobacter pylori urease and gastric colonization. *Sci.*, 287: 482–485.

Chapter 16

Optimizing Production of Antimicrobial Compounds Isolated from Endophytic *Aspergillus* sp. Obtained from An Indian Medicinal Plant: *Syzygium Cumini* (L.) Skeels

Priyanka Sharma[1,*] and Pranay Jain[2]

[1]*Department of Biotechnology Engineering, Ambala College of Engineering and Applied Research, Mithapur, Haryana, India*

[2]*Department of Biotechnology, University Institute of Engineering and Technology, Kurukshetra University, Kurukshetra, Haryana, India*

**Corresponding Author E-mail: drpriyankamailbox@gmail.com*

Abstract

Researchers all over the world have recognized the antimicrobial potential of compounds isolated from natural sources and are inclined to investigate the new and unexplored natural sources like endophytes for the discovery of novel antimicrobial agents. In the present study, an endophytic fungus Aspergillus sp. *isolated from the leaf of a medicinal plant Syzygium cumini (L.) Skeels native to the Indian Subcontinent was evaluated for its antimicrobial activity against five gram positive bacteria, two gram negative bacteria and five fungal test pathogens (including two phytopathogens). The ethyl acetate extracts of Aspergillus* sp. *inhibited the growth of all the test pathogens. The fermentation parameters were optimized to further promote the antimicrobial activity. Minimal inhibitory concentration ranged from 0.125 to 250µg/ml.*

Keywords: *Endophytic fungi, antimicrobial activity, minimal inhibitory concentration, Aspergillus* sp., *Syzygium cumini (L.) Skeels.*

1. Introduction

Unsupervised, uncontrolled and improper usage of antibiotics has caused the emergence of antimicrobial resistance among the microorganisms. There is an unending need for new and improved antibiotics. The drug discovery was traditionally focused on natural products only. But later on, the priority changed from natural to synthetic and semi synthetic route of drug development. Recently, the emphasis of the pharmaceutical industry has shifted again on microbial metabolites and other natural products. Reverting back to the natural resources for the commercial drug discovery was mainly the result of attainment of less promising leads with synthetic drugs and steeped research costs. Also, the fact that the natural products have been successfully validated as potential antimicrobial agents promoted the interest of researchers and pharmaceutical industry in natural sources of antimicrobial compounds (Strobel *et al.*, 2004; Be´rdy, 2012). Endophytes therefore, present a relatively underexplored and inexhaustible source of antimicrobial compounds (Strobel and Daisy, 2003; Kumar *et al.*, 2014).

The present study deals with the isolation of endophytic fungi capable of producing antimicrobial compound from the leaf of a medicinal plant *Syzygium cumini* (L.) Skeels (also known as jambul, jambolan, jamblang or jamun) which is native to the Indian Subcontinent and the adjoining regions of Southeast Asia. The work included purification of endophytic fungal isolate, its identification, optimization of fermentation conditions for maximum antimicrobial compound productivity, its purification and determination of minimal inhibitory concentration.

2. Materials and Methods

2.1 Collection of Plant Samples

Plant samples from *Syzygium cumini* (L.) Skeels were collected from different locations of Kurukshetra, (lying between 29 52′ and 30 12′ North latitudes and 76 25′ to 77 04′ East longitudes), Haryana, India

2.2 Isolation of Endophytic Fungi from Plant Sample

Sample (leaf/ stem) was cut and surface sterilized to remove any epiphytes.3-4 Segments of each sample were placed on Potato Dextrose Agar (PDA) amended with ciprofloxacin (100 mg/l) and incubated in a Biochemical Oxygen Demand (BOD) incubator with humidity for 5-7 days at 12-h light/dark cycle at 25 ±2°C (Sen and Batra, 2012).

2.3 Procurement of Microbial Cultures

The various test pathogens were procured from Microbial Type Culture Collection (MTCC), Institute of Microbial Technology (IMTECH), Chandigarh, which included Gram positive bacteria, *Streptococcus mutans* (MTCC 497), *Staphylococcus aureus* (MTCC 7443), *Streptococcus pyogenes* (MTCC 1924), *Bacillus megaterim* (MTCC 428) and *Bacillus subtilis* (MTCC 121); Gram negative bacteria *Escherichia coli* (MTCC 40) and *Pseudomonas fluorescens* (MTCC 1748); fungal pathogens *Candida albicans* (MTCC 227), *Candida glabrata* (MTCC 3814) and *Candida tropicalis* (MTCC 3421). Two phytopathogenic fungi *Alternaria solani* (MTCC 10690) and *Fusarium graminearum* (MTCC 2089) were also obtained. The slants of Mueller Hinton agar were made to preserve the cultures. All the slants were kept at 4°C in the refrigerator for further studies.

2.4 Fermentation

2-3 dics of the pure endophytic fungal isolates were inoculated into Potato Dextrose Broth and incubated at 25 ±2°C for 7-10 days in a BOD incubator. The broths were filtered and the culture filtrate was used to assess the antimicrobial activity against the test pathogens (Jain and Sharma, 2014).

2.5. Evaluation of Antimicrobial Potential of Endophytic Fungal Isolates by using Agar Well Diffusion Method

The inoculum of different test pathogens was adjusted according to 0.5 McFarland Standard. Test pathogens were spread aseptically using cotton Swabs on the surface of Muller Hilton agar plates. Wells of about 6.0mm diameter were aseptically punched in the agar plates using a sterile cock borer. 100µl of the crude metabolite was transferred into each well. Plates were incubated at 37°C for 24 hours. All the experiments were performed in triplicates and after 24 hours, the diameter of resulting zone of inhibition were measured in terms of millimeter (mm) using a Hi-media zone scale (Boney *et al.*, 2008). On the basis of zone of inhibition, SYZ. C was selected as the most promising isolate.

2.6. Morphological and Biochemical Characterization of Promising Isolate Exhibiting Antimicrobial Potential

The fungal endophyte isolated from the leaf sample of *Syzygium cumini* (L.) Skeels was identified as *'Aspergillus'* sp. on the basis of sporulation by staining with lactophenol cotton blue (Ainsworth *et al.*, 1973; Barnett *et al.*, 1998).

2.7. Assay for Mode of Antimicrobial Metabolite Production

To determine the extracellular and/or intracellular mode of production of antimicrobial metabolite by *Aspergillus* sp., mycelium was extracted with (1:1) ethyl acetate (100%). The mycelial extract and the culture filtrate were tested for antimicrobial activity against test pathogens by using agar well diffusion assay (Gogoi *et al.*, 2008).

2.8. Optimization of Fermentation Conditions for Antimicrobial Metabolite Production

The production of antimicrobial metabolite by the selected *Aspergillus* sp. was optimized for fermentation broths (Potato Dextrose Broth, Brain Heart Infusion Broth, Sabouraud Dextrose Broth, Tryptone Soy Broth and Czepak Dox Broth), Agitation speed (120, 150, 200, 220 and 250 rpm), pH (4.0, 5.0, 6.0, 7.0 and 8.0), Incubation temperature (20, 25, 30 and 35 C, Fermentation vessel (100, 250, 500, and 1000 ml), Inoculum size (20, 30, 35 and 40mm in diameter), Carbon source (2%) (Xylose, Fructose, Maltose, Sucrose and Galactose) and Nitrogen source (1%) [Organic: Peptone And Yeast Extract and Inorganic: Ammonium Sulphate

{$(NH_4)_2S0_4$}; Sodium Nitrate {$NaN0_3$} and Ammonium Chloride, {NH_4Cl)] (Gogoi *et al.*, 2008).

2.9. Purification and MIC Determination of Antimicrobial Compound/s

2.9.1. Solvent Extraction of Antimicrobial Metabolite

Four different organic solvents namely petroleum ether (100%) (Ranbaxy), ethyl acetate (100%) (Ranbaxy), benzene (100%) (Ranbaxy) and chloroform (100%) (Ranbaxy) were tested for extraction of antimicrobial metabolite from *Aspergillus* sp. The maximum antimicrobial potential was observed with ethyl acetate extract. The ethyl acetate extract was dried and concentrated (Atta *et al.*, 2009).

2.9.2. Determination of Minimum Inhibitory Concentration (Mic) of Powdered Metabolite

Two fold serial dilution of powdered metabolite was made in DMSO. MIC of the powdered antimicrobial metabolite against all the test pathogens was determined by using agar well diffusion assay (Boney *et al.*, 2008).

3. Results and Discussion

In the present study, antimicrobial activity of endophytic fungi isolated from *Syzygium cumini* (L.) Skeels was evaluated against twelve test pathogens; five gram+ve bacteria, two gram-ve bacteria, three opportunistic fungi and two phytopathogenic fungi. The endophytic fungi; *Aspergillus* sp. produced broad spectrum antimicrobial compounds which displayed both antibacterial as well as antifungal properties. The antimicrobial activity of *Aspergillus* sp. was found associated with both the filtered culture broth as well as the fungal mycelium extracted with solvent ethyl acetate. This suggested that the mode of antimicrobial metabolite production by *Aspergillus* sp. isolated from the leaf sample of *Syzygium cumini* (L.) Skeels is intercellular as well as extracellular. The crude culture filtrate of *Aspergillus* sp. inhibited the growth of ten out of twelve test pathogens (Fig. 16.1). The ethyl acetate extract of the culture filtrate and the mycelia ethyl acetate extract of the culture filtrate exhibited antagonism against all the twelve test pathogens in the study (Fig. 16.2, Table 16.1).

Although the culture filtrate, its ethyl acetate extract and the ethyl acetate mycelial extract, all exhibited antimicrobial potential but the antimicrobial activity of the ethyl acetate extract of the culture filtrate was maximum. Minimal inhibitory concentration ranged from 0.125 to 250µg/ml.

The growth and the antimicrobial potential of endophytic *Aspergillus* sp. was best when cultivated on potato dextrose broth as compared to other four broths in study (Fig. 16.3).

Table 16.1: Antimicrobial activity exhibited by endophytic Aspergillus sp. against test pathogens.

Code of the sample	Zone of inhibition (in mm.) against test pathogens											
	Gram-positive bacteria					Gram-negative bacteria		Opportunistic fungal pathogens			Phytopathogenic fungi	
	Bm	Bs	Sa	Sm	Sp	Ec	Pf	Ca	Cg	Ct	As	Fg
SYZ. C. culture broth (filtered)	12 ± 0.57	18±0.33	23±0.57	12±0.89	17±0.51	14±0.16	12±0.89	18	NA	NA	12±0.33	15±0.33
SYZ. C. mycelial extract (in ethyl acetate)	14±0.33	15±0.51	15±0.66	12±0.33	12±0.33	11±0.57	14±0.16	13±0.79	15±0.33	15±0.66	11±0.57	13±0.88
Positive control	48	35	NA	39	33	40	50	NA	NA	NA	NA	NA
Minimal inhibitory concentration (in µg/ml.)												
Aspergillus sp.	0.50	62.5	125	0.25	15.625	0.125	250	15.625	62.5	1.0	0.50	31.25

Values are mean inhibition zone (mm) ± S.D of three replicates

NA: No Antimicrobial activity; negative control: Dimethyl sulphoxide; positive control: Ciprofloxacin as antibacterial agent (1mg/ml) and Fluconazol as antifungal agent (1mg/ml); B: Bean; L: Leaf; S: Stem; Sm: *Streptococcus mutans* (MTCC 497); Sa: *Staphylococcus aureus* (MTCC 7443); Sp: *Streptococcus pyogenes* (MTCC 1924); Bm: *Bacillus megaterim* (MTCC 428); Bs: *Bacillus subtilis* (MTCC 121); Ec: *Escherichia coli* (MTCC 40); Pf: *Pseudomonas fluorescens* (MTCC 1748); Ca: *Candida albicans* (MTCC 227); Cg: *Candida glabrata* (MTCC 3814); Ct: *Candida tropicalis* (MTCC 3421); As: *Alternaria solani* (MTCC 10690); Fg: *Fusarium graminearum* (MTCC 2089)

The antimicrobial activity aggravated on the supplementation with carbon sources galactose and fructose. The supplementation with 2% fructose in the potato dextrose broth used for cultivation of *Aspergillus* sp. led to the inhibition of all the twelve test pathogens and also led to the increase in the diameter of halos (Fig. 16.5).

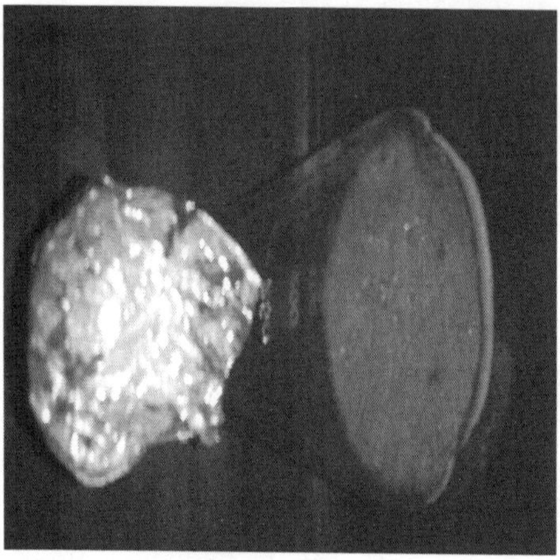

Fig.16.1: Fungal broth culture.

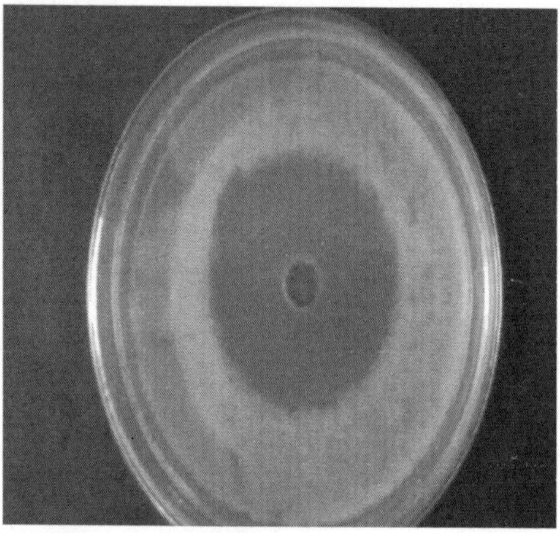

Fig. 16.2: Analyzing antimicrobial activity by Agar well diffusion technique.

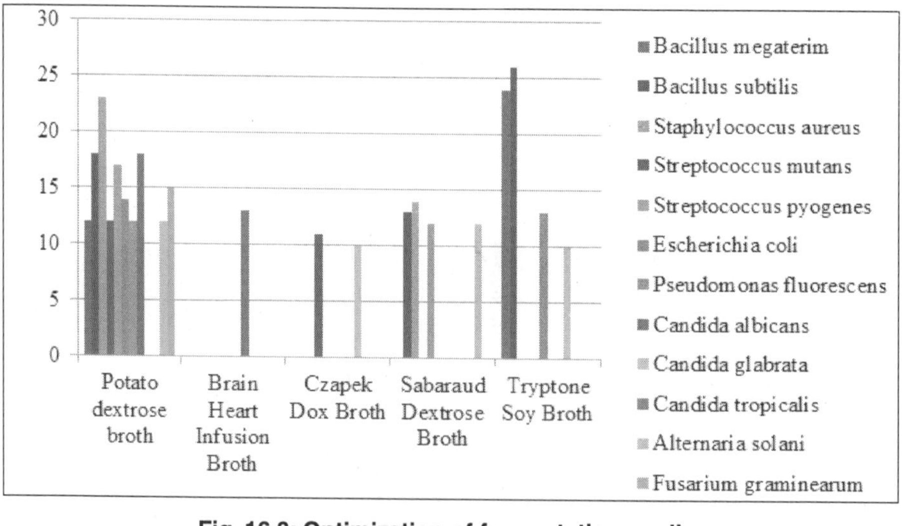

Fig. 16.3: Optimization of fermentation media.

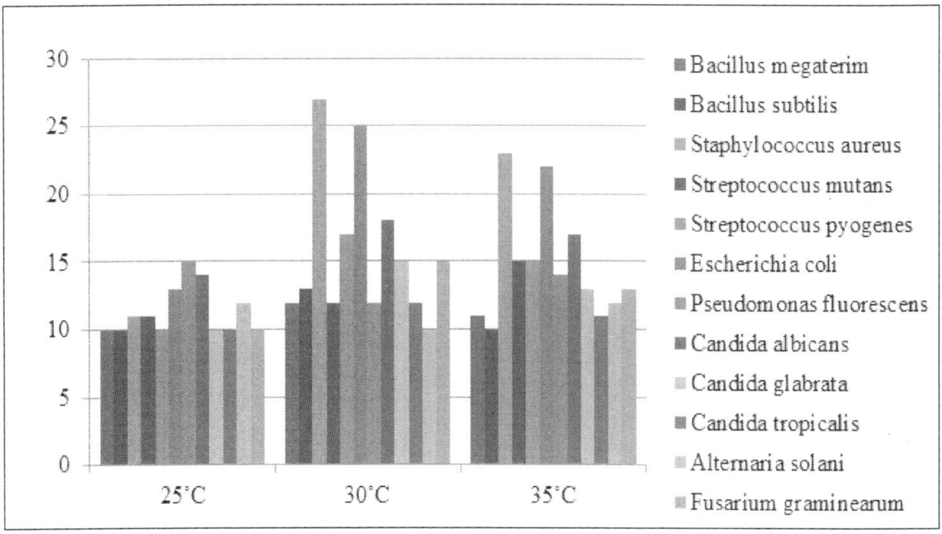

Fig. 16.4: Optimization of incubation temperature.

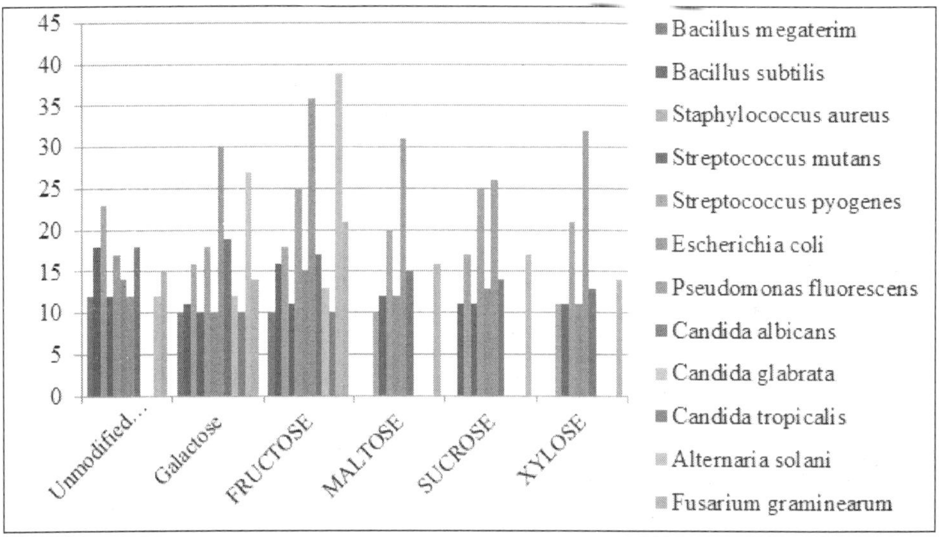

Fig. 16.5: Optimization of carbon source.

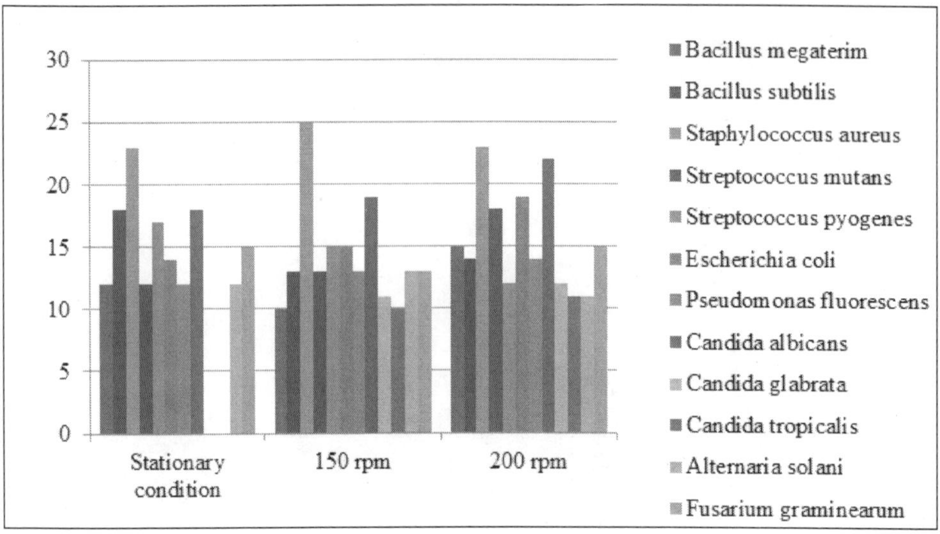

Fig. 16.6: Agitation speed optimization.

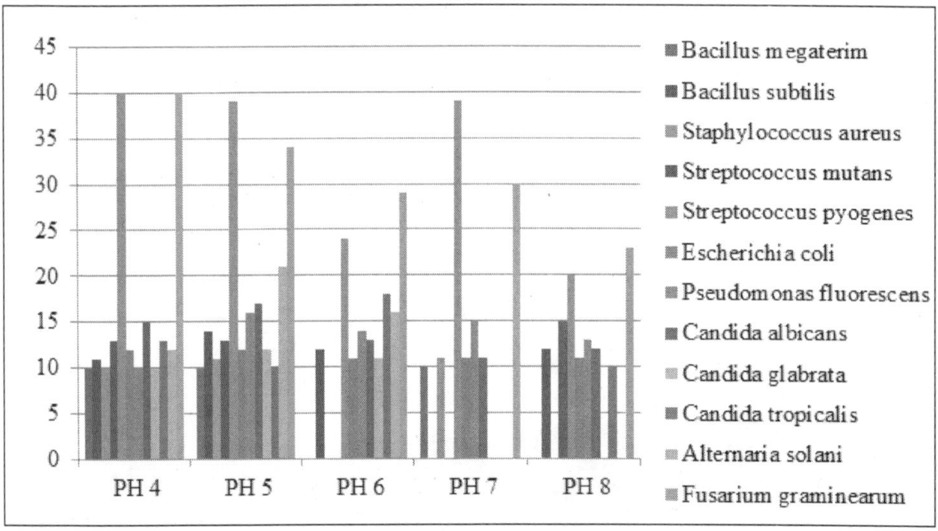

Fig. 16.7: pH optimization.

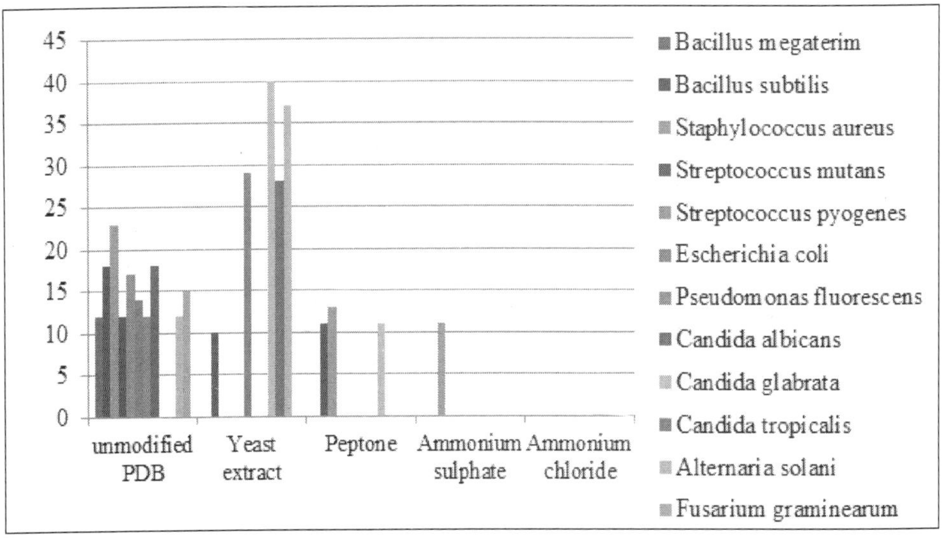

Fig. 16.8: Optimization of nitrogen source.

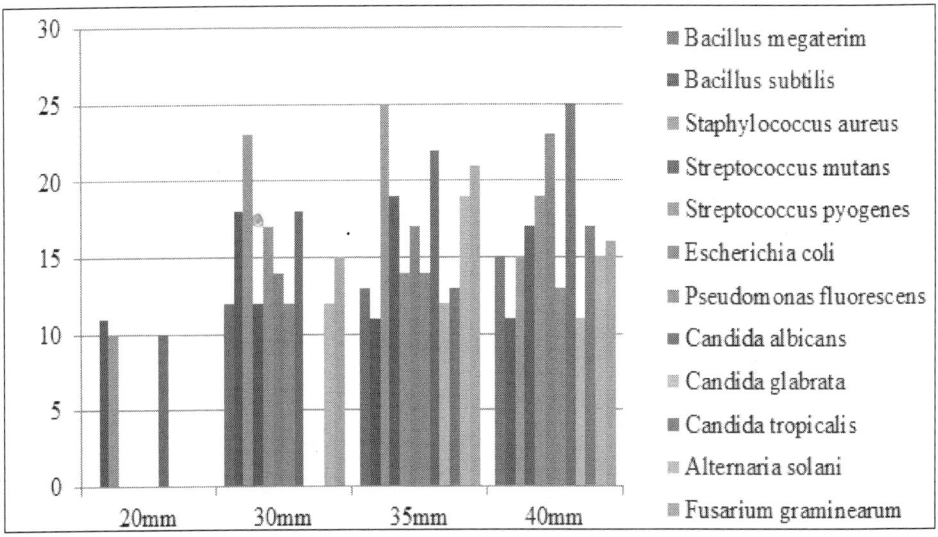

Fig. 16.9: Optimization of inoculum size.

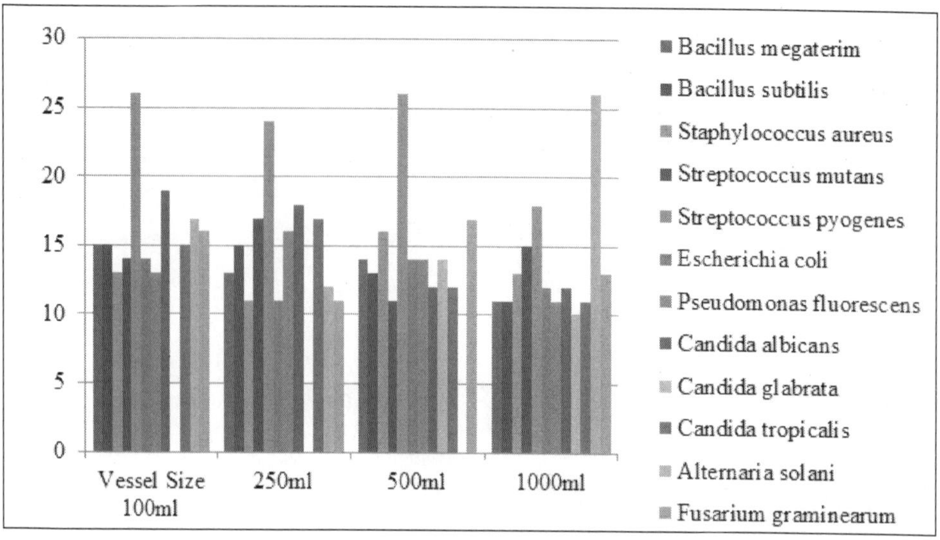

Fig. 16.10: Optimization of vessel size.

The supplementation with nitrogen sources did not have any positive effect on the antimicrobial activity as compared to the unmodified potato dextrose media (Fig. 16.8). The ph value of 5, temperature 35°C, agitation speed 200rpm, vessel size 1000ml and the inoculum size of 35mm (in diameter) led to the production of maximum extracellular antimicrobial compound (Fig. 16.4, 16.6, 16.7, 16.9, 16.10).

The present study led to the isolation of antibiotic compounds from the endophytic *Aspergillus* sp. obtained from the leaf sample of a medicinal plant *Syzygium cumini* (L.) Skeels. Inhibition of all the test pathogens was demonstrated with large halos. The results evidenced the huge potential of endophytic fungi in the isolation of pharmacologically important compounds. There are many studies on the antimicrobial or antibacterial activities of the leaf extracts or leaf essential oils of *Syzygium cumini* (Hosmat *et al.*, 2013; Mohamed *et al.*, 2013). The results of the present study suggests the possibility of isolation of bioactive compounds of the host plant from their endophotes which being microbial in nature as much more easier to grow, multiply, manipulate, maintain, saves time and space. Endophytes can used to provide high yields of the desired product.

Acknowledgement

The authors are grateful to the Hon'ble Vice-Chancellor, Kurukshetra University, Kurukshetra and the Director, University Institute of Engineering and Technology, Kurukshetra University, Kurukshetra for providing infrastructure facilities to carry out the research work.

References

Ainsworth, G.C., Sparrow, F.K., Sussman, A.S. 1973. The Fungi: advanced Treaties, Taxonomic Review with Keys (Vol. IV A). Academic Press, New York, USA. pp.1-621

Atta, H.M., Dabour, S.M., Desoukey, S.G. 2009. Sparsomycin Antibiotic Production by *Streptomyces* sp. AZ-NIOFD1: Taxonomy, Fermentation, Purification and Biological Activities. *American-Eurasian J. Agric. & Environ. Sci.*, 5(3): 368-377

Barnett, H.L., Hunter, B.B. 1998. Illustrated Genera of Imperfect Fungi. 4th ed. Mac Millan Publ. Co., New York. pp.1-218

Be´rdy, J. 2012. Thoughts and facts about antibiotics: Where we are now and where we are heading. *J. Antibiotic.*, 65: 385–395

Boney, B., Hooper, J., Parisot, J. 2008. Principles of assessing bacterial susceptibility to antibiotics using the agar diffusion method. *J. Antimicrob. Chemother.*, 61(6): 1295-1301.

Gogoi, D.K., Boruah, H.P.D., Saikia, R., Bora, T.C. 2008. Optimization of process parameters for improved production of bioactive metabolite by a novel endophytic fungus *Fusarium* sp. DF2 isolated from *Taxus wallichiana* of North East India. *World J. Microbiol. Biotechnol.*, 24:79–87. DOI 10.1007/s11274-007-9442-3

Hosmat, P.B., Sonawane, K.D., Waghmare, S.R. 2013. Antibacterial Activity of Vulgarol A extracted from the leaves of *Syzygium Cumini. Asian J. Pharm. Clin. Res.*, 6 (4): 100-102

Jain, P., Sharma, P. 2014. Antagonistic activity of endophytic fungi isolated from *Syzygium cumini* (L.) Skeels. *Int. J. Life Sci. Res.*, 3(2): 59-63.

Kumar, G.A., Antony, A.R., Kannan, V.R. 2015. Exploration of endophytic microorganisms from selected medicinal plants and their control potential to multi drug resistant pathogens. *J. Med. Plants Studies.*, 3(2): 49-57

Kumar, K., Aharwal, R.P., Shukla, H., Rajak, R.C., Sandhu, S.S. 2014. Endophytic fungi: as a source of antimicrobials bioactive compounds. *World J. Pharm. Pharmaceut. Sci.*, 3(2):1179-1197

Manonmani, R., Catherin, S.S. 2015. GC-MS Analysis of Bioactive Components of an Important Medicinal Fern *Actiniopteris Radiate* (Swartz) Link. *World J. Pharm. Res.*, 4(4): 1860-1869

Mohamed A.A., Ali S.I., El-Baz F.K. 2013. Antioxidant and Antibacterial Activities of Crude Extracts and Essential Oils of *Syzygium cumini* Leaves. *PLoS ONE.*, 8(4): e60269. DOI:10.1371/journal.pone.0060269

Sanjenbam, P., Gopal, J.V., Kannabiran, K. 2014. Isolation and Identification of anticandidal Compound from *Streptomyces* sp. VITPK91. *Appl. Biochem. Microbiol.*, 50(5): 492–499.

Sen, A., Batra, A. 2012. Determination of antimicrobial potentialities of different solvent extracts of the medicinal plant: *Phyllanthus amarus* Schum. And Thonn. *Int. J. Green Pharm.*, 6(1): 50-56.

Strobel, G., Daisy, B. 2003. Bioprospecting for microbial endophytes and their natural products. *Microbiol. Mol. Biol. Rev.*, 67:491–502.

Strobel, G., Daisy, B., Castillo, U., Harper, J. 2004. Natural products from endophytic microorganisms. *J. Nat. Prod.*, 67: 257-268.

Chapter 17

Production and Analysis of Various Attributes of Newly Synthesized Synbiotic Yoghurt *Via* Incorporation of An Indigenous Isolate of *Lactobacillus Acidophilus* and Inulin

Suman Kandola[1,*], Meenu Rathi[2]

[1]Department of Biotechnology, Deenbandhu Chhotu Ram University of Sciences and Technology (DCRUST), Murthal-132710, India

[2]Department of Botany, GMN College, Ambala Cantt., Haryana

**Corresponding Author E-mail: sumankandola84@yahoo.com*

Abstract

Modern consumers are increasingly interested in their personal health and expect the food that they eat to be healthy or even capable of preventing illness. Yoghurt is considered as a health promoting food worldwide and the most widely used method to increase the number of advantageous bacteria in intestinal tract. Probiotic foods should contain specific probiotic strains and maintain a suitable level of viable cells during the product's shelf life. Before probiotic strains can be delivered to consumers, they must first be able to be manufactured under industrial conditions, and then survive and retain their functionality during storage as frozen or freeze-dried cultures, and also in the food products into which they are finally formulated. Synbiotic yoghurt was prepared from an indigenous isolate Lactobacillus acidophilus NCDC 13 and prebiotic inulin and analyzed for various attributes like titrable acidity, viable counts and sensory evaluation. The flavor, body, texture and appearance score for synbiotic yoghurt was much higher than the plain yoghurt and the probiotic yoghurt. Also, the viable counts of L. acidophilus were higher in synbiotic yoghurt than probiotic yoghurt throughout the storage period. This study suggests the importance of incorporation of inulin during preparation of synbiotic yoghurt with Lactobacillus acidophilus as it increases the viability of probiotic strain during storage of food product and also improves the product characteristics.

Keywords: *Probiotics, Prebiotics, Synbiotics, Titrable acidity, Viable counts, Sensory evaluation*

Abbreviations: CFU- *Colony Forming Units;* **ADSA-** *American Dairy Science Association;* **NCDC-** *National Collection of Dairy Cultures.*

1. Introduction

Gibson introduced the concept of prebiotics and he also speculated as to the additional benefits one might see if prebiotics were combined with probiotics to form what he called a synbiotic. He defined synbiotics as "a mixture of probiotics and prebiotics that beneficially affects the host by improving the survival and implantation of live microbial dietary supplements in the gastrointestinal tract by selectively stimulating the growth and/or by activating the metabolism of one or a limited number of health-promoting bacteria, and thus improving host well fare". The prebiotic should offer a selective advantage for the growth of the probiotic it is combined with to provide a synergistic effect. This is because in addition to the action of probiotics that promote the growth of existing strains of beneficial bacteria in the colon, prebiotics such as inulin and oligofructose also act to improve the survival, implantation and growth of newly added probiotics strains. Inulin is a mixture of polymers consisting mainly of fructose unit; its partial enzymatic hydrolysis yields oligofructose. These prebiotic (dietary) fibers, help in stabilization of foams, assist in digestion, increase stool volume, stimulate lactobacilli and are used for formulation of low fat preparations as fat and sugar replacers, respectively. Synbiotics indicate a realistic way of using biological preparations in the prevention of gastrointestinal diseases in humans and animals. Yoghurt is produced worldwide. Hence, incorporation of probiotic strains and prebiotic in yoghurt could produce good results.

Modern consumers are increasingly interested in their personal health and expect the food that they eat to be healthy or even capable of preventing illness. Gut health in general has shown to be the key sector for functional foods in Europe. The viability and stability of probiotics has been both a marketing and technological challenge for industrial producers. Probiotic foods should contain specific probiotic strains and maintain a suitable level of viable cells during the product's shelf life. Unless strict demands are set on probiotic product definition and labeling their regulatory definition will remain obscure. The technological demands placed on probiotic strains are great and new manufacturing process and formulation technologies may often be required for bacteria primarily selected for their functional health properties. Before probiotic strains can be delivered to consumers, they must first be able to be manufactured under industrial conditions, and then survive and retain their functionality during storage as frozen or freeze-dried cultures, and also in the food products into which they are finally formulated. The probiotic strains should also survive the gastrointestinal stress factors and maintain their functionality within the host. Additionally, they must be able to be incorporated into foods without producing off-flavors or textures- they should be viable but not growing. The packaging materials used and the conditions under which the products are stored are also important for the quality of products. The sensory property is essential because it influences consumer's choice and performance. To produce a new product, if formulation of product is achieved automatically, production speed and efficiency can be improved in addition to the increased evaluation accuracy, with an accompanying reduction in production costs. Thus, aim of this study was the development of synbiotic yoghurt and evaluating its sensory properties.

2. Materials and Methods

2.1 Bacterial Strains and Growth Conditions

The probiotic strain *Lactobacillus acidophilus* NCDC 13 (LA-13) and starter strains *Streptococcus thermophilus* NCDC 74 (ST-74) and *Lactobacillus delbrueckii* ssp. *bulgaricus* NCDC 09 (LB-09) used in the present investigation were obtained from the National Collection of Dairy Cultures (NCDC), Dairy Microbiology Division, National Dairy Research Institute (NDRI), Karnal, India. Freeze dried cultures were activated in chalk litmus milk at 37 C for 24 hr and sub-cultured monthly. Before use, the cultures were subcultured twice in de Man Rogosa Sharpe (MRS) broth (Himedia, Mumbai, India).

2.2 Prebiotic Used

Inulin (Raftline ST) manufactured by ORAFTI (Tienan, Belgium) was procured form S.A. Pharmachem. Pvt. Ltd., Goregaon, Mumbai.

2.2.1 Preparation of Synbiotic Yoghurt

Fresh buffalo milk was standardized (3% fat, 12% SNF) by using skim milk. The prebiotic inulin was added at the rate of 3%. The milk was heated to 90 C for 10 min and cooled to 45 C without exposing to the atmosphere. Yoghurt starters (*S. thermophilus* NCDC 74 and *L. bulgaricus* NCDC 09) were added at 2% in the ratio 1:1 and probiotic *L. acidophilus* NCDC 13 was added at 1% as adjunct starter. The inoculated milk was distributed into clean polyethene terepthalate (PET) jars and incubated at 42 C for 3 to 4 hr. Probiotic yoghurt was prepared from standardized milk without adding inulin and following the same procedure and incubation conditions. Plain yoghurt was also prepared by inoculating standardized milk with only yoghurt cultures (2%) and incubating under similar conditions.

2.2.2 Titrable Acidity

The titrable acidity of the yoghurt sample was determined in terms of percent lactic acid as per the method described in IS: 1479 (1960).

2.2.3 Viable Count of Organisms

The viable counts of yoghurt starters and probiotic *L. acidophilus* were determined in each product on each day of storage. A sample of 11g of yoghurt was drawn aseptically after mixing the contents thoroughly. It was then transferred to 99 ml of sterile 0.1% peptone and mixed uniformly with a vortex mixture. Subsequent serial dilutions were prepared and viable numbers enumerated using pour plate technique. The counts of *S. thermophilus* were enumerated on M17 agar after incubation aerobically at 37 C for 48 hr (Terzaghi and Sandine, 1975). RCA agar adjusted to pH 5.3 and anaerobic incubation at 37 C for 72 hr was used for the selective enumeration of *L. bulgaricus* (Dave and Shah, 1997 b). Maltose-MRS agar was used for the enumeration of *L. acidophilus*, as *L. bulgaricus* did not utilize maltose. Incubation was carried out anaerobically at 37 C for 48 hr (Hull and Roberts, 1984).

2.2.4 Sensory Evaluation

Synbiotic yoghurt, probiotic yoghurt and plain yoghurt were evaluated for sensory characteristics on day 1st, 3rd, 5th and 7th of storage by using the American Dairy Science Association (ADSA) score card for yoghurt with slight modification (Bodyfelt *et al.*, 1988).

3. Results and Discussion

3.1 Sensory Evaluation of Synbiotic Yoghurt

The synbiotic yoghurt was compared with the plain and the probiotic yoghurt with respect to characteristics such as flavor, body and texture, and appearance. (Table 17.1) present the results of the sensory evaluation of different types of yoghurt using the ADSA score card. Although the flavor scores do not differ much within the different types of yoghurt, the synbiotic yoghurt scored higher (7.97/10) than the other types of yoghurt. The inulin present in the synbiotic yoghurt imparted its delicate sweetness, which was liked by most of the consumers.

Table 17.1: Sensory Evaluation of Different Types of Yoghurt Using ADSA Score Card.

Characteristics	Plain Yoghurt	Probiotic Yoghurt	Synbiotic Yoghurt
Flavour (10)	7.11	7.28	7.97
Body and texture (5)	3.75	3.86	4.44
Appearance (5)	3.86	4.03	4.28
Total (20)	14.69	15.06	16.51

The body and texture score for synbiotic yoghurt was found to be 4.44/5.00, which was much higher than the plain yoghurt (3.75/5), and the probiotic yoghurt (3.86/5). The results are in agreement with Franck (2000), who suggested that inulin improves the body texture of the product by improving the gel network resulting in a creamy texture. The synbiotic yoghurt also scored better (4.28/5) with regard to appearance also. The probiotic yoghurt resulted in some wheying off, which was reduced by incorporating inulin. Overall synbiotic yoghurt scored much higher value (16.51/20) than the plain and probiotic yoghurt.

3.2 Storage Effect on Acidity

The effect of storage (4 to 7 C) on the developed acidity and viable counts of probiotic and yoghurt bacteria was studied in all three types of yoghurt for 7 days. The changes in the titrable acidity (% lactic acid) during the storage period are illustrated in (Fig. 17.1). The initial acidity of milk (0.15%) increased to about 0.9 after the fermentation in all types of yoghurt. Thereafter, it increased gradually upto the 7th day of storage as seen in the graph. The developed acidity did not differ much in plain, probiotic and synbiotic yoghurt up to the 3rd day of storage. But after that, the acidity in synbiotic yoghurt and probiotic yoghurt increased markedly and reached above 1.2 percent on the 7th day, whereas plain yoghurt showed minimum increase with 1.08 percent. The increase in acidity of probiotic and synbiotic yoghurt can be attributed to the growth and acid production by *L. acidophilus* in addition to the yoghurt starters. On the contrary, the titrable acidity of probiotic yoghurt samples did not show much increase during refrigerated storage as reported by Dave and Shah (1997a).

3.3 Effect of Storage on Viable Counts of S. *Thermophilus*, L. *Bulgaricus* and L. *Acidophilus*

The effect of storage on the viable counts of yoghurt starters and the probiotic bacteria during a storage period of 7 days is presented in (Fig 17.2 to 17.4). The counts of *S. thermophilus* were higher in comparison to *L. bulgaricus* and *L. acidophilus* during the storage, as already reported by Vinderola *et al.* (2002) and Oliveira *et al.* (2009, 2011). The *S. thermophilus* predominates on the mixed starter cultures used for yoghurt preparation. The highest count of *S. thermophilus* was obtained on the 3rd day of storage (8.67 log cfu/g) and, thereafter, it declined slightly to 8.37 log cfu/g on the 7th day of storage. There was not much difference in counts of *S. thermophilus* in the plain, probiotic and synbiotic yoghurt. The decrease in the counts after the 3rd day can be attributed to the increased acidity as reported by Dave and Shah (1997a). As shown in Figure 3, the counts of *L. bulgaricus*, followed similar pattern in all the three types of yoghurt. The viable count increased from 6.13 log cfu/g (0 day) to 6.97 log cfu/g on the 7th day. Since there was no decrease in the counts, it can be concluded that the *L. bulgaricus* strain used in this study was more acid tolerant. Which are contrary to the results obtained by Oliveira *et al.* (2009). However, in comparison to *S. thermophilus* and *L. acidophilus*, the counts of *L. bulgaricus* were lower. This can be explained by the fact that the selective agar (RCA at pH 5.3) reduced the recovery of the organism by 1-2 log cycles as reported by Dave and Shah (1997b).

As illustrated in Figure 4, the counts of *L. acidophilus* ranged from 6.89 to 7.94 log cfu/g, which was well above the recommended level of 10^6 log cfu/g for producing the desired therapeutic benefits (Robinson, 1987). But after an increase to 7.81 log cfu/g in probiotic yoghurt, it decreased slightly to 7.72 log cfu/g after the 5th day of storage. The results are in accordance with John (2000), who reported a decline in count of *L. acidophilus* NCDC 13 after 5th day of storage in acidophilus-bifidobacteria (AB) dahi. Dave and Shah (1997a) reported that the viable counts of *L. acidophilus* were found to be affected by the oxygen content that the inulin acted as substrate for *L. acidophilus* leading to an increase in its count. Similar findings were also shown by Chick *et al.* (2001) and Bruno *et al.* (2002). They reported an increase in the viability of bifidobacteria in presence of inulin during the refrigerated storage.

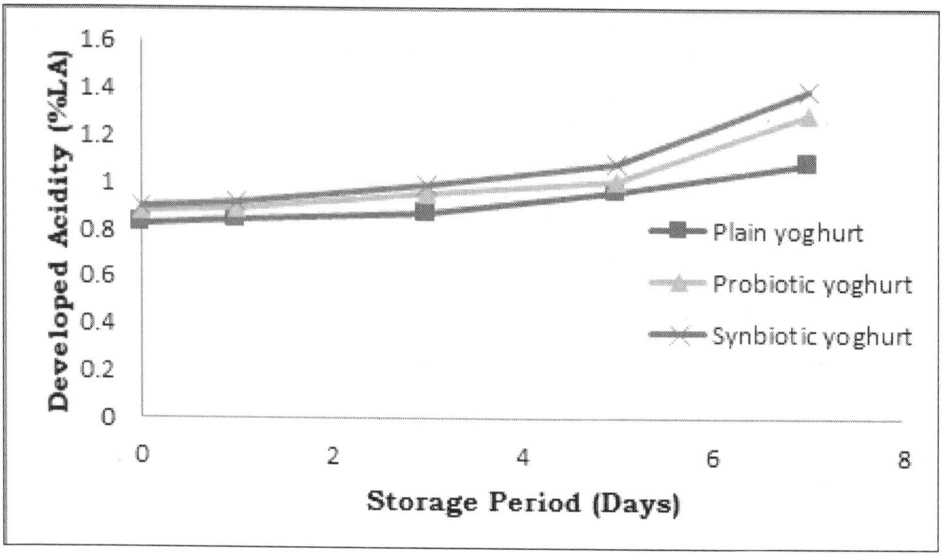

Fig. 17.1: Effect of storage on developed acidity.

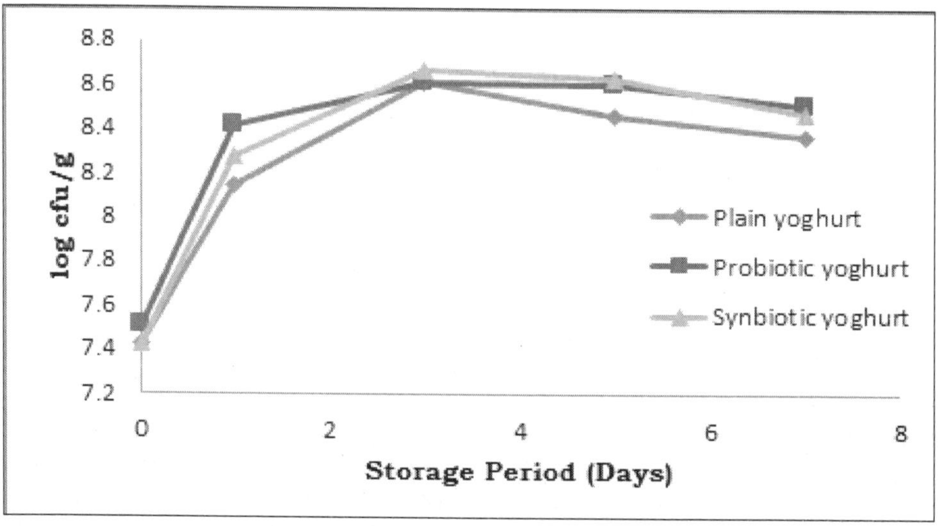

Fig. 17.2: Effect of storage on counts of *S. thermophilus*.

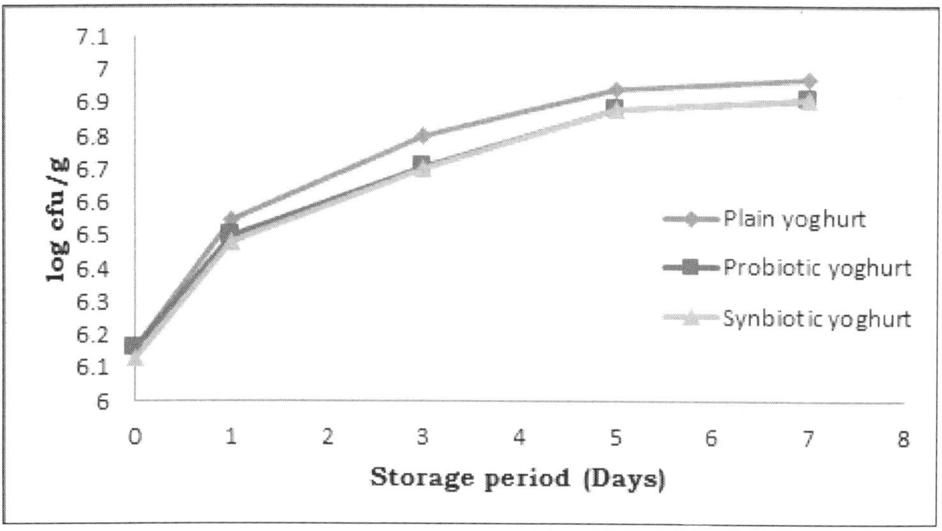

Fig. 17.3: Effect of storage on counts of *L. bulgaricus*.

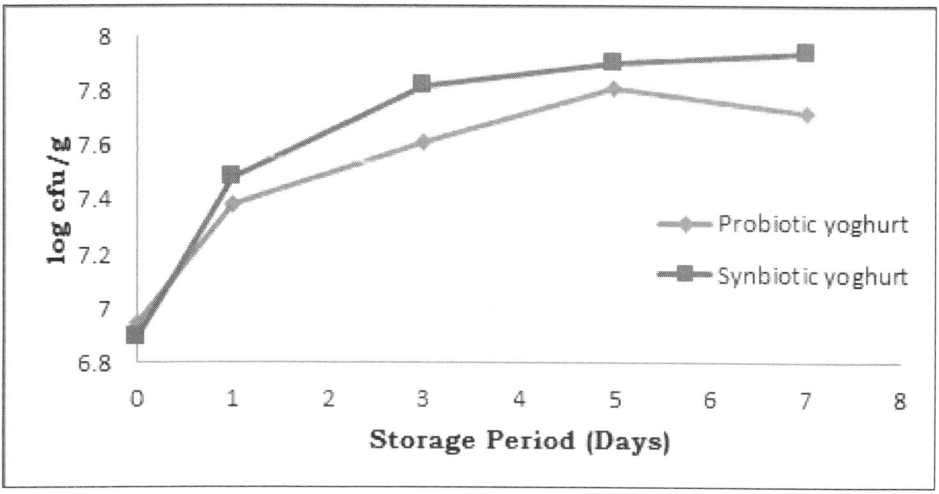

Fig. 17.4: Effect of storage on counts of *L. acidophilus*.

4. Conclusion

Results obtained in our study have shown that inulin in conjunction with *Lactobacillus acidophilus* increases its viability during storage of food product and also improves the product characteristics. Therefore, present study suggests inulin and *Lactobacillus acidophilus* can be a good combination for any synbiotic preparation.

Acknowledgment

The CSIR (Council of Scientific and Industrial Research), New Delhi, India is greatly acknowledged for financial assistance. Authors are also thankful to Dairy

Microbiology Division, National Dairy Research Institute, Karnal, India, for providing the bacterial strains and the facilities for carrying out the work.

References

Bodyfelt, F.W., Tobias, J. and Trout, G.M. 1988. The sensory evaluation of dairy products. *AVI Publ. New York.*, 48-67.

Bruno, F.A., Lankaputhra, W.E. and Shah, N.P. 2002. Growth, viability and activity of Bifidobacterium spp. in milk containing prebiotics. *J Food Sci.*, 67: 2740-744.

Chick, H., Shin, H.S. and Ustunol, Z. 2001. Growth and acid production by lactic acid bacteria and bifidobacteria grown in skim milk containing honey. *J Food Sci.*, 66: 478-481.

Dave, R.I. and Shah, N.P. 1997a. Effect of level of starter culture on viability of yoghurt and probiotic bacteria in yoghurts. *Food Aus.*, 49: 164-168.

Dave, R.I. and Shah, N.P. 1997b. Viability of yoghurt and probiotic bacteria in yoghurts made from commercial starter cultures. *Int. Dairy J.*, 7: 31-41

Franck, A.M. 2000. Inulin and oligofructose. In: Gibson, G. and Angus, F., (Eds) Surrey: Leatherhead Publishing, p. 1-18.

Hull, R.R., Roberts, A.V. 1984. Differential enumeration of *Lactobacillus acidophilus* **in yoghurt.** *Aust. J. Dairy Technol.*, 39:160–163.

John, S. 2000. Studies on incorporation of bifidobacteria and *L. acidophilus* into dahi. *Ph.D Thesis*, National Dairy Research Institute (Deemed University), Karnal, India.

Oliveira, R.P.S., Perego, P., Converti, A. and Oliveira, M.N. 2009. Effect of inulin on growth and acidification performance of different probiotic bacteria in cocultures and mixed culture with Streptococcus thermophilus. *J Food Eng.*, 91: 133-139.

Oliveira, R.P.S., Perego, P., Oliveira, M.N., and Converti, A. 2011. Effect of inulin as a prebiotic to improve growth and counts of a probiotic cocktail in fermented skim milk. *LWT-Food Sci Technol.*, 44: 520-523.

Robinson, R.K. 1987. Survival of Lactobacillus acidophilus in fermented products. *SA Tyd Sui.*, 19: 25-27.

Terzaghi, B.E. and Sandine, W. 1975. Improved medium for lactic acid streptococci and their bacteriophages. *J Appl Microbiol.*, 29: 807-813.

Vinderola, C.G., Mocchiutti, P. and Reinheimer, J.A. 2002. Interaction among lactic acid starter and probiotic bacteria used for fermented dairy products. *J Dairy Sci.*, 85: 721-729.

Chapter 18

Translation Initiation Regulation and Cancer: A Review

Harnek Singh Saini[1*] and Ritu Saini[2], Rajesh Kumar[1], Anjali Dahiya[1]

[1*]*Department of Biotechnology, University Institute of Engineering and Technology, Kurukshetra University, Kurukshetra*

[2]*Department of Chemistry and Biochemistry, CCS Haryana Agricultural University, Hisar*

*Corresponding Author Email: harneksingh89@gmail.com

Abstract

Protein synthesis is the most energy consuming process in the cell and it has a major role in the regulation of gene expression. Translation, which involves the orchestrated interaction of tRNAs, ribosomes, auxiliary factors and mRNA, is tightly controlled. Dysregulation of this process can be considered a hallmark of cancer and is linked to aberrant proliferation, survival, angiogenesis, alterations in immune response and cancer energetics. Although all steps of mRNA translation are highly regulated, most of the translational control occurs at the rate limiting initiation step. Translation initiation can be regulated by alterations in the expression or phosphorylation status of the various factors involved like eIF4E, S6 kinases, eIF2 phosphorylation, mTOR signaling pathway, polyadenylation, RNA packaging. Three main alterations at the translational level occur in cancer: (1) variations in mRNA sequences that increase or decrease translational efficiency by mutations in c-myc, BRCA1, Cyclin-dependent Kinase Inhibitor 2A, ATM etc. (2) changes in the expression or availability of components of the translational machinery like alterations in eIF4E in cancer, overexpression of S6K, Polyadenylate polymerase(PAP). (3) activation of translation through aberrantly activated signal transduction pathways. Therapeutic agents that targets the translation machinery in cancer is mainly divided into two categories-(1) Inhibitors of eIF4F upstream regulators includes- mTOR inhibitors (Rapamycin and rapalogues), dual PI3K/mTOR Inhibitors, MAPK-interacting kinases (MNKs) inhibitors and (2) Direct inhibitors of translational machinery includes- 4E- antisense oligonucleotides, inhibitors of eIF4E–cap interaction, inhibitors of eIF4E–eIF4G interaction, inhibitors of eIF4A. Therefore, therapeutic agents that target components of the protein synthesis apparatus hold promise as novel anticancer drugs that can overcome intra-tumour heterogeneity.

Keywords: *Translation, cancer, eIF4E, mTOR, therapeutic inhibitors.*

1. Introduction

Cancer, medically known as 'neoplasm' is a class of disorder characterized by out of control cell growth. According to an estimate, about 13% of the human deaths worldwide occur due to cancer. Rates are rising due to changing lifestyle in the developing world. It is well established that defective processes of cellular homeostasis, like functions involved in DNA repair and genomic stability, result in a predisposition toward malignant transformation (Hanahan and Weinberg, 2000). All tumor cells show the six hallmarks of cancer. These characteristics are required to produce a malignant tumor. They include, aberrant cell growth and division absent the proper signals, Continuous growth and division even given contrary signals, avoidance of programmed cell death, limitless number of cell divisions, promoting blood vessel construction, invasion of tissue and formation of metastases, The progression from normal cells to cells that can form a detectable mass to outright cancer involves multiple steps known as malignant progression. Protein synthesis is a vital cellular process that regulates growth and metabolism. It is controlled via signaling networks in response to environmental changes, including the presence of nutrients, mitogens, or starvation. The phosphorylation state of proteins involved in translation initiation is a limiting factor that regulates the formation or activity of translational complexes. In cancer cells, hyperactivated signaling pathways influence translation, allowing uncontrolled growth and survival (Grzmil and Hemmings, 2012).

2. Initiation of Translation in Eukaryotes

Translation is divided into three distinct phases: initiation, elongation and termination (Mathews *et al.*, 2007). Although all three phases are subject to regulatory mechanisms, under most circumstances translation initiation is considered the rate-limiting step (Sonenberg and Hinnebusch, 2009; Jackson *et al.*, 2010; Meric and Hunt, 2009) Translation initiation is a complex process that begins with the interaction of the cap-binding protein complex, eukaryotic initiationfactor-4F (eIF4F), with the mRNA 5′-end cap structure(m7GpppN; where N is any nucleotide). eIF4F comprises three subunits: eIF4E, which is the cap-binding protein;eIF4A,which is an RNA helicase; and the scaffolding protein eIF4G that bridges the mRNA and the ribosomethrough eIF3, which binds the ribosome directly10 (Fig. 18.1). The 48S complex then scans the 5′ untranslated region(UTR) of mRNA until the initiation codon AUG is recognized. This triggers eIF5 to hydrolyze GTP, the eIFs then dissociate and the 60S ribosomal subunit joins in assembling the fully functional 80S ribosome that is ready to begin peptide synthesis - elongation phase (Gebauer and Hentze, 2004; Fraser and Doudna, 2007).

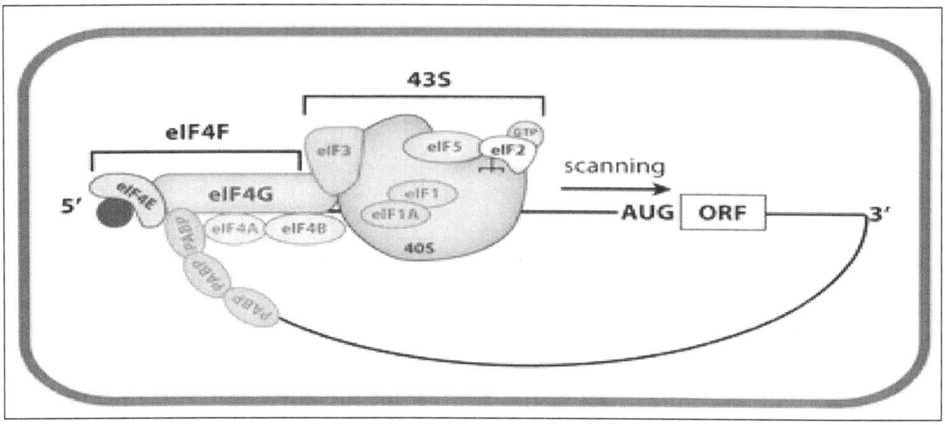

Fig. 18.1: Cap dependent translation initiation.

3. Regulation of Translation Initiation

Translation initiation can be regulated by alterations in the expression or phosphorylation status of the various factors involved like eIF4E, S6 kinases, eIF2 phosphorylation, mTOR signaling pathway, polyadenylation, RNA packaging (Bhat *et al.*, 2015). These Key components in translational regulation that may provide potential therapeutic targets are as follow:

3.1 eIF4E

eIF4E plays a central role in translation regulation. It is the least abundant of the initiation factors and is considered the rate-limiting component for initiation of cap dependent translation (Sonenberg and Hinnebusch, 2009). eIF4E may also be involved in mRNA splicing, mRNA processing, and mRNA nucleocytoplasmic transport (Rosenwald, 1995). Overexpression of eIF4E leads to the deregulation of translational and cellular homeostasis (Avdulov *et al.*, 2004). eIF4E levels in the cell are tightly controlled at several levels:

3.1.1 Transcription

eIF4E expression can be increased at the transcriptional level in response to serum or growth factors. eIF4E overexpression may cause preferential translation of mRNAs containing excessive secondary structure in their 5′ UTR that are normally discriminated against by the translational machinery and thus are inefficiently translated (Koromilas *et al.*, 1992). As examples of this, overexpression of eIF4E promotes increased translation of vascular endothelial growth factor, fibroblast growth factor-2, myc and cyclin D1 (Nathan *et al.*,1999).

3.1.2 Phosphorylation

eIF4E undergoes phosphorylation at Ser209 by MAP kinase signaling integration kinases 1 and 2 (Mnk1/2). eIF4E phosphorylation appears to enhance its transformation potential Several mitogens, such as serum, plateletderived growth factor, epidermal growth factor, insulin, angiotensin II, src kinase overexpression, and ras overexpression, lead to eIF4E phosphorylation (Kleijn *et al.*, 1998).

3.1.3 The translational Repressors, 4E-BPs

Assembly of the eIF4F complex is inhibited by members of a family of repressors termed the eIF4E-binding proteins (4E-BPs). 4E-BPs compete with eIF4G for a binding site in eIF4E. The binding of eIF4E to the best characterized eIF4E-binding protein, 4E-BP1, is regulated by 4E-BP1 phosphorylation. Hypophosphorylated 4EBP1 binds to eIF4E, whereas 4E-BP1 hyperphosphorylation decreases this binding. Insulin, angiotensin, epidermal growth factor, platelet-derived growth factor, hepatocyte growth factor, nerve growth factor, insulin-like growth factors I and II, interleukin 3, granulocyte-macrophage colony-stimulating factor - steel factor, gastrin, and the adenovirus have all been reported to induce phosphorylation of 4E-BP1 and to decrease the ability of 4E-BP1 to bind eIF4E (Takeuchi *et al.*, 2001).

3.2 eIF2 Phosphorylation

The binding of the initiator tRNA to the small ribosomal unit is mediated by translation initiation factor eIF2. Phosphorylation of the -subunit of eIF2 prevents formation of the eIF2/GTP/Met-tRNA complex and inhibits global protein synthesis (Kimball, 1999). eIF2 is phosphorylated under a variety of conditions, such as viral infection, nutrient deprivation, heme deprivation, and apoptosis. eIF2 is phosphorylated by heme-regulated inhibitor, nutrient regulated protein kinase, and the interferon-induced double stranded RNA-activated protein kinase (PKR).

3.3 S6Ks

Mammalian cells contain two ribosomal protein S6 kinases (S6K1 and S6K2) encoded by two separate genes. Both S6Ks are downstream targets of mTOR ((Ma and Blenis 2009; Zoncu *et al.*, 2011). In the absence of extracellular stimuli such as growth factors and nutrients, S6K1 is associated with eIF3. Upon addition of growth factors or nutrients, the mTOR/raptor is recruited to eIF3 to phosphorylate S6K1 and 4E-BP1. Phosphorylation of S6K1 at T389 leads to its dissociation from eIF3. T389-phosphorylated S6K1 binds to phosphoinositide- dependent kinase 1 (PDK1), which phosphorylates S6K1 at T229 (Fenton and Gout, 2011). The fully activated S6K1 then phosphorylates eIF4B and S6. Phosphorylation of eIF4B by S6K1 at S422 promotes its association with eIF3 (Holz *et al.*, 2005). eIF4B undergoes phosphorylation not only by the PI3K/mTOR pathway, but also by the ERK 1/2 MAPK pathway. All of these molecular interactions cooperate to enhance translation initiation complex formation and cap-dependent translation.

3.4 mTOR

Mammalian target of rapamycin (mTOR) is a serine/ threonine kinase that exists as two functionally and structurally divergent complexes: mTOR complex 1 (mTORC1) and mTORC2 (Laplante and Sabatini, 2012). mTOR is the mammalian homologue of the yeast TOR proteins that regulate G1 progression and translation in response to nutrient availability. mTOR is a serine-threonine kinase that modulates translation initiation by altering the phosphorylation status of 4E-BP1 and S6K (Roux and Topisirovic, 2012; Brown *et al.*, 1995). 4E-BP1 is phosphorylated on multiple residues. mTOR phosphorylates the Thr-37 and Thr-46 residues of 4E-BP1. However, phosphorylation at these sites is not associated with a loss of eIF4E binding. Phosphorylation of Thr-37 and Thr-46 is required for subsequent phosphorylation at several COOH-terminal, serum-sensitive sites; a combination of these phosphorylation events appears to be needed to inhibit the binding of 4E-BP1 to eIF4E. (Fig. 18.2)

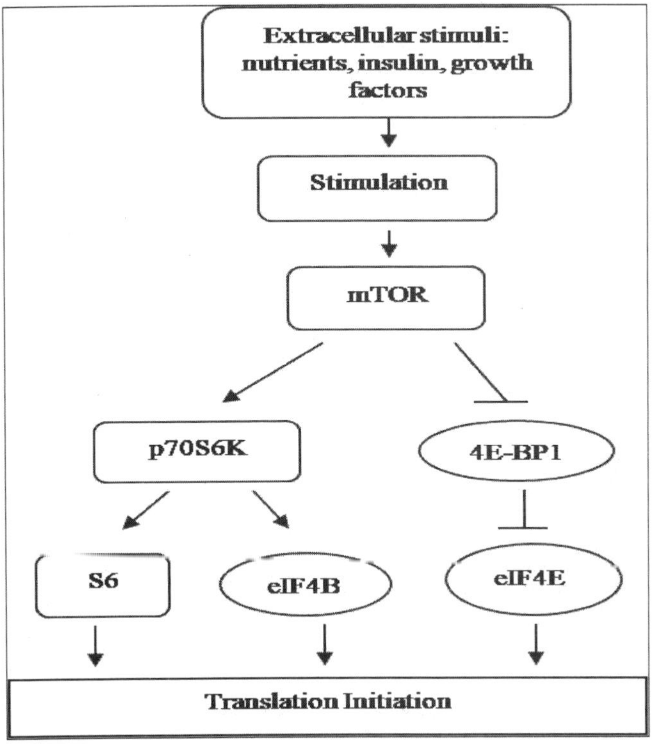

Fig. 18.2: Activation of translation initation by mTOR.

The product of the *ATM gene*, p38/MSK1 pathway, and protein kinase C also play a role in 4E-BP1 phosphorylation. Three mammalian homologs of (4E-BPs) were identified *i.e* 4E-BP1, 4E-BP2 and 4E-BP3. All three isoforms harbor a conserved eIF4E-binding motif Tyr-X-X-X-X-Leu-F (where X is variable and F is a hydrophobic amino acid) and phosphorylation sites critical to the regulation of cap-dependent translation. 4E-BPs and eIF4G share the same eIF4E-binding motif. Structural studies have demonstrated that the eIF4Gs and the 4E-BPs interact through mutually exclusive binding sites on the surface of eIF4E. Hypophosphorylated 4E-BPs interact strongly with eIF4E, whereas the hyperphosphorylated forms bind weakly. Rapamycin treatment inhibits 4E-BP phosphorylation, increases the interaction between eIF4E and 4E-BP, and consequently inhibits cap-dependent translation.

3.5 Polyadenylation

The poly(A) tail in eukaryotic mRNA is important in enhancing translation initiation and mRNA stability. The 5′ cap and poly(A) tail are thought to function synergistically to regulate mRNA translational efficiency (Sachs and Varani, 2000) .

3.6 RNA Packaging

Most RNA-binding proteins are assembled on a transcript at the time of transcription, thus determining the translational fate of the transcript (Wolffe and Meric, 1996). A highly conserved family of Y-box proteins is found in cytoplasmic messenger ribonucleoprotein particles, where the proteins are thought to play a role in restricting the recruitment of mRNA to the translational machinery (Evdokimova *et al.*, 1995). The major mRNA-associated protein, YB-1, destabilizes the interaction of eIF4E and the 5′ mRNA cap *in vitro*, and overexpression of YB-1 results in translational repression *in vivo* (Evdokimova *et al.*, 2001). Thus, alterations in RNA packaging can also play an important role in translational regulation.

4. Translation Alterations Encountered in Cancer

4.1 Mutations in the mRNA Sequence can Alter its Translational Efficiency

Mutations in the mRNA sequence, especially in the 5′ UTR, can alter its translational efficiency, as seen in the following examples.

4.1.1 c-myc

The protein encoded by this gene is a multifunctional, nuclear phosphoprotein that plays a role in cell cycle progression, apoptosis and cellular transformation. In several Burkitt lymphomas that have deletions of the mRNA 5′ UTR, translation of c-myc is more efficient (Stoneley *et al.*, 1998). In patients with multiple myeloma, a C→T mutation in the c-myc IRES was identified and found to cause an enhanced initiation of translation via internal ribosomal entry (Chappell *et al.*, 2000).

4.1.2 BRCA1

A tumor suppressor gene that, when inherited in a mutated state, is associated w ith the development of various cancers, especially breast and ovarian cancer. A somatic point mutation (G→C) in position 3 with respect to the start codon of the BRCA1 gene was identified in a highly aggressive breast cancer (Signori *et al.*, 2001). Chimeric constructs consisting of the wild-type or mutated BRCA1 5′ UTR and a downstream luciferase reporter demonstrated a decrease in the translational efficiency with the 5′ UTR mutation.

4.1.3 Cyclin-dependent Kinase Inhibitor 2A

It is a tumor suppressor protin that slows down the cell cycle by prohibiting progression from G1 phase to S phase. Some inherited melanoma have a G→T transversion at base 34 of cyclin-dependent kinase inhibitor-2A, which encodes a cyclin-dependent kinase 4/cyclin-dependent kinase 6 inhibitor important in G1 checkpoint regulation (Liu *et al.*, 1999). This mutation gives rise to a novel AUG translation initiation codon, creating an upstream open reading frame that competes for scanning ribosomes and decreases translation from the wild-type AUG.

4.1.4 Alternate Splicing and Alternate Transcription Start Sites

Alterations in splicing and alternate transcription sites can lead to variations in 5 UTR sequence, length, and secondary structure, ultimately impacting translational efficiency.

4.1.5 ATM

Ataxia telangiectasia mutated (ATM) is a serine/threonine protein kinase that is recruited and activated by DNA double-strand breaks. It phosphorylates several key proteins that initiate activation of the DNA damage checkpoint, leading to cell cycle arrest, DNA repair or apoptosis. A defect in ATM has severe consequences in repairing certain types of damage to DNA, and cancer may result from improper repair (Savitsky *et al.*, 1997).

4.1.6. Alternate Polyadenylation Sites

Multiple polyadenylation signals leading to the generation of several transcripts with differing 3′ UTR have been described for several mRNA species, such as the RET proto-oncogene (Myers *et al.*, 1995), ATM gene, tissue inhibitor of metalloproteinases-3 and calmodulin-I (Senterre-Lesenfants *et al.*, 1995). The role of these alterations in cancer development and progression is unknown. (Fig. 18.3)

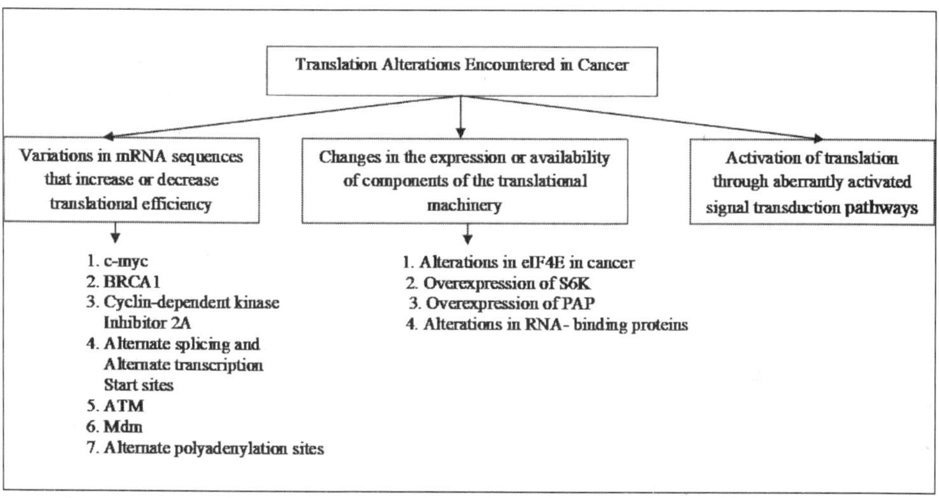

Fig. 18.3: Alterations in the Components of the Translation Machinery in Cancer.

4.2 Alterations in the Components of the Translation Machinery

4.2.1 Alterations in eIF4E in Cancer

eIF4E is the least abundant of the initiation factors and is considered the rate-limiting component for initiation of cap dependent translation. eIF4E is overexpressed by

~3-10 fold in many human cancers, including those of head and neck, bladder, colon, breast, prostate, lung and blood (Silvera *et al.*, 2010). eIF4E overexpression does not strongly promote global protein synthesis but rather selectively stimulates translation of a subset of mRNAs referred to as "eIF4E-sensitive" (Miyagi *et al.*, 1995). In general, these mRNAs contain long, highly structured 5′ UTRs and are thus expected to be highly dependent on the unwinding activity of the eIF4A subunit of eIF4F for efficient translation "eIF4E-sensitive" mRNAs encode proliferation and survival promoting factors and proto-oncogenic proteins, such as cyclins, c-myc, and BclxL. Thus, the proto-oncogenic activity of eIF4E stems from its ability to selectively stimulate translation of mRNAs, which encode tumor-promoting factors. Phosphorylation of eIF4E is also implicated in cancer development and progression (Furic *et al.*, 2010; Topisirovic *et al.*, 2004).

Alterations in Other Initiation Factors: Other translation initiation factors might be implicated in tumorigenesis: Increased eIF4G expression is associated with aggressive stages in lung cancer and inflammatory breast cancer (Comtesse *et al.*, 2007). Moreover, eIF4G is overexpressed in cervical cancer (Silvera *et al.*, 2009), and its overexpression correlates with poor prognosis in nasopharyngeal carcinoma. eIF4A is overexpressed in human melanoma cells and in primary hepatocellular carcinomas (Liang *et al.*, 2014) .eIF3 has also been implicated in cellular transformation in breast, liver and prostate cancers (Rothe *et al.*, 2000). eIF-2 is found in increased levels in bronchioloalveolar carcinomas of the lung (Rosenwald *et al.*, 2001).

4.2.2 Overexpression of S6K

S6K is amplified and highly overexpressed in the MCF7 breast cancer cell line, compared with normal mammary epithelium (Barlund *et al.*, 2000).

4.2.3 Overexpression of PAP

Polyadenylate polymerase catalyzes 3 poly (A) synthesis. PAP is overexpressed in human cancer cells compared with normal and virally transformed cells (Topalian *et al.*, 2001).

4.2.4 Alterations in RNA-binding Proteins

Even less is known about alterations in RNA packaging in cancer. Increased expression and nuclear localization of the RNA-binding protein YB-1 are indicators of a poor prognosis for breast cancer (Janz *et al.*, 2002), ovarian cancer and non-small cell lung cancer (Shibahara *et al.*, 2001).

4.2.3 Activation of Signal Transduction Pathways

Activation of signal transduction pathways by loss of tumor suppressor genes or overexpression of certain tyrosine kinases can contribute to the growth and aggressiveness of tumors. An important mutant in human cancers is the tumor suppressor gene PTEN, which leads to the activation of the PI3K/Akt pathway. Activation of PI3K and Akt induces the oncogenic transformation of chicken embryo fibroblasts. The transformed cells show constitutive phosphorylation of S6K and of 4E-BP1 (Aoki *et al.*, 2001). A mutant Akt that retains kinase activity but does not phosphorylate S6K or 4E-BP1 does not transform fibroblasts, which suggests a correlation between the oncogenicity of PI3K and Akt and the phosphorylation of S6K and 4E-BP1. Several tyrosine kinases such as platelet-derived growth factor, insulin-like growth factor, HER2/neu, and epidermal growth factor receptor are overexpressed in cancer. Because these kinases activate downstream signal transduction pathways known to alter translation initiation, activation of translation is likely to contribute to the growth and aggressiveness of these tumors (Child *et al.*, 1999).

5. Therapeutic Agents that Targets the Translation Machinery in Cancer

Important drug targets in the translation machinery and compounds that target them are shown in Table 18.1.

Table 18.1: Therapeutic inhibitors of translation in cancer.

Inhibitors of eIF4F upstream regulators	
mTOR inhibitors-Rapamycin and rapalogues.	Rapamycin (also known as sirolimus) is a naturally occurring macrolide produced by Streptomyces hygroscopicus, works as an anticancer compound that inhibits cell growth and proliferation as well as a potent immunosuppressant that effectively prevents allograft rejection (Vezina *et al.*, 1975).
	Rapamycin was the first mTOR inhibitor but despite its antitumor activity in preclinical models it was not successful as an anticancer drug (Yip *et al.*, 2010).
	Subsequently several analogs of rapamycin, nowcalled rapalogs, with better solubility and pharmacokinetic properties were synthesized such as temsirolimus, everolimus, and deforolimus.
	Like rapamycin these rapalogs form a complex with intracellular receptor FKBP12 which binds to mTOR and inhibits mTORC1 downstream signaling (Sarbassov *et al.*, 2006).
Dual PI3K/mTOR Inhibitors	These molecules inhibit mTORC1, mTORC2, and PI3K, thus inhibiting the phosphorylation of AKT, S6K1, and 4E-BP1, and are therefore attractive drugs for targeting cancers driven by PI3K activation (Fan *et al.*, 2006).
	This was made possible because of the high homology that is shared by the kinase domains of PI3K and mTOR. These inhibitors include XL-765, PI-103, and NVP-BEZ235.
	PI-103 and NVP-BEZ235 have been found to suppress AKT as well as S6K1 in breast tumors and leukemia cells (Fruman and Rommel, 2014).
MNK inhibitors	Several small-molecule inhibitors targeting MNKs, including CGP57380, cercosporamide, Resorcylic acid lactones, Retinamides and 2-anilino-pyrimidines (Knauf *et al.*, 2001).
	MNK inhibitors inhibit eIF4E phosphorylation, colony formation, epithelial-to-mesenchymal transition, invasion and MCL1 expression in cancer cell lines (Konicek *et al.*, 2011).
Direct inhibitors of the translation machinery	
Reduction of eIF4E expression by using Antisense oligonucleotides	It inhibits tumour growth and angiogenesis with low toxicity in mice. Reduces eIF4E protein levels in human tumours; Phase I/II clinical trials in combination with established chemotherapies in non-small-cell lung cancer and prostate cancer are in progress (Lin *et al.*, 2012).

Inhibitors of eIF4E–cap interaction	The ability of eIF4E to cause neoplastic transformation depends on its cap-binding activity, as overexpression of an eIF4E mutant defective in cap binding is not tumorigenic (Wendel *et al.*, 2007).
	N-7-benzyl guanosine monophosphate tryptamine phosphoramidate pronucleotide 4Ei-1 was reported to inhibit cap-dependent translation and the epithelial-to-mesenchymal transition in zebrafish (Ghosh *et al.*, 2009).
	Ribavirin (1-β-D-ribofuranosyl-1,2,4-triazole-3-carboxamide) acts as a physical mimic of the m7G cap and subsequently, inhibits eIF4E activity (Malina *et al.*, 2011).
Ribavirin targets eIF4E dependent Akt signaling	ribavirin (1-β-D-ribofuranosyl-1,2,4-triazole-3-carboxamide) acts as a physical mimic of the m7G cap and subsequently, inhibits eIF4E activity (Kentsis *et al.*, 2004; Tan *et al.*, 2008)
Inhibitors of eIF4E–eIF4G interaction	This is accomplished through interaction of the eIF4E-binding motif YXXXLΦ (X is any amino acid and Φ is a hydrophobic amino acid) of eIF4G, with Val69 and Trp73 residues on the dorsal surface of eIF4E (Marcotrigiano *et al.*, 1999). In addition, eIF4G binds directly to the mRNA, and this interaction stabilizes the 5′-mRNA-cap–eIF4E association (O'Leary *et al.*, 2013). 4E-BPs bind to the same region as eIF4G on the dorsal surface of eIF4E, thereby blocking eIF4E–eIF4G association (Altmann *et al.*, 1997).
Inhibitors of eIF4A	Hippuristanol, pateamine A and silvestrol are eIF4A inhibitors that suppress translation. (Malina *et al.*, 2012).
	Hippuristanol belongs to a family of polyoxygenated steroids, which bind to the C-terminal domain of eIF4A. It allosterically prevents eIF4A from interacting with RNA and blocks its helicase activity, both in its free form and as part of eIF4F (Bordeleau *et al.*, 2006).
	Pateamine A and silvestrol increase the ATPase, RNA-binding and helicase activities of eIF4A. Both compounds seem to function as chemical inducers of dimerization, and they increase the RNA-binding affinity of eIF4A in a non-sequence dependent manner, resulting in its depletion from the eIF4F complex (Bordeleau *et al.*, 2005).

6. Conclusion

Dysregulation of mRNA translation is a frequent feature of neoplasia. Many oncogenes and tumour suppressors affect the translation machinery, making aberrant translation a widespread characteristic of tumour cells, independent of the genetic make-up of the cancer. Therefore, therapeutic agents that target components of the protein synthesis apparatus hold promise as novel anticancer drugs that can overcome intra-tumour heterogeneity.

References

Altmann, M., Schmitz, N., Berset, C. and Trachsel, H. 1997. A novel inhibitor of cap-dependent translation initiation in yeast: p20 competes with eIF4G for binding to eIF4E. *EMBO J.* 16: 1114-1121.

Aoki, M., Blazek, E., and Vogt, P. K. 2001. A role of the kinase mTOR in cellular transformation induced by the oncoproteins P3k and Akt. *Proc. Natl.* Acad. *Sci. USA*, **98:** 136-141.

Avdulov, S., Li, S., Michalek, V., Burrichter, D., Peterson, M., Perlman, D.M., Manivel, J.C., Sonenberg, N., Yee, D., Bitterman, P.B. and Polunovsky, V.A. 2004. Activation of translation complex eIF4F is essential for the genesis and maintenance of the malignant phenotype in human mammary epithelial cells. *Cancer Cell*. 5: 553-563.

Barlund, M., Forozan, F., Kononen, J., Bubendorf, L., Chen, Y., Bittner, M. L., Torhorst, J., Haas, P., Bucher, C., Sauter, G., Kallioniemi, O. P., and Kallioniemi, A. 2000. Detecting activation of ribosomal protein S6 kinase by complementary DNA and tissue microarray analysis. *J. Natl. Cancer Inst. (Bethesda)*. **92:** 1252-1259.

Bhat, M., Robichaud, N., Hulea, L., Sonenberg, N., Pelletier J., and Topisirovic, I. 2015. Targeting the translation machinery in cancer. *Nature Reviews Drug Discovery*. 14(4): 261-278

Bordeleau, M. E., Matthews, J., Wojnar, J.M., Lindqvist, L., Novac, O., Jankowsky, E., Sonenberg, N., Northcote, P., Teesdale-Spittle, P. and Pelletier, J. 2005. Stimulation of mammalian translation initiation factor eIF4A activity by a small molecule inhibitor of eukaryotic translation. *Proc. Natl Acad. Sci. USA*. 102: 10460-10465.

Bordeleau, M.E., Mori, A., Oberer, M., Lindqvist, L., Chard, .LS., Higa, T., Belsham, G.J., Wagner, G., Tanaka, J. and Pelletier, J.2006, Functional characterization of IRESes by an inhibitor of the RNA helicase eIF4A. *Nature Chem. Biol*. 2: 213-220.

Brown, E. J., Beal, P.A., Keith, C.T., Chen, J., Shin, T.B. and Schreiber, S.L. 1995, Control of p70 S6 kinase by kinase activity of FRAP *in vivo*. *Nature*. 377: 441-446.

Chappell, S. A., LeQuesne, J. P., Paulin, F. E., de Schoolmeester, M. L., Stoneley, M., Soutar, R. L., Ralston, S. H., Helfrich, M. H., and Willis, A. E. 2000. A mutation in the c-myc-IRES leads to enhanced internal ribosome entry in multiple myeloma: *A novel mechanism of oncogene de-regulation. Oncogene*. 19: 4437-4440.

Child, S. J., Miller, M. K., and Geballe, A. P. 1999. Cell type-dependent and -independent control of HER-2/neu translation. *Int. J. Biochem. Cell Biol*. 31: 201-213.

Comtesse, N., Keller, A., Diesinger, I., Bauer, C., Kayser, K., Huwer, H., Lenhof, H.P. and Meese, E. 2007. Frequent overexpression of the genes *FXR1*, *CLAPM1* and *EIF4G* located on amplicon 3q26–27 in squamous cell carcinoma of the lung. *Int. J. Cancer*. 120: 2538-2544.

Evdokimova, V. M., Wei, C. L., Sitikov, A. S., Simonenko, P. N., Lazarev, O. A., Vasilenko, K. S., Ustinov, V. A., Hershey, J. W., and Ovchinnikov, L. P. 1995. The major protein of messenger ribonucleoprotein particles in somatic cells is a member of the Y-box binding transcription factor family. *J. Biol. Chem*. 270: 3186-3192.

Evdokimova, V., Ruzanov, P., Imataka, H., Raught, B., Svitkin, Y., Ovchinnikov, L. P., and Sonenberg, N. 2001. The major mRNA-associated protein YB-1 is a potent 5_ cap-dependent mRNA stabilizer. *EMBO J*. 20: 5491-5502.

Fan, Q. W., Knight, Z.A., Goldenberg, D.D., Yu, W., Mostov, K.E., Stokoe, D., Shokat, K.M. and Weiss, W.A.2006 A dual PI3 kinase/mTOR inhibitor reveals emergent efficacy in glioma. *Cancer Cell*. 9: 341-349.

Fenton, T.R. and Gout, I.T. 2011. Functions and regulation of the 70kDa ribosomal S6 kinases. 43: 47-59.

Fraser, C.S. and Doudna, J.A. 2007. Structural and mechanistic insights into hepatitis C viral translation initiation. *Nature Reviews Microbiology*. 5: 29-38.

Fruman, D. A. and Rommel, C. 2014. PI3K and cancer: lessons, challenges and opportunities. *Nature Rev. Drug Discov.* 13: 140–156.

Furic, L., Rong, L., Larsson, O., Koumakpayi, I.H., Yoshida, K., Brueschke, A., Petroulakis, E., Robichaud, N., Pollak, M., Gaboury, L.A., Pandolfi, P.P., Saad, F. and Sonenberg, N. 2010. eIF4E phosphorylation promotes tumorigenesis and is associated with prostate cancer progression. *Proc. Natl Acad. Sci. USA* 107: 14134-14139.

Gebauer, F. and Hentze, M. W. 2004. Molecular mechanisms of translational control. *Nature Rev. Mol. Cell Biol.* 5: 827-835.

Ghosh, B., Benyumov, A.O., Ghosh, P., Jia, Y., Avdulov, S., Dahlberg, P.S., Peterson, M., Smith, K., Polunovsky, V.A., Bitterman, P.B. and Wagner, C.R. 2009. Nontoxic chemical interdiction of the epithelial-tomesenchymal transition by targeting cap-dependent translation. *ACS Chem. Biol.* 4: 367-377.

Grzmil, M. and Hemmings, B. A. 2012. Translation Regulation as a Therapeutic Target in Cancer. *Cancer res.* 72(16): 3891-3900.

Hanahan, D. and Weinberg, R.A. 2000. "The hallmarks of cancer". *Cell*. 100 (1): 57-70.

Holz, M.K., Ballif, B.A., Gygi, S.P., and Blenis, J. 2005. mTOR and S6K1 mediate assembly of the translation preinitiation complex through dynamic protein interchange and ordered phosphorylation events. *Cell*. 123: 569-580.

Jackson, R.J., Hellen, C.U. and Pestova, T.V. 2010. The mechanism ofeukaryotic translation initiation and principles of its regulation. *Nat Rev Mol Cell Biol*. 11: 113-127.

Janz, M., Harbeck, N., Dettmar, P., Berger, U., Schmidt, A., Jurchott, K., Schmitt, M., and Royer, H. D. 2002. Y-box factor YB-1 predicts drug resistance and patient outcome in breast cancer independent of clinically relevant tumor biologic factors HER2, uPA and PAI-1. *Int. J. Cancer*. 97: 278-282.

Kentsis A, Topisirovic I, Culjkovic B, Shao L, Borden KL. 2004. Ribavirin suppresses eIF4E-mediated oncogenic transformation by physical mimicry of the 7-methyl guanosine mRNA cap. Proc Natl. Acad. Sci. USA. 101:18105-18110.

Kimball, S.R. 1999. Eukaryotic initiation factor eIF2. *The international journal of biochemistry and cell biology*. 31(1): 25-29.

Kleijn, M., Scheper, G. C., Voorma, H. O., and Thomas, A. A. 1998. Regulation of translation initiation factors by signal transduction. *Eur. J. Biochem.* 253: 531-544.

Knauf, U., Tschopp, C. and Gram, H. 2001. Negative regulation of protein translation by mitogen-activated protein kinase-interacting kinases 1 and 2. *Mol. Cell. Biol.* 21: 5500-5511.

Konicek, B. W., Stephens, J.R., McNulty, A.M., Robichaud, N., Peery, R.B., Dumstorf, C.A., Dowless, M.S., Iversen, P.W., Parsons, S., Ellis, K.E., McCann, D.J., Pelletier, J., Furic, L., Yingling, J.M., Stancato, L.F., Sonenberg, N. and Graff, J.R. 2011. Therapeutic inhibition of MAP kinase interacting kinase blocks eukaryotic initiation factor 4E phosphorylation and suppresses outgrowth of experimental lung metastases. *Cancer Res.* 71:1849-1857.

Koromilas, A. E., Lazaris-Karatzas, A., and Sonenberg, N. 1992. mRNAs containing extensive secondary structure in their 5′ non-coding region translate efficiently in cells overexpressing initiation factor eIF-4E. *EMBO J.* 11: 4153-4158.

Laplante, M. and Sabatini, D.M. 2012. mTOR signaling in growth control and disease. *Cell.* 149: 274-293.

Liang, S., Zhou, Y., Chen, Y., Ke, G., Wen, H. and Wu, X. 2014. Decreased expression of EIF4A1 after preoperative brachytherapy predicts better tumor-specific survival in cervical cancer. *Int. J. Gynecol. Cancer* 24: 908-915.

Lin, C. J., Nasr, Z., Premsrirut, P.K., Jr. J.A.P., Hippo, Y., Lowe, S.W. and Pelletier, J. 2012. Targeting synthetic lethal interactions between Myc and the eIF4F complex impedes tumorigenesis. *Cell Rep.* 1: 325-333.

Liu, L., Dilworth, D., Gao, L., Monzon, J., Summers, A., Lassam, N., and Hogg, D. 1999. Mutation of the CDKN2A 5_ UTR creates an aberrant initiation codon and predisposes to melanoma. *Nat. Genet.* 21:128-132.

Ma, X.M. and Blenis, J. 2009. Molecular mechanisms of mTOR-mediated translational control. *Nat Rev Mol Cell Biol.* 10: 307-318.

Malina, A., Cencic, R. and Pelletier, J. 2011. Targeting translation dependence in cancer. *Oncotarget* . 2: 76-88.

Malina, A., Mills, J. R. and Pelletier, J. 2012. Emerging therapeutics targeting mRNA transltion. *Cold Spring Harb. Perspect. Biol.* 4: 12377.

Marcotrigiano, J., Gingras, A.C., Sonenberg, N. and Burley, S.K. 1999. Cap-dependent translation initiation in eukaryotes is regulated by a molecular mimic of eIF4G. *Mol. Cell.* 3: 707-716.

Mathews, M.B., Sonenberg, N. and Hershey, J.W.B. 2007. Translationalcontrol in biology and medicine (ed. Sonenberg N, Mathews MB, Hershey JWB). *Cold Spring Harbor Laboratory Press,Cold Spring Harbor, NY.*

Meric, F. and Hunt, K. K. 2009. Translation initation in cancer: A novel target for therapy. *Molecular cancer therapeutics.*1: 971-979.

Grzmil, M., and Hemmings, B. A. 2012. Translation Regulation as a Therapeutic Target in Cancer. *Cancer res.* 72(16): 3891-3900.

Miyagi, Y., Sugiyama, A., Asai, A., Okazaki, T., Kuchino, Y., and Kerr, S. J. 1995. Elevated levels of eukaryotic translation initiation factor eIF-4E, mRNA in a broad spectrum of transformed cell lines. *Cancer Lett.* 91: 247-252.

Myers, S. M., Eng, C., Ponder, B. A., and Mulligan, L. M. 1995. Characterization of RET proto-oncogene 3′ splicing variants and polyadenylation sites: a novel C-terminus for RET. *Oncogene*. 11: 2039-2045.

Nathan, C. A., Franklin, S., Abreo, F. W., Nassar, R., De Benedetti, A., and Glass, J. 1999. Analysis of surgical margins with the molecular marker eIF4E: a prognostic factor in patients with head and neck cancer. *J. Clin. Oncol*. 17: 2909-2914.

O'Leary, S. E., Petrov, A., Chen, J. and Puglisi, J.D. 2013. Dynamic recognition of the mRNA cap by *Saccharomyces cerevisiae* eIF4E. *Structure*. 21: 2197-2207.

Rosenwald, I. B., Hutzler, M. J., Wang, S., Savas, L., and Fraire, A. E. 2001. Expression of eukaryotic translation initiation factors 4E and 2 is increased frequently in bronchioloalveolar but not in squamous cell carcinomas of the lung. *Cancer (Phila.)*. 92: 2164-2171.

Rosenwald, I. B., Kaspar, R., Rousseau, D., Gehrke, L., Leboulch, P., Chen, J. J., Schmidt, E. V., Sonenberg, N., and London, I. M. 1995. Eukaryotic translation initiation factor 4E regulates expression of cyclin D1 at transcriptional and post-transcriptional levels. *J. Biol. Chem*. 270: 21176-21180.

Rothe, M., Ko, Y., Albers, P., and Wernert, N. 2000. Eukaryotic initiation factor 3 p110 mRNA is overexpressed in testicular seminomas. *Am. J. Pathol*. 157: 1597-1604.

Roux, P. P. and Topisirovic, I. 2012. Regulation of mRNA translation by signaling pathways. *Cold Spring Harb. Perspect. Biol*. 4:12252.

Sachs, A. B., and Varani, G. 2000. Eukaryotic translation initiation: there are (at least) two sides to every story. *Nat. Struct. Biol*. 7: 356-361.

Sarbassov, D.D., Ali, S.M., Sengupta, S., Sheen, J.H., Hsu, P.P., Bagley, A.F., Markhard, A.L. and Sabatini, D.M. 2006. Prolonged rapamycin treatment inhibits mTORC2 assembly and Akt/PKB. *Mol. Cell*. 22: 159-168.

Savitsky, K., Platzer, M., Uziel, T., Gilad, S., Sartiel, A., Rosenthal, A., Elroy-Stein, O., Shiloh, Y., and Rotman, G. 1997. Ataxia-telangiectasia: structural diversity of untranslated sequences suggests complex post-transcriptional regulation of *ATM* gene expression. *Nucleic Acids Res*. 25: 1678-1684.

Senterre-Lesenfants, S., Alag, A. S., and Sobel, M. E. 1995. Multiple mRNA, species are generated by alternate polyadenylation from the human *calmodulin- I* gene. *J. Cell. Biochem*. 58: 445-454.

Shibahara, K., Sugio, K., Osaki, T., Uchiumi, T., Maehara, Y., Kohno, K., Yasumoto, K., Sugimachi, K., and Kuwano, M. 2001. Nuclear expression of the Y-box binding protein, YB-1, as a novel marker of disease progression in non-small cell lung cancer. Clin. *Cancer Res*. 7: 3151-3155.

Signori, E., Bagni, C., Papa, S., Primerano, B., Rinaldi, M., Amaldi, F., and Fazio, V.M.A. 2001. A somatic mutation in the 5′ UTR of *BRCA1* gene in sporadic breast cancer causes down-modulation of translation efficiency. *Oncogene*, 20: 4596-4600.

Silvera, D., Arju, R., Darvishian, F., Levine, P.H., Zolfaghari, L., Goldberg, J., Hochman, T., Formenti, S.C. and Schneider, R.J. 2009. Essential role for eIF4GI overexpression in the pathogenesis of inflammatory breast cancer. *Nature Cell Biol.* 11: 903-908.

Silvera, D. S., Formenti, C. and Schneider, R. J. 2010. Translational control in cancer. *Nature Rev. Cancer.* 10: 254-266.

Sonenberg, N. and Hinnebusch, A.G. 2009. Regulation of translation initiation in eukaryotes: Mechanisms and biological targets. *Cell.* 136: 731-745.

Stoneley, M., Paulin, F. E., Le Quesne, J. P., Chappell, S. A., and Willis, A. E. 1998. c-Myc 5′ untranslated region contains an internal ribosome entry segment. *Oncogene,* 16: 423-428.

Takeuchi, K., Shibamoto, S., Nagamine, K., Shigemori, I., Omura, S.,Kitamura, N., and Ito, F. 2001. Signaling pathways leading to transcription and translation cooperatively regulate the transient increase in expression of c-Fos protein. *J. Biol. Chem.* 276: 26077-26083.

Tan, K., Culjkovic, B., Amri, A. and Borden, K.L.B. 2008. Ribavirin targets eIF4E dependent Akt survival signaling. *Biochem Biophys Res Commun.* 375(3): 341-345.

Topalian, S. L., Kaneko, S., Gonzales, M. I., Bond, G. L., Ward, Y., and Manley, J. L. 2001. Identification and functional characterization of neo-poly(A) polymerase, an RNA processing enzyme overexpressed in human tumors. *Mol. Cell. Biol.* 21: 5614-5623.

Topisirovic, I., Ruiz-Gutierrez, M. and Borden, K.L. 2004. Phosphorylation of the eukaryotic translation initiation factor eIF4E contributes to its transformation and mRNA transport activities. *Cancer Res.* 64: 8639-8642.

Vezina, C., Kudelski, A. and Sehgal, S.N. 1975. Rapamycin (AY22,989), a new antifungal antibiotic. I. Taxonomy of the producing streptomycete and isolation of the active principle. *J. Antibiot. (Tokyo)* 28: 721-726.

Wendel, H. G., Silva, R.L., Malina, A., Mills, J.R., Zhu, H., Ueda, T., Watanabe-Fukunaga, R., Fukunaga, R., Teruya-Feldstein, .J, Pelletier, J. and Lowe, S.W. 2007. Dissecting eIF4E action in tumorigenesis. *Genes Dev.* 21: 3232-3237.

Wolffe, A.P. and Meric, F. 1996. Coupling transcription to translation: a novel site for the regulation of eukaryotic gene expression. *Int. J. Biochem. Cell Biol.* 28, 247-257.

Yip, C. K., Murata, K., Walz, T., Sabatini, D.M., and Kang, S.A. 2010. Structure of the human mTOR complex I and its implications for rapamycin inhibition. *Mol. Cell.* 38: 768-774.

Zoncu, R., Efeyan, A., and Sabatini, D.M. 2011. mTOR: From growth signal integration to cancer, diabetes and ageing. *Nat Rev Mowl Cell Biol.* 12: 21-35.

Chapter 19

Assessing the Effect of GA_3 Pretreatment and Postharvest Treatments on Quality of Spikes of *Polianthes Tuberosa* cv. Prajwal

Pooja Rani[1*] and Narender Singh[2]

[1]*Department in Botany, D.B.G. Govt. P.G. College, Panipat, Haryana- 132103, India*

[2]*Department of Botany, Kurukshetra University, Kurukshetra, Haryana -136119, India*

**Corresponding Author E-mail-poojabishan@gmail.com*

Abstract

In the present investigation, Gibberellic acid (GA_3) pretreatment was given by dipping the Polianthes tuberosa cv. Prajwal bulbs at three different concentrations viz. 0, 50, 100 and 150 ppm before sowing for about 12 hours. To assess the vaselife of these spikes, postharvest treatments used were T0- Control (DDW); T_1- 8% Sucrose; T_2- 8% Sucrose + 200 ppm Citric Acid ; T_3- 8% Sucrose + 400 ppm 8-Hydroxyquinoline; T_4- 200 ppm Citric Acid; T_5- 4% Sucrose; T_6- 4% Sucrose+ 200 ppm Citric acid; T_7- 4% Sucrose+ 400 ppm 8-Hydroxyquinoline.The results showed that moisture content and fresh weight was higher in tuberose spikes obtained from GA_3 pretreated bulbs placed in 4 % Sucrose + 400 ppm 8-Hydroxyquinoline solution as postharvest treatment followed by 4 % Sucrose + 200 ppm Citric acid. 4% Sucrose+ 400 ppm Hydroxyquinoline followed by 4% Sucrose + 200 ppm Citric acid also showed the best results in enhancing water uptake and vase life which was extended 10 days more in spikes obtained from GA_3 pretreated bulbs.

Keywords: *GA_3, 8-hydroxyquinoline, citric acid, vaselife.*

1. Introduction

Quality of cut flowers is an important factor contributing to marketing and profitability of the tuberose which is a perennial bulbous plant belonging to the family Agavaceae. Extending the vase life after harvest is a significant issue that has a major impact on efficient trading of cut flowers and can increase the economic benefits (Weaver *et al.*, 1998). The application of gibberellins has brought a sort of revolution for the floriculture industry. Gibberellins are the plant growth regulators that are known to stimulate physiological responses in plants and alter the source-sink metabolism through their effect on photosynthesis and sink formation (Iqbal *et al.*, 2011) which results in improving important vegetative as well as floral traits in floricultural crops. Various factors such as bacterial contamination, vascular blockage, ethylene sensitivity and oxidative stress cause petal burning or browning, and floret wilting or abscission in cut flowers (Rubinstein, 2000; Rogers, 2006; Zhang *et al.*, 2011; Shahri & Tahir, 2011). Moreover, various agents such as microbial activities, air obstruction and physiological response to cutting wound, gum aggregation in xylem, latex leakage, and other processes in the wound lead to vascular blockage and prevent water uptake, which reduces the freshness of cut flowers (Rattanawisalanon *et al.*, 2003; van Ieperen *et al.*, 2002; van Meeteren *et al.*, 2005; van Meeteren *et al.*, 2006; Imsabai *et al.*, 2013). Additionally, recent studies on some cut flowers demonstrated that with the decrease in the freshness of cut flowers is due to some bacteria that aggregate in the vase solution and lead to senescence of cut flowers (Macnish *et al.*, 2008). To cope up with this problem of cut flowers, postharvest solutions such as 8-Hydroxyquinoline and citric acid were used in our study. Therefore, vase life of various cut flowers significantly increased, which is supposedly due to the reduction in vascular blockage (Alaey *et al.*, 2011; Shimizu-Yumoto & Ichimura 2010; Mansouri, 2012).

2. Materials and Methods

The bulbs of tuberose were raised in the Botanical Garden, Kurukshetra University, Kurukshetra, India based on a randomized block design under open field conditions. Healthy bulbs of tuberose cv. Prajwal having a diameter of about 3 cm were used. Bulbs were procured from the Indian Agricultural Research Institute, New Delhi. The sowing of tuberose was carried out on raised beds during the month of April. Gibberellic acid treatment was given by dipping the bulbs at three different GA_3 concentrations viz. 0, 50, 100 and 150 ppm before sowing. The bulbs were dipped for 12 hours. After treatment, the solution was drained off. Thereafter, all the bulbs were taken out and were kept under shade for two hours. Sowing was carried out at a distance of 20-25 cm apart on raised beds in the field. Plants were irrigated at regular intervals. Standard package and practices were followed for raising the crop under field conditions.

The spikes were harvested only from control and best-found concentrations of GA_3 from both the crops at the time of flowering. The spikes were harvested in the morning hours with a sharp blade or scalpel. The spikes were cut under water to prevent cavitation and immediately brought to laboratory. The basal few centimeters of spikes were re-cut under water to a uniform length. Then, spikes were transferred to the following solutions containing biocides in laboratory. The holding solutions used were T_0- Control (DDW); T_1- 8% Sucrose; T_2- 8% Sucrose + 200 ppm Citric Acid; T_3- 8% Sucrose + 400 ppm 8-Hydroxyquinoline; T_4- 200 ppm Citric Acid; T_5- 4% Sucrose; T_6- 4% Sucrose+ 200 ppm Citric acid; T_7- 4% Sucrose+ 400 ppm Hydroxyquinoline. Only healthy spikes were selected for experimentation.

All the spikes were washed before putting in holding solutions. Conical flasks (Borosil-make) of 100 ml capacity were used to have concentrations of different holding solutions.

For each concentration, ten conical flasks were used and 100 ml test solution was poured in each flask in case of tuberose and 200 ml in case of gladiolus depending upon spike length. Conical flask containing double distilled water (DDW) served as control. Spikes were kept in each flask. All experiment was set up at room temperature under day and night intensity of 2.24 $\mu mol\ m^{-2}s^{-1}$ and 1.13 $\mu mol\ m^{-2}s^{-1}$ respectively.

Sampling stages: Samples was collected at- 0 day; 3rd day & 6th day.

Observations were made and samples were collected at different time intervals (0, 3 and 6- day) to analyze the different parameters. 0-day in the text and tables denotes initial day when experiment was started. Petals of floret on spikes were plucked with the help of foreceps from different post-harvest treatments. Samples were collected in such a manner that 3 replicates were available for each biochemical extraction and analyses.

3 Results

3.1 Visible Effect

Florets of spikes exhibited shrinkage at 3-day and the process intensified up to 6-day stage. The shrinkage was noticed maximum in spikes held in DDW in both untreated and GA_3-pretreated spikes in tuberose and least in 4 % Sucrose and its combination with 400 ppm 8-Hydroxyquinoline followed by 200 ppm Citric acid.

3.2 Volume of Absorbed Solution

Data in (Table 19.1) presented differences in water uptake of tuberose when subjected to different holding solutions. It was observed that all treatments increased the solution uptake which ranged between 49.16 to 71.84 ml in untreated spikes and 55.24 to 75.40 ml in GA_3 pretreated spikes. Furthermore, the significantly highest values of water uptake were obtained in the spikes obtained from pretreated GA_3 bulbs when kept in 4 % Sucrose + 400 ppm 8-Hydroxyquinoline followed by 4 % Sucrose + 200 ppm Citric acid as post harvest treatment and the differences were significant compared with the other post harvest treatments.

Table 19.1: Effect of pretreatment and holding solutions on volume of holding solution absorbed by spikes in *Polianthes tuberosa* cv. Prajwal.

Pretreatments Post harvest treatments	Volume of holding solution (ml) Control (0 ppm)	GA_3 (150 ppm)
T_0	48.51 ± 0.75	52.31 ± 1.44
T_1	49.16 ± 0.77	55.24 ± 0.65
T_2	50.20 ± 0.45	58.41 ± 0.658
T_3	53.44 ± 0.48	58.62 ± 1.50
T_4	60.34 ± 0.48	68.44 ± 1.04
T_5	62.70 ± 0.65	69.52 ± 1.71

Pretreatments Post harvest treatments	Volume of holding solution (ml) Control (0 ppm)	GA_3 (150 ppm)
T_6	69.64 ± 0.77	72.31 ± 1.11
T_7	71.84 ± 0.51	75.40 ± 0.75

*Initial Volume = 100 ml

*Each value indicates mean of 10 replicates

T_0- Control (DDW); T_1- 8% Sucrose; T_2- 8% Sucrose + 200 ppm Citric Acid ; T_3- 8% Sucrose + 400 ppm 8-Hydroxyquinoline; T_4- 200 ppm Citric Acid; T_5- 4% Sucrose; T_6- 4% Sucrose+ 200 ppm Citric acid; T_7- 4% Sucrose+ 400 ppm Hydroxyquinoline

3.3 Moisture Content

Moisture content was maximum at 0-day, then it decreases in all the treatments. Data in (Table 19.2) showed difference in moisture content in tuberose when subjected to various postharvest treatments in spikes of both GA_3 pretreated and untreated bulbs in tuberose. Moisture content was maximum in spikes obtained from GA_3 pretreated bulbs in tuberose. Furthermore, the significantly highest value of moisture content was obtained in the GA_3 pretreated spikes placed in 4 % Sucrose + 400 ppm 8-Hydroxyquinoline solution as postharvest treatment followed by 4 % Sucrose + 200 ppm Citric acid in tuberose.

Table 19.2: Effect of pretreatment and holding solutions on moisture content (%) of petals in *Polianthes tuberosa* cv. Prajwal.

Pretreatments	Postharvest treatments	0 day	Days after treatment 3 day	6 day
Control	T_0	91.7 ± 0.152	78.2 ± 0.208	51.1 ± 0.556
(0 ppm)	T_1		82.7 ± 0.152	57.4 ± 0.264
	T_2		83.3 ± 0.153	58.2 ± 0.251
	T_3		83.8 ± 0.111	58.5 ± 0.264
	T_4		85.2 ± 0.173	60.2 ± 0.208
	T_5		87.5 ± 0.173	68.2 ± 0.208
	T_6		89.2 ± 0.153	68.8 ± 0.253
	T_7		89.9 ± 0.153	69.2 ± 0.416
GA3 (150 ppm)	T_0	93.2 ± 0.053	79.2 ± 0.484	54.4 ± 0.208
	T_1		83.5 ± 0.088	59.1 ± 0.152
	T_2		86.4 ± 0.057	60.1 ± 0.208
	T_3		88.4 ± 0.100	60.7 ± 0.360
	T_4		92.1 ± 0.234	64.5 ± 0.264
	T_5		91.3 ± 0.173	70.5 ± 0.264
	T_6		92.2 ± 0.165	71.7 ± 0.305
	T_7		92.7 ± 0.76	72.2 ± 0.173

*Each value indicate mean of 10 replicates

T_0- Control (DDW); T_1- 8% Sucrose; T_2- 8% Sucrose + 200 ppm Citric Acid ; T_3- 8% Sucrose + 400 ppm 8-Hydroxyquinoline; T_4- 200 ppm Citric Acid; T_5- 4% Sucrose; T_6- 4% Sucrose+ 200 ppm Citric acid; T_7- 4% Sucrose+ 400 ppm Hydroxyquinoline

3.4 Vase Life

It is interesting to note that among all the treatments, 4% Sucrose + 400 ppm 8-Hydroxyquinoline was most effective as far as vase life is concerned followed by 4 % Sucrose + 200 ppm Citric acid and 4% Sucrose over control (Table 19.3).

Table 19.3: Effect of pretreatment and holding solutions on vase life of spikes in *Polianthes tuberosa* cv. Prajwal.

Pretreatments Post harvest treatments	Control (0 ppm)	Vase life (days) GA_3 (150 ppm)
T_0	4.82 ± 0.408	5.89 ± 0.71
T_1	5.81 ± 0.707	6.04 ± 0.68
T_2	6.33 ± 0.516	6.34 ± 0.82
T_3	6.51 ± 0.316	7.87 ± 0.64
T_4	7.03 ± 0.316	8.12 ± 0.85
T_5	7.45 ± 0.483	8.23 ± 1.14
T_6	8.07 ± 0.258	8.98 ± 0.65
T_7	8.99 ± 0.258	10.92 ± 0.34

*Each value indicates mean of 10 replicates

T_0- Control (DDW); T_1- 8% Sucrose; T_2- 8% Sucrose + 200 ppm Citric Acid; T_3- 8% Sucrose + 400 ppm 8-Hydroxyquinoline; T_4- 200 ppm Citric Acid; T_5- 4% Sucrose; T_6- 4% Sucrose+ 200 ppm Citric acid; T_7- 4% Sucrose+ 400 ppm Hydroxyquinoline

3.5 Fresh Weight

The pre- and postharvest treatments increased the fresh weight which proved its superiority as compared to control (Table 19.4). It is elucidated from results that a gentle increase in weight of spike was observed up to the 3-day in spikes of both pretreated and untreated bulbs in tuberose. Thereafter, reduction in weight of spikes was observed in spikes of pretreated and untreated bulbs. But, increasing trend continued up to 6 days in spikes obtained from pretreated spikes kept in 4% sucrose+400 ppm 8-Hydroxyquinoline followed by 4% sucrose+ 200 ppm citric acid. The maximum increase was observed in GA_3 pretreated spikes held in 4 % Sucrose + 400 ppm 8-Hydroxyquinoline solution and 4 % Sucrose + 200 Citric acid as postharvest treatments in tuberose.

Table 19.4: Effect of pretreatment and holding solutions on fresh weight of spikes (gm) in *Polianthes tuberosa* cv. Prajwal.

Pretreatments	Post harvest treatments	0 day	Days after treatment 3 day	6 day
Control (0 ppm)	T_0	34.82 ± 0.721	40.21 ± 0.354	30.36 ± 0.481
	T_1		42.45 ± 0.456	31.86 ± 0.621
	T_2		43.43 ± 0.564	32.98 ± 0.500

Pretreatments	Post harvest treatments	0 day	Days after treatment 3 day	6 day
	T_3		43.98 ± 0.223	33.56 ± 0.234
	T_4		45.63 ± 0.446	38.75 ± 0.456
	T_5		47.62 ± 0.408	40.76 ± 0.308
	T_6		48.43 ± 0.623	45.44 ± 0.548
	T_7		50.88 ± 0.762	47.55 ± 0.658
GA_3 (150 ppm)	T_0	43.89 ± 0.412	53.27 ± 0.230	45.34 ± 0.348
	T_1		55.32 ± 0.407	46.22 ± 0.408
	T_2		57.34 ± 0.512	47.38 ± 0.401
	T_3		57.34 ± 0.733	47.88 ± 0.465
	T_4		59.23 ± 0.812	50.43 ± 0.189
	T_5		60.32 ± 0.489	51.67 ± 0.876
	T_6		61.45 ± 0.287	65.32 ± 0.223
	T_7		65.65 ± 1.652	69.33 ± 0.543

Each value indicate mean of 10 replicates

T_0- Control (DDW); T_1- 8% Sucrose; T_2- 8% Sucrose + 200 ppm Citric Acid ; T_3- 8% Sucrose + 400 ppm 8-Hydroxyquinoline; T_4- 200 ppm Citric Acid; T_5- 4% Sucrose; T_6- 4% Sucrose+ 200 ppm Citric acid; T_7- 4% Sucrose+ 400 ppm Hydroxyquinoline

4. Discussion

In the present study, the fresh weight of cut spikes placed in different holding solutions, there was gain in fresh weight of spikes of both GA_3-pretreated and non-treated bulbs in tuberose initially up to 3-day. After progressive days of vase life, there was a significant decrease in fresh weight of petals of *Polianthes tuberosa* cv. Prajwal. The reduction in fresh weight and rate of senescence was comparatively less pronounced in spikes of GA_3-pretreated bulbs than those of non-treated bulbs. Fresh weight of spikes increased with the application of GA_3 comparative to control plants of tuberose. However, there is a limited water uptake in untreated flowers and unbalance between water uptake and transpiration that finally leads to an unrecoverable situation and the premature end of its vase life (Van Meeteren *et al.*, 2001). GA_3 application increased the photosynthetic rate by enhancing chlorophyll content, leaf area and accumulation of larger amount of assimilates for growth of spikes with increased length when placed in different preservatives in 4% sucrose and its combination with citric acid and 8-hydroxyquinoline. However, the other holding solutions could retain appreciable fresh weight of spikes. Sucrose proved to be most promising in delaying shrinkage, freshness and fresh weight and also in combination, it gave better results.

The reduction in fresh weight after 3-days could be due to decreased water uptake or increase in water loss and decrease in membrane stability index. Loss in fresh weight during senescence has also been reported by other workers like Bieleski and Reid (1992); Singh *et al.* (2007) and Serek *et al.* (1994) in lilly, rose and petunia. It was noticed that chemicals used for enhancing vase life also improved water uptake and moisture content of spikes by reducing vascular blockage and ultimately fresh weight. The 8-HQ has been known to possess strong anti-microbial properties that eliminate vascular blockage and enhance water uptake in cut flowers (Burdett, 1970). Treatment of antimicrobial compounds inhibits bacterial proliferation and suppresses decrease in hydraulic conductance of cut roses (Van Doorn *et al.*, 1989). In our study, 4 % Sucrose + 400 ppm 8 Hydroxyquinoline followed by 4 % Sucrose + 200 ppm Citric acid has been found to be the best concentration in making spikes fresh for a long time in tuberose by enhancing water uptake and inhibited bacterial growth. The addition of sugars in vase solutions is essential for a good flower development and facilitates movement and utilization of substances with GA_3. The effect of GA_3 with sucrose was significant on fresh weight of spikes due to the vase solution uptake and energy source. GA_3 along with sucrose has also been suggested to increase fresh weight in cut flowers of gerbera and gladiolus (Emongor, 2004, Singh *et al.*, 2008). Sucrose is source of energy and good respiratory substrate for maintenance of osmotic potential by improving ability to absorb water and maintain turgidity (Varu and Barad, 2008). Therefore, different retentive solutions were used in present study to increase flower's post harvest longevity. Hydoxyquinoline and citric acid act as most potent biocide by inhibiting growth of bacteria in xylem vessels resulting in enhanced water uptake and by reducing pH of vase solution and carbohydrates energy especially sucrose act as important source of energy especially sucrose in plant tissues (Marousky, 1971, Ali and Hassan, 2014 and Jafarpour *et al.*, 2015).

As far as other parameters like absorbed solution by the spikes, vase life and moisture content are concerned, GA_3 as pretreatment was found to have significant effect on all the parameters during postharvest life of spikes of both the crops. Spikes obtained from GA_3 pretreated bulbs absorbed maximum volume of vase solutions. Interaction of GA_3 + 4 % sucrose + 8 Hydroxyquinoline (400 ppm) showed best results in enhancing water uptake in comparison to other post harvest treatments. GA3 along with sucrose has been suggested to induce water uptake in cut flowers of gerbera (Emongor, 2004).

Application of GA_3 along with holding solutions used in present study extended the display or vase life of tuberose cut flowers during the vase period. Since cut flowers are prone to desiccation and oxidative stress during vase life; therefore, GA_3 being growth regulator, could regulate anti-oxidative activity and membrane stability in the petals of spikes under stress. The reduced membrane leakage with GA_3 supply protects the phospholipids and proteins from per oxidation due to the combined effect of GA_3 and preservatives like sucrose, biocides viz. 8-HQ and Citric acid. This phenomenon resulted in reduced membrane leakage and enhanced the vase life of cut flowers. Mackay *et al.*, (2005) showed that gibberellic acid with sucrose increased longevity of cut lupine, increased flower size and opening rate and also improve the quality of petal color. Mutui *et al.*, (2006) showed that gibberellic acid delayed the start of leaf senescence up to 7 days and falling petals by 2 days.

Application of biocides with sugars has become very important preservatives for many cut flowers (Pun and Ichimura, 2003). Application of HQS significantly increased the vase life and fresh weight of rose. But this treatment become more effective when sucrose was added to HQS resulted in longest vase life in *Antirrhinum majus* (Asrar, 2012).

It is concluded that spikes obtained from best concentration of GA_3 when placed in 4% Sucrose+ 400 ppm Hydroxyquinoline followed by 4% Sucrose+ 200 ppm Citric acid, these preservatives performed better in maintaining fresh weight as well as moisture content in tuberose. Vase life was also found to be the maximum in spikes obtained from best concentration of GA_3 placed in 4% Sucrose+ 400 ppm Hydroxyquinoline followed by 4% Sucrose+ 200 ppm Citric acid.

References

Alaey, M., Babalar, M., Naderi, R. and Kafi, M. 2011. Effect of pre- and postharvest salicylic acid treatment on physio-chemical attributes in relation to vase-life of rose cut flowers. *Postharvest Biol. Technol.*, 61(1): 91-94.

Ali, E. and Hassan, F. 2014. Postharvest quality of *Strelitzia reginae* Ait. cut flowers in relation to 8-hydroxyquinoline sulphate and gibberellic acid treatments. *Sci. Agri.*, 5 (3): 97-102.

Asrar, A.W. 2012. Effects of some preservative solutions on vase life and keeping quality of snapdragon (*Antirrhinum majus* L.) cut flowers. *J. Saudi Soc. Agri. Sci.*, 11(1): 29-35.

Bieleski, R.L. and Reid, M.S. 1992. Physiological changes accompanying senescence in ephemeral daylily flower. *Plant Physiol.*, 98: 1042-1049.

Burdett, A.N. 1970. The cause of bent neck in cut roses. *J. American. Soc. Hort. Sci.*, 95: 427-431.

Emongor, V. E. 2004. Effect of gibberellic acid on postharvest quality and vase life of gerbera cut flowers (*Gerbera jamesonii*). *J. Agrn.*, 3: 191-195.

Imsabai, W., Leethiti, P., Netlak, P. and van Doorn, W.G. 2013. Petal blackening and lack of bud opening in cut lotus flowers (*Nelumbo nucifera*): role of adverse water relations. *Postharvest Biol. Technol.*, 79: 32-38.

Iqbal, N., Nazar, R., Iqbal, M., Khan, R., Masood, A. and Khan, N.A. 2011. Role of gibberellins in regulation of source–sink relations under optimal and limiting environmental conditions. *Curr. Sci.*, 100(7): 998-1007.

Jafarpour, M., Golparvar, A.R., Askarikhorasgani, O. and Amini, S. 2015. Improving postharvest vase-life and quality of cut gerbera flowers using natural and chemical preservatives. *J. Central Eurp. Agri.*, 16(2): 199-211.

Mackay, W.A., Sankhla, N. and Davis, T. D. 2004 Gibberellic acid and sucrose delay senescence of cut *Lupinus densiflorus* benth flowers. Proc., pp 104–106.

Macnish, A.J., Leonard, R.T. and Nell, T.A. 2008. Treatment with chlorine dioxide extends the vase life of selected cut flowers. *Postharvest Biol. Technol.*, 50: 197-207.

Mansouri, H. 2012. Salicylic acid and sodium nitroprus-side improve postharvest life of *Chrysanthemums*. *Sci. Hortic.*, 145: 29-33.

Marousky, F.J. 1971.Inhibition of vascular blockage and increased moisture retention in cut roses induced by pH, 8-HQC and sucrose. *J. Amer. Soc. Hort. Sci.*, 96: 38-41.

Muthui, T.M., Emongor, V.E. and Hutchinson, M.J. 2006. The effects of gibberellin$_{4+7}$ on the vase life and flower quality of *Alstroemeria* cut flowers. *Plant Growth Regul.*, 48: 207–214.

Pun, U.K. and Ichimura, K. 2003. Role of sugars in senescence and biosynthesis of ethylene in cut flowers. *Jpn Agri. Res. Quart.*, 37: 219-224.

Rattanawisalanon, C., Ketsa, S. and van Doorn, W.G. 2003. Effect of aminooxyacetic acid and sugars on the vase life of *Dendrobium* flowers. *Post harvest Biol. Technol.*, 29: 93-100.

Rogers, H.J. 2006. Programmed cell death in floral organs: how and why do flowers die? *Ann. Bot.*, 97: 309-315.

Rubinstein, B. 2000. Regulation of cell death in flower petals. *Plant Mol. Biol.*, 44: 303-318.

Serek, M., Jones, R.B. and Reid, M.S. 1994. Role of ethylene in opening and senescence of *Gladiolus sp.* flowers. *J. Am. Soc. Hort. Sci.*, 119: 1014-1019.

Shahri. W. and Tahir, I. 2011. Flower senescence-strategies and some associated events. *Bot. Rev.*, 77: 152-184.

Shimizu-Yumoto, H. and Ichimura, K. 2010. Combination pulse treatment of 1-naphthaleneacetic acid and aminoethoxyvinylglycine greatly improves post-harvest life in cut *Eustoma* flowers. *Postharvest Biol and Technol.*, 56: 104-107.

Singh, A., Kumar, J., Kumar, P. and Singh, V. 2007. Membrane stability and post harvest keeping quality of gladiolus cut spikes as influenced by certain chemicals with sucrose in vase solution. *Indian J. Plant Physiol.*, 12(1): 63-71.

Singh, A., Kumar, J. and Kumar, P. 2008. Effects of plant growth regulators and sucrose on postharvest physiology, membrane stability and vase life of cut spikes of gladiolus. *Plant Growth Regul.*, 55: 221–229.

Van Meeteren, U., van Gelder, H., van Ieperen, W. and Slootweg, C. 2001. Should we reconsider the use of deionised water as control vase solutions? *Acta Hort.*, 543: 257-262.

Van Doorn, W.G., Schurer, K. and De Witte, Y. 1989. Role of endogenous bacteria in vascular blockage of cut rose flowers. *J. Plant Physiol.* 134: 371-381.

Van Ieperen, W., van Meeteren U. and Nijsse, J. 2002. Embo-lism repair in cut flower stems: a physical ap-proach. *Postharvest Biol. Technol.* 25: 1-14.

Van Meeteren, U., van Gelder, H., van Ieperen, W. and Slootweg, C. 2001. Should we reconsider the use of deionised water as control vase solutions? *Acta Hort.*, 543: 257-262.

Van Meeteren, U., Arévalo-Galarza, L. and van Doorn, W.G. 2006. Inhibition of water uptake after dry storage of cut flowers: role of aspired and wound-induced processes in *Chrysanthemum*. *Postharvest Biol. Technol,*. 41(1): 70-77.

Van Meeteren, U., van Gelde, A. and van Ieperen, W. 2005. Effect of growth conditions on postharvest rehydration ability of cut chrysanthemum flowers. *Acta Hortic.*, 669: 287-295.

Varu, D.K. and Barad, A.V. 2008. Effect of harvesting date and floral preservatives on vase life of tuberose (*Polianthes tuberosa* L.) cut flowers during winter season. *J. Orna. Hort.*, 11: 181-187.

Weaver, L.M., Gan S., Quirino B. and Amasino R.M. 1998. A comparison of the expression patterns of several senescence-associated genes in response to stress and hormone treatment. *Plant Mol. Biol.*, 37: 455-469.

Zhang, H., Hu, S.L., Zhang, Z.J., Hu, L.Y., Jiang, C.X. and Wei, S.T. 2011. Hydrogen sulfide acts as a regulator of flower senescence in plants. *Postharvest Biol. Technol.*, 60: 251-257.

Chapter 20

Nano-puerarin Disrupts Decidualization and Migratory Response of Human Endometrial Stromal Cells

Ghungroo Saraswat[1*], Aditya Konar[2], S.N. Kabir[2]

[1]*S.D. (P.G.) College, Panipat, Haryana, India*

[2]*CSIR-Indian Institute of Chemical Biology, Kolkata, India*

**Corresponding Author Email: ghungroo3789@gmail.com*

Abstract

Puerarin, a selective estrogen receptor modulator (SERM), provokes uninterrupted proliferative action of estrogen on endometrial luminal epithelia and prevents implantation in rats, albeit at unacceptably higher doses. We, therefore, developed poly lactic-co-glycolic acid-puerarin nanoparticles (PN), and explored its contragestative potential in rats. Oral administration of PN at 75mg/kg body weight for day 1-2 of pregnancy resulted in complete implantation failure. Decidualization of uterine stromal cells is a critical prerequisite for successful implantation. To further explore the precise mechanism of PN action, decidualization process was studied in human endometrial stromal cell lines (hESCs) by directed migration of the cells in presence and absence of platelet derived growth factor (PDGF-BB) together with immunofluorescence detection of desmin; RT-PCR analysis of IGF binding protein-1 (IGFBP-1), decidual prolactin related protein (dPRL), FOS and STAT3; western blot analysis of extracellular-signal-regulated kinases (ERKs), pERK, myosin light chain (MLC), Ras homolog gene family, member A (RhoA), and Rho-associated protein kinase. PN adversely alters the ERK1/2 and RhoA signalling pathways, and attenuated ESC migration is perhaps one mechanism underlying the PN-inhibited decidualization process. In Conclusion, PN exerts its anti-implantation activity by way of prohibitory effect on endometrial bed preparation and decidualization process. PN is perhaps a prospective candidate molecule that may be further explored for future development of non-steroidal emergency or post-coital contraceptive formulation.

Keywords: *Decidualization, Human Endometrial Stromal Cells, Nano-puerarin*

1. Introduction

Puerarin, a naturally occurring isoflavone C-glycoside extracted from the n-BuOH fraction of the plant *Pueraria tuberose*, possesses significant antifertility activity in rats, mice and hamsters (Shukla, 1996). Our previous work reported that puerarin is a SERM (selective estrogen receptor modulator) that counteracts the proliferative action of oestrogen and disrupts embryo-uterine communication to inhibit implantation in the rats. The future prospect of puerarin, however, seemed limited by its high minimum effective dose to induce 100% implantation failure (MED100) (Saha *et al.*, 2012), reduced solubility, short elimination half-life, and poor bioavailability (Yu *et al.*, 2011). We, therefore, developed poly lactic-co-glycolic acid-puerarin nanoparticles (PN), and explored its contragestative potential in rats. Oral administration of PN at 75mg/kg body weight for day 1-2 of pregnancy resulted in complete implantation failure.

In the rats, the uterus responds to an implanting blastocyst by transforming the endometrium into deciduas (Wooding and Flint, 1994), which is a critical prerequisite for successful implantation. Thus, implantation necessarily requires endometrial decidualization to occur, while induction of decidualization primarily needs the mechanical trauma, which in pregnancy is exerted by the active blastocyst.

Human implantation involves extensive tissue remodelling at the fetal-maternal interface. This is evident that decidualizing endometrial stromal cell (ESC) is inherently motile and invasive, and likely contributes to the highly dynamic processes at the implantation site (Schwenke *et al.*, 2013). Migration can either be random (chemokinesis; triggered by a uniformly present stimulus) or directed (chemotaxis; following an external cue in a concentration gradient) (Petrie *et al.*, 2009). Multiple lines of evidence suggest that chemotactic response of ESC is mediated equally through the Ras/Raf/MEK/ERK and PI3K/AKT (Schwenke *et al.*, 2013) pathways. Transformation of ESC into decidual cells require changes in cellular architecture controlled by Rho family small guanosine triphosphate (GTP)-binding proteins (GTPases). In their active GTP-bound state, Rho GTPases regulate the extent of MLC phosphorylation through its effector protein Rho-associated serine/threonine kinase (ROCK) (Schwenke *et al.*, 2013). In addition to AKT, the Rho GTPase Rac1 is a downstream effector of PI3K (Schiller, 2006) which antagonizes RhoA (Sander *et al.*, 1999) and results in ROCK (Rho Kinase) inhibition. ROCK inhibition in ESC causes a rapid decrease in MLC phosphorylation (Schwenke *et al.*, 2013) to destabilize the stroma and favour ESC migration and decidualization.

Our previous results explained that PN attenuated the signalling pathway involved in the stromal-epithelial interaction that effectively hinders the implantation process to block the establishment of pregnancy in rats. For further insight into the mechanism of PN action, we evaluated the effect of PN on human ESC migration and the underlying pathways.

2. Materials and Methods

2.1 Chemicals

Dulbecco's Modified Eagle's Medium and Ham's F-12 (1:1, v/v; DMEM/Ham's F-12; D2906), MPA, db-cAMP, E_2 and HBEGF (E4643) were purchased from Sigma-

Aldrich Co (St. Louis, MO, USA). ITS + Premix (Cat# 354352) was purchased from BD Biosciences, NJ, USA. Heat-inactivated FBS (cat # 1008214), penicillin–streptomycin solution (cat #15140- 122) were obtained from Thermofisher Scientific, (MA, USA). PI3-kinase activator (Sc 3036),Goat polyclonal desmin (sc-7559), IGFBP1(sc- 6072) , mouse monoclonal β-Actin (Sc-47778), mouse anti-goat IgG secondary antibody FITC (sc-2356), donkey anti-goat IgG secondary antibody-HRP (Sc-2020), donkey anti-rabbit IgG secondary antibody-HRP (Sc- 2313), and goat anti-mouse IgG secondary antibody-HRP (Sc-2031) were obtained from Santa Cruz Biotechnology, Santa Cruz (CA, USA). Rabbit polyclonal p44/42 MAPK (ERK½) (#9102), p-p44/42 MAPK (pERK½) (#4730p), p-p44/42 MAPK (pERK½) (#4695p), Myosin light chain (MLC) (# 3672p), phospho-MLC (pMLC) (# 3671p), rabbit monoclonal Ras homolog gene family, member A (RHOA), (#2117p) and ROCK (#4035s) were obtained from Cell Signaling Technology (MA, USA). Rac activator (CN02-A) was obtained from cytoskeleton, (Denver, USA). All other common chemicals including epidermal growth factor receptor (EGFR) activator (cat# 530536), CHCl3, Tris-HCl, ethyleneglycol tetraacetic acid, KCl, MgCl2, DTT and picric acid were obtained from Merck Specialities Pvt Ltd. (Darmstadt, Germany).

2.2 Culture of Human Endometrial Cell Lines

Human ESC (hESC) (ATCC® CRL-4003Tm) (Krikun *et al.*, 2004) were grown in phenol red free DMEM)/ F12 supplemented with 1% ITS+ Premix, 1% penicillin/ streptomycin solution and 10% heat-inactivated FBS. Steps while working with adherent cell lines:-

Medium change: Occasional change of medium was required in every 2 to 3 days.

Harvesting of cell: Since the cell line was adherent, trypsinization (or breaking of intercellular peptide linkages) was required before any further process to be carried out. For harvesting the cells, 1.0 ml of 0.25% Trypsin-0.53mM EDTA (TE) solution was added to the flask, put in CO_2 incubator for 5 min, and the cells were monitored under inverted microscope until the cell layer dispersed (a long term incubation of cells to TE will damage the intracellular proteins, thereby bringing about death of cells). One ml medium was then added to the flask (medium contains FBS which inhibited the action of TE).

Centrifugation: After harvesting, 1 ml medium was further added and the cells in the medium were transferred into 15 ml sterile centrifuge tubes. Centrifugation was done at 125 x g at 26° C for 5 min. The pellet was then washed with 1 ml medium by gentle pipetting. For further experiment, ESC were cultured in the presence or absence of PN (1nM and 5nM) for 24 h followed by culture for a further period of 72 h in the presence of MPA (100 nM), db-cAMP (0.5 mM), and E2 (10 nM) in phenol red-free DMEM/F12 supplemented with 2% FBS. For the western analysis, in the presence of activator, cells were treated with PI3K (100μg/ml for 24h along with PN treatment), EGFR (100μM for 15 min before PN treatment) and Rac (0.25 unit/ml for 2 min before PN treatment).The cultured cells were used to evaluate the effect of PN on the decidualization, migration potential and the underlying signalling pathway.

2.3 Reverse Transcription Polymerase Chain Reaction (RT-PCR)

Total RNA isolated from pregnant rat uterus using Trizol reagent. RNA (5μg) was reverse transcribed to synthesize single-stranded cDNA using cDNA synthesis for PCR kit from Fermentas (#K1621, Maryland, USA). Amplification of the desirable genes were carried out using specific forward and reverse primers for human PRL, IGFBP1, FOS, STAT3 and β- Actin and a PCR premix solution, according to the manufacturer's instructions (PCR Reagents, Thermo Fisher Scientific Inc.). The primer sequences used were PRL 5'- CATCAACAGCTGCCACACTT -3' (Forward), 5'-CGTTTGGTTTGCTCCTCAAT- 3'(Reverse), IGFBP1 5'-CTATGATGGCTCGAAGGCTC-3' (Forward),5'-TTCTTGTTGCAGTTTGGCAG-3' (Reverse), FOS 5'-AAGGAGAATCCGAAGGGAAA-3'(Forward), 5'-CTTCTCCTTCAGCAGGTTGG-3' (Reverse), STAT3 5'-TCACTTGGGTGGAGAAGGAC -3' (Forward), 5'-GCTACCTGGGTCAGCTTCAG-3'(Reverse) and β-Actin 5'-GCA CCA CAC CTT CTA CAA TG-3'(Forward), 5'-TGC TTG CTG ATC CAC ATC TG-3'(Reverse).

Following an initial heat denaturation at 95°C for 4 min, the PCR cycle was repeated 35 times under the following conditions: 95°C/30 s, 60°C/30 s, 72°C/45s for PRL; 95°C/30s, 60°C/30s, 72°C/30s for IGFBP1; 95°C/30s, 58°C/30s, 72°C/45s for FOS; 94°C/30s, 60°C/45s, 72°C/45s for STAT3 and 94°C/30s, 56°C/45s, 72°C/45s for β-Actin with a final extension at 72°C for an additional 10 min. The amplified PCR products were separated by 1.5% agarose gel electrophoresis and stained with ethidium bromide. The gels were viewed using Gel Doc™ XR+ System (BioRad Laboratories, CA, USA) and quantified using Image J Software, version 1.45. The intensity of the PCR products was normalized by comparison with β-Actin and was calculated as relative amounts.

2.4 Indirect Immunofluorescence

Immunofluorescence assessment of desmin, IGFBP1, ERK1/2, pERK1/2, pMLC and RHOA was done in human endometrial stromal cell lines. Cells were fixed with 4% paraformaldehyde in PBS and permeabilized with 0.2% Triton X-100 in PBS. Then, cells were blocked with blocking solution (4% BSA in PBS). Primary goat polyclonal desmin and IGFBP1 (1:200 dilution in PBS) and rabbit polyclonal ERK1/2, pERK1/2, pMLC and RHOA antibodies were added and incubated overnight at 4°C. Primary antibody was removed with 6x rinses in PBS and incubated with rabbit anti-goat IgG secondary antibody conjugated to FITC and goat anti-rabbit IgG secondary antibody-FITC for 90 min at room temperature in dark. The sections were 6x rinsed in PBS and mounted in anti-fade Vectashield (Vector Laboratories, Inc., Burlingame, CA) mounting medium (Cai *et al.*, 2000). The sections were visualized and photographed (x200) with an Andor spinning disk confocal microscope.

2.5 Western Blot Analysis

hESC were lysed in RIPA buffer (50 mmol/L Tris–Cl pH 7.4, 150 mmol/L NaCl, 1% NP-40, 0.1% SDS, and 0.5% sodium deoxycholate) supplemented with 1 mmol/L PMSF. Protein concentration was determined by the method of Lowry and coworkers (1951). 50 μg protein samples were resolved by SDS-PAGE and transferred to a polyvinylidenedifluoride membrane (Millipore, MA, USA) at 50V for 2 h. After blocking the non-binding sites with skimmed milk in TBST buffer (5mmol/L Tris-HCl, pH-7.4, 136 mmol/L NaCl, and 0.1% Tween-20) for 1 h at room temperature, the blots were incubated overnight at 40°C with respective primary antibodies (dilution 1:2000) against decidualization and migration mediating molecules. The blots were then washed three times with TBST buffer, incubated for 2 h at room temperature with HRP-linked secondary antibody at a dilution of 1:3000. The antibodies were detected with the SuperSignal West Pico Chemiluminescence Substrate (Thermo Scientific, Rockford, USA). The target bands were quantified using Image J software, version 1.45. Comparison with β-Actin normalized the intensity of a specific protein band, and the results have been calculated and expressed as relative amounts.

2.6 Chemotactic Migration Assay

2.6.1 Principle

Cell migration, the movement of cells from one area to another generally in response to a chemical signal, is central to achieving functions such as wound repair, cell differentiation and embryonic development. The transwell migration assay, also called Boyden chamber assay is a commonly used test to study the migratory response of cells. During this assay, cells are placed on the upper layer of a cell permeable membrane and a solution containing the test agent is placed below the membrane. Following an incubation period (3–18 hours), the cells that have migrated through the membrane are stained and counted.

2.6.2 Protocol

Directed migration was monitored by transwell migration assay using non-coated cell culture inserts with 8 mm pores for 24-well recipient plates (BD Biosciences, NJ, USA).

Treated hESC were plated at a density of 3 104 cells/insert in 300 μl DMEM without serum. The lower reservoir received 750 μl DMEM with or without HBEGF (10ng/ml). Each treatment was performed in quadruplicate wells. Migration was allowed to proceed for 18 h. After the incubation, cells were fixed in 3.7% formaldehyde for 2 min and permeabilized in methanol for 20 min. Cell culture inserts were then washed with PBS and cells were stained with Giemsa stain for 15 min. After staining, the non-migrated cells on the upper surface of the filter were removed by scraping. The cells that had migrated to the lower side of the filter were counted in 10 unit field per filter at 20 1.6 X magnification using a Zeiss microscope.

2.7 Statistical Analysis

Data were expressed as mean ± S.E.M. The results were analysed by *t*-test and one-way ANOVA, followed by Newman-Keuls test as applicable, using the GraphPad Prism 3.0 software (GraphPad Software). A P value <0.05 was considered to be statistically significant.

3. Results

3.1 Effect of PN on the Decidualization of hESC

As a marker of decidualization, hESC were evaluated for the expression of IGFBP1, PRL and desmin. Cells treated with PN shows dose-dependent reduction in mRNA expression of IGFBP1and PRL (Fig. 20.1a). Immunofluoresence analysis of desmin and IGFBP1 in hESC is presented in (Fig. 20.1b-c). Compared with higher expression in the decidualized cells, PN treated cells exhibited dose-dependent reduction in the expression of desmin (b) and IGFBP1(c).

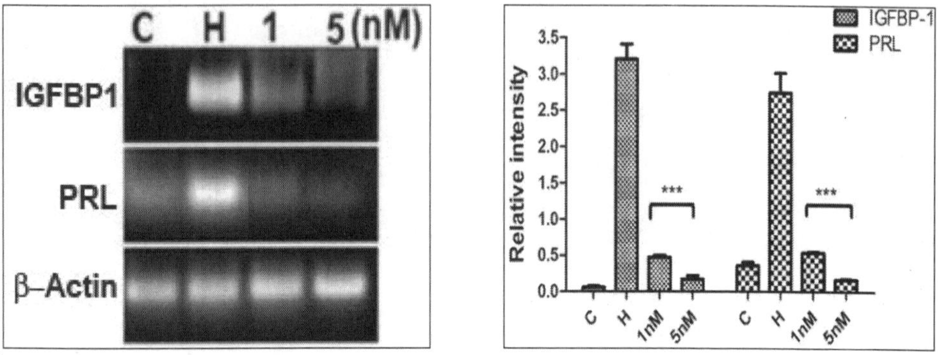

(a)

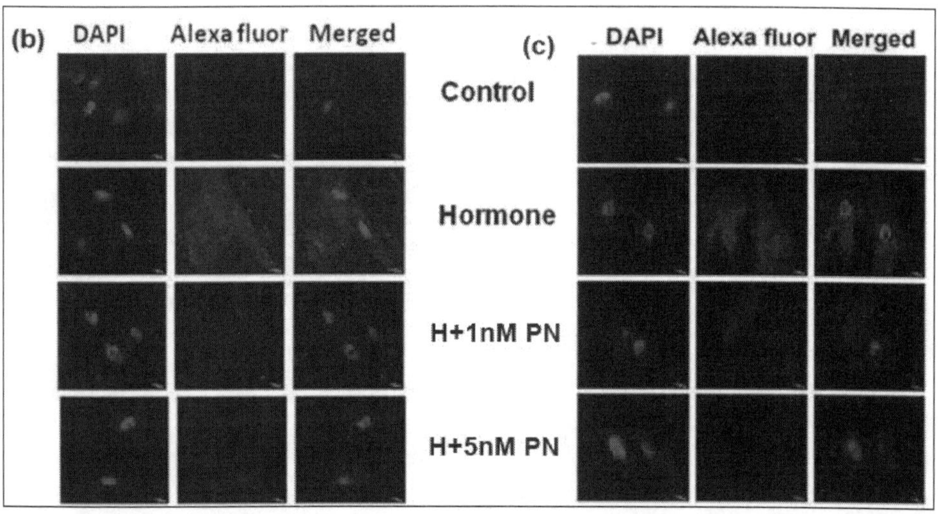

(b)

Fig. 20.1: Effect of PN on the decidualization of hESC. (a) presents the mRNA expression level of PRL and IGFBP1 in hESC as evaluated by RT-PCR. PN treatment dose-dependently inhibited the hormone-induced expression of PRL and IGFBP1 in the cultured stromal cell; (b-c) represents the overlaid immunofluorescence images of cultured stromal cells analyzed for the expression of desmin (b) and IGFBP1(c),

respectively. As compared to the control, PN treatment reduced the expression of desmin and IGFBP1. Each immunofluorescence image is one representative of four to five replicates of the concerned group (x400). The experiments were repeated three times with similar results and data were analyzed by ANOVA. Values represent the mean±S.E.M. of three individual experiments. Figure bars with asterisk vary significantly (*P<0.001).C: Control, H: Hormone (MPA, E2 and db-cAMP).**

3.2 Effect of PN on Expression of ERK

Immunofluoresence analysis of ERK1/2 and pERK1/2 in hESC treated with and without PN are presented in (Fig. 20.2a and 20.2b). Compared to the hormone-induced decidualized cells, PN treated cells exhibited dose-dependent reduction in the expression of pERK1/2(b); however, the total ERK1/2 expression (a) remained identical in all the groups.

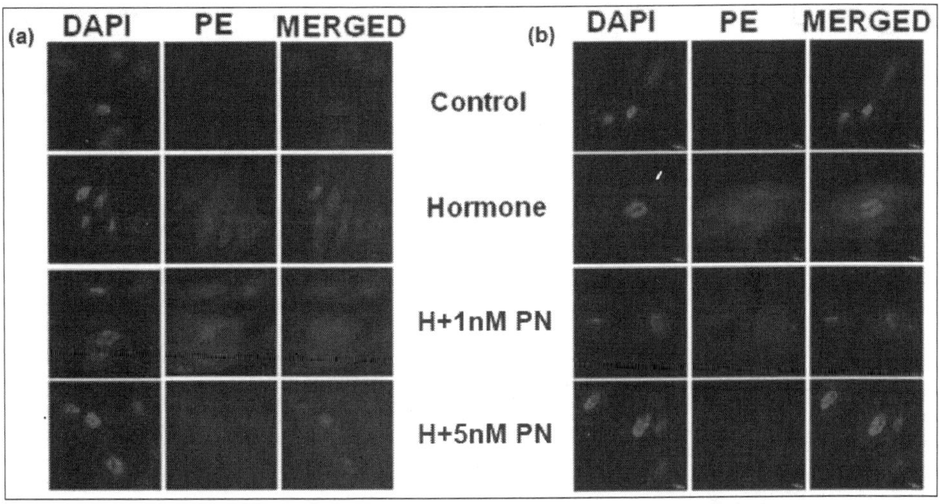

Fig. 20.2: Effect of PN on the expression of ERK1/2 and pERK1/2. The overlaid immunofluorescence Images viewed under confocal microscope represent control, hormone, and PN (at 1 nM and 5 nM) treated hESC cells analysed for the expression of ERK1/2 (a) and pERK1/2(b). Total ERK1/2 was consistently expressed in decidualized cells, however, hormone induced expression of pERK1/2 was downregulated in PN treated hESC. Each image is one representative of four to five replicates of the concerned group (x400). H: Hormone (MPA, E2 and db-cAMP).

3.3 Effect of PN on Expression of FOS and STAT3

To determine if downstream target genes of ERK1/2 are altered by inhibition of ERK1/2 phosphorylation during *in vitro* decidualization of hESC, we examined the expression profiles of the ERK1/2 target genes FOS and STAT3 by RT-PCR. As shown in (Fig. 20.3), expression of FOS and STAT3 were significantly reduced in PN treated decidualized hESC.

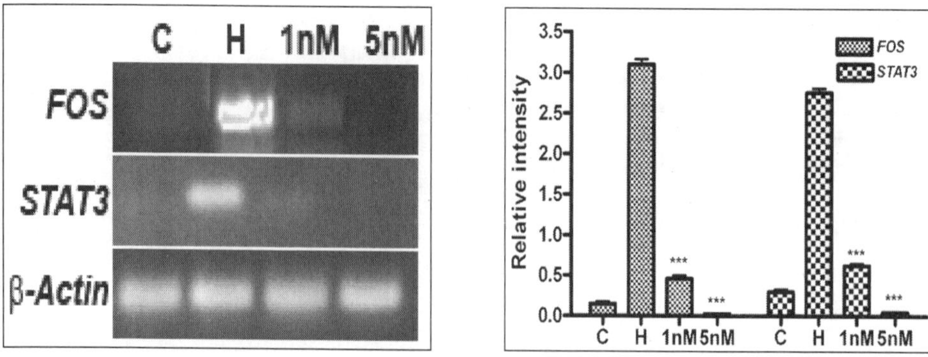

Fig. 20.3: PN effect on the expression of ERK1/2 target genes. The hESC expression of ERK 1/2 target genes, FOS and STAT3 were significantly decreased by treatment with PN during hormone-induced in vitro decidualization. The experiments were repeated three times with similar results and data were analyzed by ANOVA. Values represent the mean±S.E.M. of three individual experiments. Figure bars with asterisk (***) vary significantly ($P<0.001$).C: Control, H: Hormone (MPA, E2 and db-cAMP), 1nM PN+H treated cells, 5nM+H PN treated cells.

3.4 Effect of PN on Migration of Human Endometrial Stromal Cells

We investigated the role of PN on the chemotactic response of the decidualizing endometrial stromal cells (Fig. 20.4) with or without the growth factor, HBEGF. hESC exhibited marked chemotactic response to HBEGF by a significant increase in the migration of the decidualizing cells as compared to the migratory response in decidualized cells without HBEGF. In both condition, pre-exposure to PN dose-dependently inhibited the hormone-induced hESC migration.

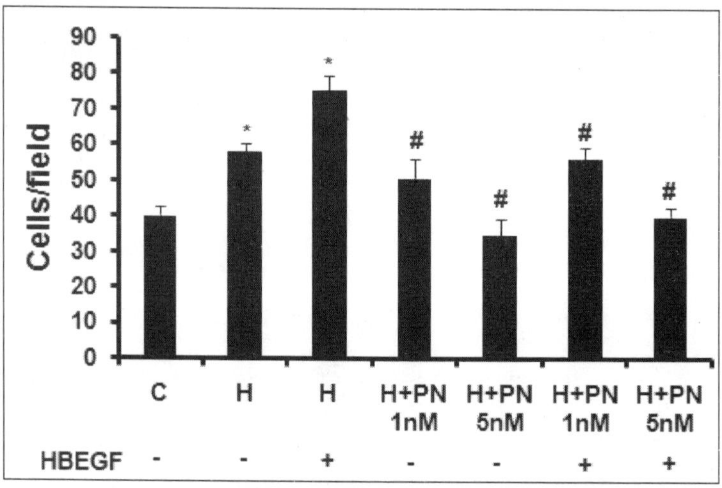

Fig. 20.4: Chemotactic response of hESC. Cells treated with PN (1nM and 5nM) for 24h followed by hormonal treatment (MPA, E2 and db-cAMP) for 72h, were subjected to transwell migration assay. The histogram reflects that PN dose-dependently inhibits hESC migration as induced by decidualizing hormone both in the presence or absence

of HBEGF. The results are expressed as number of migrated cells at the end of 18 h migration period and values represent the mean±S.E.M. of three individual experiments. Data were analyzed by ANOVA. Figure bars with asterisk vary significantly *P<0.05 (C vs. H), #P<0.05 (H vs. PN-treatment). H: Hormone (MPA, E2 and db-cAMP).

3.5 Effect of PN on Migration-Mediating Signalling Pathway

To investigate the PN-induced inhibition of decidual cell migration, we analyzed the expression of ERK, pERK, RHOA, ROCK, MLC and pMLC (Fig. 20.5) at protein levels. Decidualization-inducing hormone upregulated the expression of ERK, pERK and down regulated pMLC, RHOA and ROCK. PN dose-dependently blocked the hormone mediated effects (a). HBEGF potentiated the hormone-induced expression and activation of ERK and downregulated the expression of RHOA, ROCK, and pMLC (b). PN inhibited the HBEGF mediated effects of these migratory molecules.

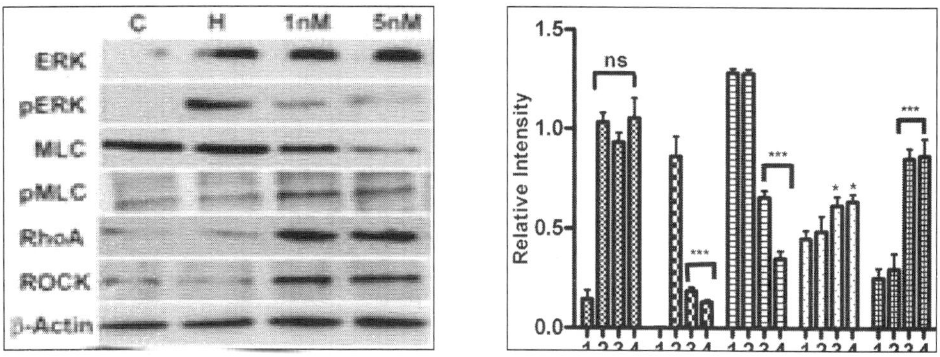

(a)

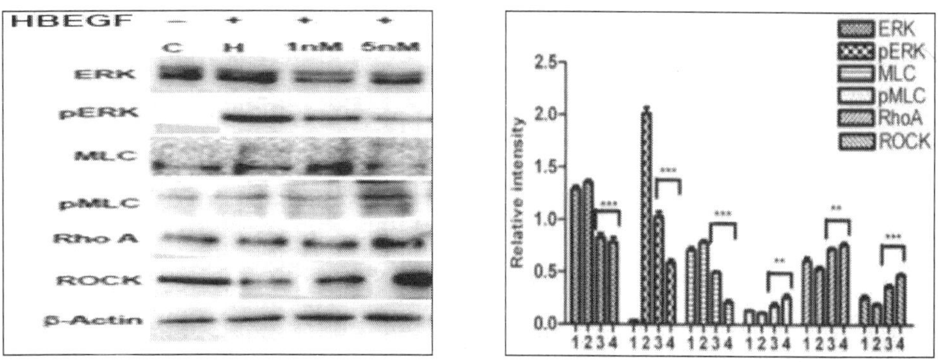

(b)

Fig. 20.5: Effect of PN on the migration pathway. hESC were treated with PN (1nM and 5nM) for 24h followed by hormonal treatment (MPA, E2 and db-cAMP) for 72h. Lysates were then immunoblotted with the relevant antibodies. Equal amounts of protein (50 µg) were loaded onto the gel and its intensity was calculated and compared after normalization with β-Actin. Fig. 5a-b present western analysis data on the stromal cell migrationregulating molecules in hormone-stimulated cultured hESC treated with/

without HBEGF. Values represent the means ± SEMs of three individual experiments analysed by ANOVA. Figure bars with asterisks vary significantly (*P<0.05, **P<0.01, ***P<0.001).The numerical in the X-axis of the graphs represent; 1: untreated control, 2: hormone-primed control, 3: pre-treated with 1nM PN, and 4: pretreated with 5 nM PN.

3.6 Effect of EGFR-, PI3K-, and RAC- activators on PN-inhibited Decidualizing Process

To further evaluate specific target of PN on migration-mediating pathway, hESC were incubated with EGFR, PI3K, and Rac activators (Fig. 20.6). Immunofluorescence analysis revealed that there was a gross increase in the phosphorylation of ERK1/2 (a and b) and decrease in the expression of RHOA (c) and pMLC (d) under the influence of hormone in hESC. Pretreatment with PN inhibited the hormone-induced activation of ERK1/2(b) and upregulated the expression of RHOA (c) and pMLC (d). These dose-dependent PN-mediated effects were reversed in the presence of EGFR activator whereas no change was observed with the PI3K and RAC activators (a, b, c, d). Western blot analysis of the migration signaling molecules in the presence of EGFR (e), PI3K (f) and Rac (g) activators also demonstrated that PN-mediated effect was reversed back in the presence of EGFR activator whereas little or no change observed with the PI3K and RAC activator.

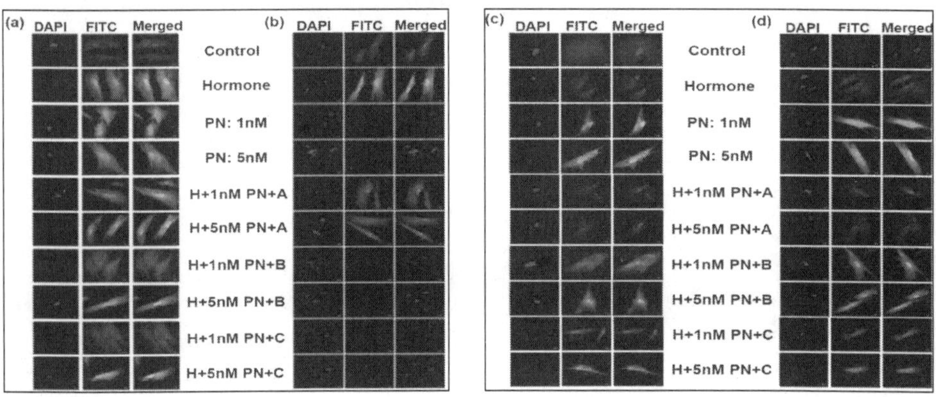

ERK pERK RHOA pMLC

(e)

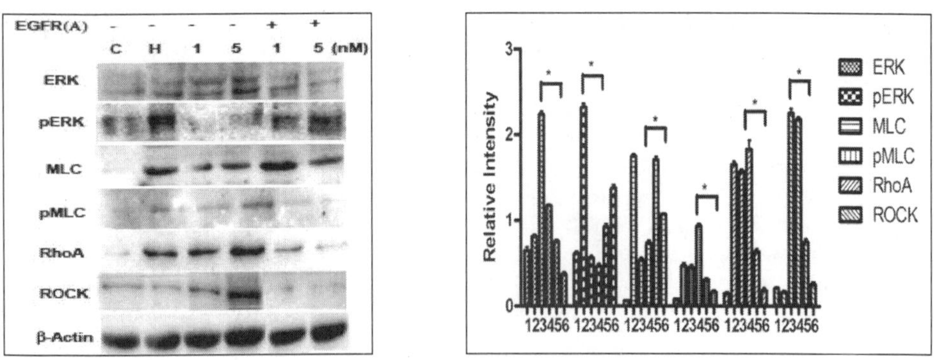

(f)

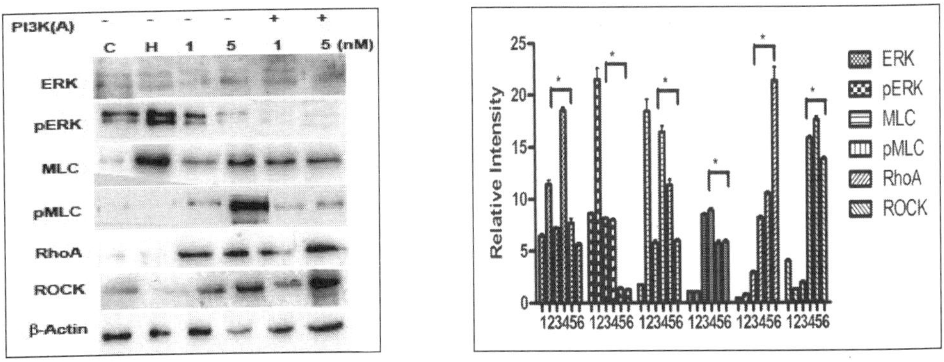

(g)

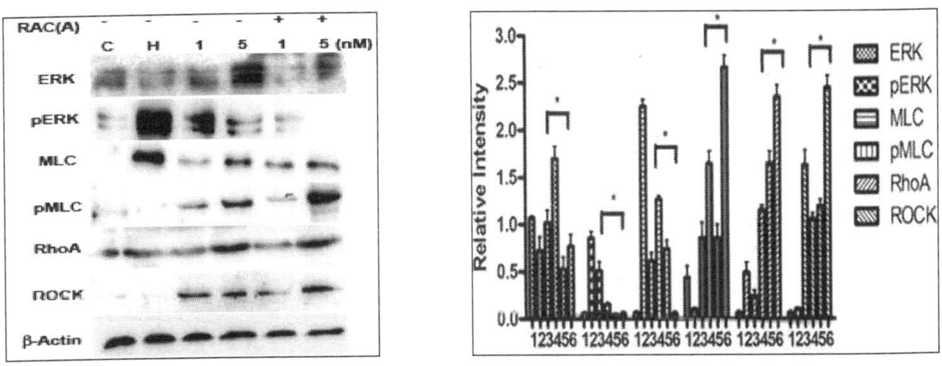

Fig. 20.6: Effect of PN in the presence of EGFR, PI3K and RAC activator. hESC were incubated with EGFR (15min), PI3K (24h) and Rac (2min) activator and treated with PN (1nM and 5nM) for 24h followed by hormonal treatment (MPA, E2 and db-cAMP) for 72h. The overlaid immunofluorescence images (a-d) viewed under confocal microscope represent expression of ERK1/2 (a), pERK1/2(b) RHOA (c) and pMLC (d) in hESC treated without (CONTROL) or with decidualization-inducing hormone (HORMONE or H) and/or PN(1 nM or 5 nM) in the presence of EGFR activator (A), PI3K activator (B), or Rac activator (C). Effect of PN is reversed in the presence of EGFR activator. Fig. 6e, f, and g represent western blots of migration regulatory molecules. EGFR activator (e) reversed the PN inhibited pERK expression and PN-stimulated RHOA, ROCK, and pMLC expression. No significant change in the PN effect was observed with PI3K (f) and Rac activator (g). Each immunofluorescence image is one representative of four to five replicates of the concerned group (x600). Values represent the means ± SEMs of three individual experiments analysed by ANOVA. Figure bars with asterisks vary significantly (*P<0.05). A: EGFR activator, B: PI3K activator, C: Rac activator. 1: untreated control, 2: hormone-primed control, 3: pre-treated with 1nM PN, 4: pre-treated with 5 nM PN, 5: pre-treated with activator and 1nM PN, 6: pre-treated with activator and 5nM PN.

4. Discussion

With a long-term objective to develop a non-steroidal EC formulation, we had earlier investigated the anti-implantation potential of puerarin, a SERM isolated from n-BuOH fraction of the plant *pueraria tuberose.* Administration of puerarin at 300 mg/kg for D1–2 or 125 mg/kg for D1–4 completely disrupted the establishment of pregnancy in rats (Saha *et al.*, 2012). The high dose of puerarin, therefore, should not have been a major key constraint in exploring its future potential as an antifertility agent. However, NP holds tremendous potential as an efficient drug delivery system (Suri *et al.*, 2007). Therefore, we prepared and characterized PLGA-encapsulated PN. Oral doses of PN at 35 mg/kg for D1–D4 or 75 mg/kg for D1–D2 are the minimum effective regimen that completely disrupts pregnancy. Thus, MED100 of PN is ~1/4th of that of puerarin, as reported by Saha *et al.* (2012).

Our previous studies demonstrated that PN disrupted stromal-epithelial cell cross-talk that effectively hindered decidualization and blocked the implantation process in rats. In this study, we have evaluated the effect of PN on the hormone-induced hESC decidualization and migration. Human endometrial stromal cells decidualize *in vitro* in the presence of certain hormones and growth factors (Tseng *et al.*, 1992; Frank *et al.*, 1994) and they are typically characterized by their secretary products, particularly prolactin (PRL), IGFBP1(Tabanelli *et al.*, 1992) and by the intermediate filament protein desmin (Can *et al.*, 1995). We adopted this *in vitro* model to examine the effects of PN. As evidenced by the decreased expression of the decidual marker genes, IGFBP1, PRL and desmin, PN inhibited the artificially-induced hESC decidualization.

ERK1/2 is a member of the well-known MAPK pathway, activated by the upstream Ras/Raf/MEK signalling cascade via phosphorylation on residues threonine 202 and tyrosine 204. This activated form of ERK1/2 (phospho-ERK1/2) can subsequently phosphorylate various transcription factors such as FOS, STAT3 to regulate cellular proliferation and differentiation (Lee *et al.*, 2013). The present study clearly shows that PN inhibits the activation of ERK and expression of its target genes FOS and STAT3.

Migration is one of many cellular functions altered upon decidualization of ESC (Gellersen *et al.*, 2010). In transwell migration assay, ESC motility was enhanced in the decidualizing compared with undifferentiated ESC. Differentiating ESC actively promotes implantation by moving around and encapsulating the conceptus (Weimar *et al.*, 2013). At the time of implantation, local growth factor, HBEGF, participates in modulating endometrial stromal cell dynamics and triggering the chemotaxis of hESC (Schwenke *et al.*, 2013). We demonstrated that PN prohibited with and without HBEGF-stimulated migratory response of the decidualizing ESC in a dose-dependent manner.

ESC migration involves changes in the endometrial cytoskeletal architecture which is determined primarily by actin-myosin II interactions. The phosphorylation of MLC governs the process under the regulation of MLC kinase. The extent of MLC phosphorylation is regulated by RhoA through its effector ROCK. Reports suggest that RHOA activity in hESC is anti-migratory; it induces ROCK activity leading to increased level of phosphorylated MLC (Weimar *et al.*, 2013). We demonstrated that in cultured ESC, PN dose-dependently increased MLC phosphorylation and up-regulated the expression of RHOA and ROCK that inhibited the migratory response of hESC.

Our results also showed that PN attenuated the migration of hESC and underlying signalling pathway; however, to investigate the exact target of PN, we further examined the effect of PN in the presence of the positive key regulators (PI3K, RAC and EGFR) of the migratory pathways. Cells incubated in the presence of activators reveals that PN-mediated antimigratory effect is reversed in the presence of EGFR while PI3K and RAC had little or no reversal effect. These findings strongly suggest that PN mediates its effect through EGFR. It may be relevant in this context that the responsiveness of ESC to HB-EGF increased upon decidualization concomitantly with enhanced expression of EGFR (Schwenke *et al.*, 2013).

In summary, PN adversely alters the ERK1/2 and RhoA signaling pathways, and attenuated ESC migration is perhaps one mechanism underlying the PN-inhibited decidualization process. This study clearly implies that PN may have identical effect on both human and rat implantation.

References

Cai, J., Nelson, K.C., Wu, M., Stemberg, P. and Jones, D.P. 2000. Oxidative damage and protection of the RPE. *Progress in Retinal and Eye Research*, 19: 205–221.

Can, A., Tekeliioglu, M. and Baltaci, A. 1995. Expression of desmin and dimentin intermediate filaments in human decidual cells during first trimester pregnancy. *Placenta*, 16: 261-275.

Frank, G.R., Brar, A.K., Cedars, M.I. and Handwerger, S. 1994. Prostaglandin E2 enhances human endometrial stromal cell differentiation. *Endocrinology*, 134: 258–263.

Gellersen, B., Reimann, K., Samalecos, A., Aupers, S. and Bamberger, A.M. 2010. Invasiveness of human endometrial stromal cells is promoted by decidualization and by trophoblast-derived signals. *Human Reproduction*, 25: 862–873.

Krikun, G., Mor, G., Alvero, A., Guller, S., Schatz, F., Sapi, E., Rahman, M., Caze, R., Qumsiyeh, M. and Lockwood, C.J. 2004. A novel immortalized human endometrial stromal cell line with normal progestational response. *Endocrinology*, 145: 2291-2296.

Lee, C.H., Kim, T.H., Lee, J.H., Oh, S.J., Yoo, J.Y., Kwon, H.S., Kim, Y.I., Ferguson, S.D., Ahn, J.Y., Ku, B.J., Fazleabas, A.T., Lim, J.M. and Jeong, J.W. 2013 Extracellular signal-regulated kinase ½ signaling pathway is required for endometrial decidualization in mice and human. *PLoS ONE*, 8: e75282.

Lowry, O.H., Rosebrough, N.J., Farr, A.L. and Randall, R.J. 1951. Protein measurement with Folin phenol reagent. *Journal of Biological Chemistry*, 193: 265-275.

Petrie, R.J., Doyle, A.D. and Yamada, K.M. 2009. Random versus directionally persistent cell migration. *Nature Reviews Molecular Cell Biology*, 10: 538–549.

Saha, P., Saraswat, G., Chakraborty, P., Banerjee, S., Pal, B.C. and Kabir, S.N. 2012. Puerarin, a selective oestrogen receptor modulator, disrupts pregnancy in rats at preimplantation stage. *Reproduction*, 144: 633-645.

Sander, E.E., Klooster, J.P., Delft S.V., Kammen, R.A.V.D. and Collard, J.G. 1999. Rac downregulates Rho activity: reciprocal balance between both GTPases determines cellular morphology and migratory behavior. *Journal of Cellular Biology*, 147: 1009–1022.

Schiller, M.R. 2006. Coupling receptor tyrosine kinases to Rho GTPases—GEFs what's the link. *Cell Signalling*, 18: 1834-1843.

Schwenke, M., Knöfler, M., Velicky, P., Weimar, C.H.E., Kruse, M., Samalecos, A., Wolf, A., Macklon, N.S., Bamberger, A.M. and Gellerson, B. 2013. Control of human endometrial stromal cell motility by PDGF-BB, HB-EGF and trophoblast-secreted factors. *PLoS ONE*, 8: e54336.

Shukla, S. 1996. Post-coital contraceptive action of *Pueraria tuberosa* DC. extract in rats. *Phytotherapy Research*, 10: 95–99.

Suri, S.S., Fenniri, H. and Singh, B. 2007. Nanotechnology-based drug delivery systems. *J Occup Med Toxicol.*, 2: 16, doi: 10.1186/1745-6673-2-16.

Tabanelli, S., Tang, B. and Gurpide, E. 1992. *In vitro* decidualization of human endometrial stromal cells. *Journal of Steroid Biochemistry and Molecular Biology*, 42: 337–344.

Tseng, L., Gao, J.G., Chen, R., Zhu, H.H., Mazella, J. and Powell, D.R. 1992. Effect of progestin, antiprogestin and relaxin on the accumulation of prolactin and insulin-like growth factor binding protein-1 messenger ribonucleic acid in human endometrial stromal cells. *Biology of Reproduction*, 47: 441–450.

Weimar, C.H., Macklon, N.S., Uiterweer, E.D., Brosens, J.J. and Gellersen, B. 2013. The motile and invasive capacity of human endometrial stromal cells: implications for normal and impaired reproductive function. *Human Reproduction Update*. 19: 542-557.

Wooding, F.B.P. and Flint, A.P.F. 1994. Placentation. In: Lamming, G.E. (Eds.), Marshall's Physiology of Reproduction. Chapman and Hall, London, pp. 233–460.

Yu, A., Wang, H., Wang, J., Cao, F., Gao, Y., Cui, J. and Zhai, G. 2011. Formulation optimization and bioavailability after oral and nasal administration in rabbits of puerarin-loaded microemulsion. *Journal of Pharmaceutical Sciences*, 100: 933–941.

Chapter 21

Green Synthesis of Copper Nanoparticles and its Application as Antimicrobial Agent

Abhishek Mehra, Archit Sharma and Deepak Kumar Malik*

Department of Biotechnology, University Institute of Engineering and Technology, Kurukshetra University Kurukshetra

**Corresponding Author E-mail: deepmolbio@rediffmail.com*

Abstract

The Ocimum sanctum extract was used for the synthesis of Cu nanoparticles. The synthesized copper nanoparticles were charactized by using UV-Vis spectrophotometer and FTIR analysis. The synthesised copper nanoparticles were found to be around 25 nm size through XRD analysis. The antimicrobial study of CuNPs were analysed against both gram negative and gram positive pathogenic bacteria. The maximum zone of inhibition was found more in gram negative bacteria as compared to gram positive bacteria. Therefore, eco-friendly synthesis is a sucessful way to produce copper nanoparticles using leaf extract of Ocimum sanctum.

Keywords: *Copper nanoparticles (CuNPs), Ocimum sanctum, UV-Vis spectrophotometer, FTIR analysis, XRD analysis.*

1. Introduction

An integration of bio and nano techniques for biological synthesis of Nanomaterials is known as bionanotechnology (Krithiga *et al.*, 2013; Sobha *et al.*, 2010; Sahayaraj *et al.*, 2011). Nano-materials show large surface area to volume ratio therefore finds various applications in industries (Abboud *et al.*, 2013; Roco *et al.*, 2003) including consumer products, energy, transportation, cosmetics, pharmaceuticals, antimicrobial agents and agriculture. The research in the field of biosynthesis mainly focused Ag Nanoparticles, Au Nps followed by ZnO and finally CuO Nps (Thombre *et al.*, 2013; Marcia *et al.*, 2013). Copper-based nanoparticles are of great interest because of low cost, availability and properties possessed are similar to that of other metallic Nps (Kulkarni *et al.*, 2013). It finds applications in heat transfer systems as super strong materials, sensors (Cerník *et al.*, 2013). Cu nanoparticles synthesis reported that it is susceptible to oxidation (Majumder *et al.*, 2012). Therefore, most successful chemical synthesis of metallic copper nanoparticles is either carried out in organic or aqueous phase synthesis to avoid potential oxidation of Cu (Varshney *et al.*, 2010). Various chemical and physical methods have been widely used to synthesize Copper nanoparticles. Today's nanotechnology needs to develop alternative method to chemical synthesis which is reliable, non-toxic, clean and eco-friendly. The *Ocimum sanctum*, a medicinal plant was used as a source of bio-reduction and stabilizers for synthesis of Nps.

2. Material and Methods

2.1 Material and Synthesis of Cu Nanoparticles by using *Ocimum Sanctum* Leaf Extracts

The *Ocimum sanctum* leaves were collected from Urban Estate, Kurukshetra, Haryana. The washed leaves (10 g) were cut and boiled with 100 ml of deionized water for 15 min. The resulting filtrates stored at 4 °C for further experiments. For Cu nanoparticles synthesis, three different concentrations of $CuSO_4.5H_2O$ solution were taken (0.5 mM, 1 mM and 2 mM). For each concentration, 100 ml of $CuSO_4.5H_2O$ solutions were taken. In every concentration of $CuSO_4.5H_2O$, 1 ml, 2 ml, 3 ml and 4 ml of *Ocimum sanctum* leaf extract was added respectively and kept overnight at room temperature. The Cu nanoparticles solution was purified by repeated centrifugation at 10,000 rpm for 15 min followed by re-dispersion of the pellet in water. The synthesised nanoparticles were stored at 4 °C for further use.

2.2 Characterisation and Antimicrobial Activity of Green Synthesised Cu Nanoparticles

The spectral analysis of synthesised Cu nanoparticles was done by using UV-Vis spectroscopy. All 12 samples were characterised for the spectral analysis. In FTIR measurements; the bio synthesised Cu nanoparticle solution was centrifuged at 10,000 rpm for 15 min. The dried sample was analyzed on Perkin Elmer Frontier FTIR spectrometer, which operated at a resolution 4 cm^{-1} in the region of 4000-400 cm^{-1}. For X- Ray Diffraction (FTIR) measurements; the dried sample was analyzed on X-ray Diffraction system X'Pert PRO Diffractometer. The antimicrobial activity of green synthesised copper nanoparticles was analysed by well diffusion assay. The antimicrobial activity was checked against nine different microbes: *E. coli* (MTCC 43), *S. aureus* (MTCC 3160), *B. subtilis* (MTCC 121), *P. fluorescens* (MTCC 103), *C. glabrata* (MTCC 3019) *and S. mutans* (MTCC 497), *B.megatherium* (MTCC 428), *C. albicans* (MTCC 183) and *P. aeruginosa* (MTCC 424).

3. Results & Discussion

3.1 Syntheses of Leaf Extract and Copper Nanoparticles

After incubating the different concentrations of copper sulphate solution with different volumes of *Ocimum sanctum* leaf extract, a definite colour change of solution was observed. It was seen that initially the colour of copper sulphate solution was light blue in colour. The colour of solution changed to pale green in colour. After incubating it for one day, the pale green colour changed to dark brown colour which indicates the copper nanoparticle formation. This result is in reference with previous results on copper nanoparticles (Renganathan *et al.*, 2014). The bioreduced Cu nanoparticle solution was centrifuged to get dark brown pellet of Cu nanoparticles. It was observed that the colour intensity of aqueous copper solution treated with 3 ml and 4 ml of leaf extract was higher than other samples with same copper solution concentration. Similarly colour intensity of 2 ml leaf extract treated samples was more as compared to 1 ml treated samples with same copper solution concentration. In case of copper sulphate different concentrations, samples with 1 mM and 2 mM concentrations showed maximum colour change intensity with 3 ml and 4 ml of extract volumes. Further, it was noticed that the amount of pellet obtained in two above mentioned samples was more than the other samples.

3.2 Characterisation of Synthesised Copper Nanoparticles

Curtis *et al.* (1988) reported that copper nanoparticles showed UV-Vis absorption spectra in the range of 510- 640 nm. The UV-Visible spectroscopy analysis of the samples is shown in (Fig. 21.1). The CuNPs formation was confirmed from the peak at 525 nm. The peak value was found to be gradually decreased with increase in particle size. Copper SPR effects decrease with the time because of the oxidation of the synthesized copper nanoparticles. The maximum peak value was observed in 1 mM of copper solution with 3 ml of *Ocimum sanctum* leaf extract. The further increase in extract volume does not significantly increase the peak value, which shows that for 1 mM copper solution with 3 ml of *Ocimum sanctum* extract volume was optimum. In case of 2 mM of copper solution, samples with 2 ml and 3 ml of extract volume showed the same peak value as in case of 1 mM copper solution with same extract volumes. The most optimum concentration of copper solution was found to be of 1 mM with 3 ml of extract because further increase in concentration or extract volume did not resulted in increased peak values. In 0.5 mM copper nanoparticle solution samples, the peak values were least as compared to 1 mM and 2 mM concentrations of copper sulphate solution tested.

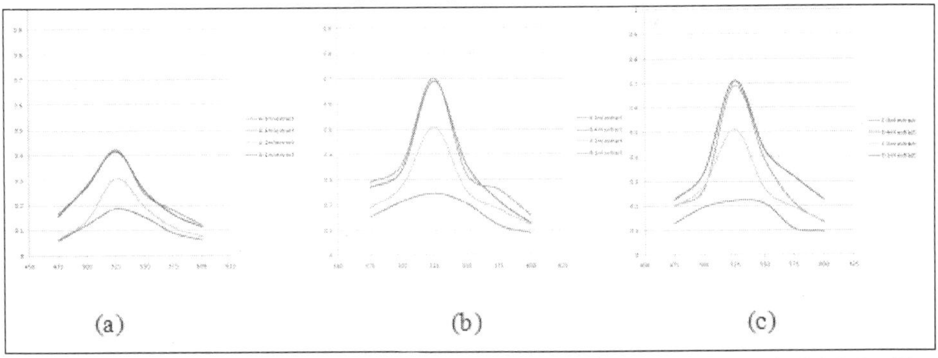

Fig. 21.1: UV-Vis spectroscopy analysis of Cu nanoparticles synthesis (a) 0.5 mM, (b)1 mM, (c) 2 mM copper sulphate solution.

3.3 FTIR Analysis and XRD Analysis

The FTIR analysis and XRD analysis of copper nanoparticles is shown in (Fig. 21.2). It shows five major peaks, i.e, at 3420 cm^{-1}, 2900 cm^{-1} , 1650 cm^{-1}, 1050 cm^{-1} and 620 cm^{-1}. It was observed that the band at 3420 cm^{-1} corresponds to O-H, as also the H-bonded alcohols and phenols. The peak at 2900 cm^{-1} indicates carboxylic acid. The band at 1650 cm^{-1} states amide. The band at 1050 cm^{-1} indicates that C-O stretching alcohols, carboxylic acids, esters and ethers. Therefore, the synthesized nanoparticles were encapsulated by some proteins and metabolites such as terpenoids having functional groups of alcohols, ketones, aldehydes and carboxylic acids. Banerjee *et al.* 2014 reported that *Ocimum sanctum* leaf extract is responsible for the formation of Ag nanopatical. The FTIR analysis confirmed the bioreduction of copper sulphate solution using *Ocimum sanctum* leaf extract. The XRD pattern synthesized Cu nanoparticles by using a 3 ml leaf extract of *Ocimum sanctum* in 1 mM copper solution shows a high crystallinity of Cu sample level with diffraction angles of 74.5 , 50.8 and 43.8 , which correspond to the characteristic face centered cubic (FCC) of copper. Kulkarni *et al.* 2013 reported the copper nanoparticle diffraction angle of 44.8 . The size of the NPs obtained were estimated to be 25 nm using Debye-Scherrer Equation, which may indicate a high surface area, and surface area to volume ratio of the nano-crystals. Average particle size has been estimated by using Debye-*Scherrer* formula: $D= K\lambda /\beta \cos\theta$. Where D is the particle diameter size, Constant [K] = 0.9, β is the full width at half maximum [FWHM], Full width at half maximum in radius [β] = FWHM x /180, $\lambda = 1.5406 \times 10^{-10}$, Cos θ = Bragg angle.

1) $2\theta = 43.64$

   $\beta = (43.80 - 43.46) \times 3.14 / 180 = 0.0059$ radians

   $$D = \frac{0.9 \times 0.1541}{0.0059 \times \text{Cos } 21.82} = 25 \text{ nm}$$

2) $2\theta = 50.80$

   $\beta = (50.92 - 50.54) \times 3.14 / 180 = 0.0066$ radians

   $$D = \frac{0.9 \times 0.1541}{0.0066 \times \text{Cos } 25.4} = 23.26 \text{ nm}$$

3) $2\theta = 74.42$

   $\beta = (74.68 - 74.26) \times 3.14 / 180 = 0.0024$ radians

   $$D = \frac{0.9 \times 0.1541}{0.0024 \times \text{Cos } 37.21} = 24.88 \text{ nm}$$

The different peak values and corresponding size of copper nanoparticles formed of XRD analysis is shown in (Table 21.1). There are three main peak diffraction angle (2θ) values, 43.64, 50.80 and 74.42.

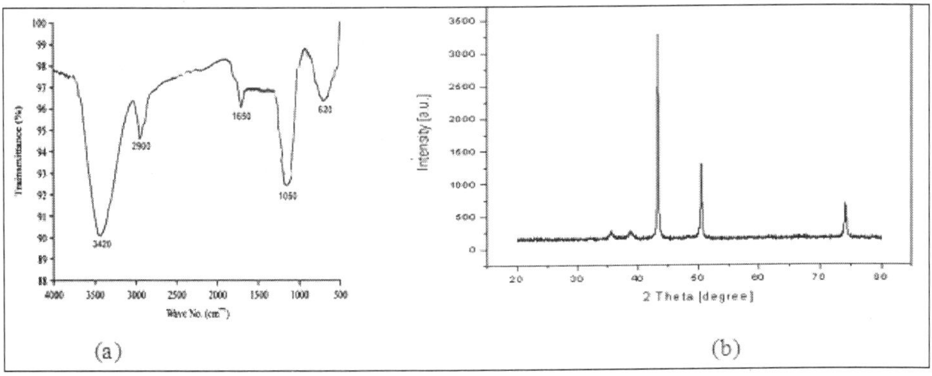

Fig. 21.1: (a) FTIR spectrum, (b) XRD of Copper nanoparticles synthesised through 1 mM copper sulphate solution with 3 ml of *Ocimum sanctum* leaf extract.

Table 21.1: XRD Data values

POSITION (2θ)	WIDTH	SIZE (nm)
43.64	0.0059	25
50.80	0.0066	23.26
74.42	0.0024	24.88

3.4 Antimicrobial Activity of Green Synthesised Copper Nanoparticles

The antimicrobial activity of green synthesised copper nanoparticles using *Ocimum sanctum* leaf extract was highest in case of *E. coli*. The zone of inhibition was moderate in case of *S. aureus, S. mutans* and *C. glabrata*. About 60 µl of copper nanoparticle solution was poured in each well. It was observed that the zone of inhibition was larger in case of well poured with nanoparticle formed from higher volume of leaf extract in 1 mM with 3 ml. All the values of zone of inhibition is given in (Table 21.2). It can be noticed that in *E. coli* the zone of inhibition was 32 ± 1 mm where 3 ml of leaf extract in 1 mM copper sulphate solution was used for Cu nanoparticle synthesis , 21 ± 1 mm for the case of 2 ml of leaf extract in 1 mM copper solution was used for Cu nanoparticle synthesis and 10 ± 1 mm for the case of 3 ml of leaf extract in 0.5 mM copper solution was used for Cu nanoparticle synthesis. It was also observed that there were no zones of inhibition in case of control (1 mM Copper sulphate solution). Renganathan *et al.* 2014 reported that Cu nanoparticle formed by the leaf extract of *Capparis zeylanica* showed antimicrobial activity.

Table 21.2: Zone of Inhibition formed by different samples of Cu nanoparticles.

Culture	Zones of Inhibition (mm)			
	Control- 1 mM CuSO4.5H20	0.5 mM + 3 ml Extract Sample 'A'	1 mM + 2 ml Extract Sample 'B'	1 mM + 3 ml Extract Sample 'C'
E. coli	Nil	10±1	21±1	32±1
S. aureus	Nil	10±1	16±1	26±1
S. mutans	Nil	9±1	14±1	22±1

Culture	Zones of Inhibition (mm)			
	Control- 1 mM CuSO4.5H20	0.5 mM + 3 ml Extract Sample 'A'	1 mM + 2 ml Extract Sample 'B'	1 mM + 3 ml Extract Sample 'C'
C. glabrata	Nil	15±1	19±1	25±1
P. fluorescens	Nil	11±1	16±1	22±1
B. subtilis	Nil	8±1	11±1	15±1
B. megatherium	Nil	7±1	15±1	18±1
Candida albicans	Nil	7±1	15±1	21±1
P. aeruginosa	Nil	9±1	14±1	17±1

4. Conclusion

The copper nanoparticles were successfully synthesized by using *Ocimum sanctum* plant leaves extract and analysed for anti microbial study. The synthesized copper nanoparticles were analyzed by using UV-Vis spectrophotometer and synthesis was confirmed by presence of peak at 525 nm. Through FTIR analysis it was confirmed that bioreduction of aqueous copper sulphate solution was mediated by leaf extract biomolecules. The synthesised copper nanoparticles were found to be around 25 nm size through XRD analysis. The antimicrobial activity of synthesised nanoparticles was analysed which showed its effectiveness in inhibiting various gram positive & negative pathogens and alongwith some fungus pathogens. In the future work, there is need to analyse the cytotoxity activity of synthesised nanoparticles.

References

Abboud, Y., Saffaj, T., Chagraouim, A., Bouari, E., Brouzi, K., Tanane, Ihssane, O. 2013. Biosynthesis, characterization and antimicrobial activity of copperoxide nanoparticles (CONPs) produced using brown alga extract (*Bifurcaria bifurcata*). *App. Nano.* 4: 571.

Banerjee, P., Satapathy, M., Mukhopahayay, A., Das, P. 2014. Leaf extract mediated green synthesis of silver nanoparticles from widely available Indian plants: synthesis, characterization, antimicrobial property and toxicity analysis, Bioresources and Bioprocessing. *Springer Open J.* 13: 14-17.

Cerník, M. 2013. Green synthesis of copper oxide nanoparticles using gum karaya as a biotemplate and their antibacterial application. *Int. J. Nano.* 8: 889–898.

Curtis, A.C., Duff, D.G., Edwards, P.P., Jefferson, D.A., Kirkland, A.I., Wallace, A.S. 1988. *J. Physical Chemistry.* 92: 2211.

Krithiga, Jayachitra, A., Rajalakshmi. 2013. Synthesis, characterization and analysis of the effect of copper oxide nanoparticles in biological systems. *An Ind. J. NanoScience.* 1: 6-15.

Kulkarni, P. 2013. Green Synthesis of Copper Nanoparticles Using *Ocimum Sanctum* Leaf Extract. *Int. J. Chem Std.* 1(3): 1-4.

Majumder, D.R. 2012. Bioremediation: Copper Nanoparticles from Electronic-waste. *Int. J. Eng. Sci. Technol.* 4(10): 4380-4389.

Marcia, R. Salvadori, Luiz, F. Lepre, A. Ro^mulo and Cla'udio A. Oller do Nascimento, 2013. Biosynthesis and Uptake of Copper Nanoparticles by Dead Biomass of *Hypocrea lixii* isolated from the Metal Mine in the Brazilian Amazon Region. *Plos One,* 8: 1-8.

Saranyaadevi, K., Subha, V., Ernest Ravindran, R.S. and Renganathan, S. 2014. Synthesis and Characterization of Copper Nanoparticle using *Capparis Zeylanica* leaf extract. *Int. J. Chem. Tech. Res.* 6(10): 4533-4541.

Roco, M.C. 2003. Broader societal issue of nanotechnology. *J. Nanoparticle Res.,* 5: 181-189.

Sahayaraj, K. and Rajesh, S. 2011. Bionanoparticles synthesis and antimicrobial applications In: Science against microbial pathogens: communicating current research and technological advances. (Ed.) A. Méndez-Vilas, FORMATEX Microbiology Series, Spain, No.3, 1: 228-244.

Sobha, D.K., Surendranath, K., Meena, V., Jwala, K.T., Swetha, N. and Latha, K.S.M. 2010. Emerging trends in nanobiotechnology, *J. Biotech. Mol. Bio. Rev.* 5(1): 001-012.

Stoytcheva, M. 2011. Pesticides Formulations, Effects, Fate. InTech Janeza Trdine, *Croatia.* 9: 1-805.

Thombre, R., Mehta, S., Mohite, J. and Jaisinghani, P. 2013. Synthesis of Silver Nanoparticles and Its Cytotoxic Effect against Thp-1 Cancer Cell Line. Int. J. Pharma Bio Sci., 4(1): 184 – 192.

Varshney, B.S., Gaur, M.S., Pasricha, R. 2010. Characterization of copper nanoparticles synthesized by a novel microbiological method. *JOM: the Journal of the Minerals, Metals & Materials Soc.,* 62: 102-104.

Chapter 22

Carcinogenicity of Nickel: A Review

Veena Vishwakarma[1], Sunil Kumar[1], Bharti Yadav[1], Ranjan Gupta[2], Neeraj K. Aggarwal[3] and Anita Yadav[1*]

[1]Department of Biotechnology, Kurukshetra University, Kurukshetra (Haryana)-136119

[2]Department of Biochemistry, Kurukshetra University, Kurukshetra (Haryana)-136119

[3]Department of Microbiology, Kurukshetra University, Kurukshetra (Haryana)-136119

**Corresponding Author E-mail: ayadav@kuk.ac.in*

Abstract

Nickel is well known for its wide industrial and chemical applications. It is naturally found in environment (air, water and soil) at low levels. But through the time increased industrialization and its commercial use have elevated its level in our environment. The workers in nickel refineries are the most exposed to nickel through inhalation of refinery dust containing nickel particles. Nickel exposure is majorly related to increased risk of lung and nasal cancers. Other organ such as kidneys, heart, spleen, liver, tested and breast are also affected at a lower extent. EPA, US (Environmental Protection Agency, United States) has classified nickel refinery dust under Group-A, human carcinogen, and nickel carbonyl under group B2, probable human carcinogen on the basis of various epidemiological studies. From various in vitro and in vivo studies, it is found that oxidative stress, DNA damage, changes in histone methylation patterns, gene expression and interference with DNA repair system are some molecular mechanisms by which nickel induces genotoxicity and carcinogenicity. Carcinogenetic assessment of nickel compounds indicates the close relation of nickel exposure and lung and nasal carcinogenesis. Molecular mechanism of nickel induced carcinogenesis is a very complex process which requires a proper pathway analysis for the benefit of research on nickel induced carcinogenesis and diseases.

Keywords: *Nickel, carcinogenesis, lung cancer, nasal cancer*

1. Introduction

Nickel is a transition metal and naturally found in environment at low level. It is widely used for industrial and commercial applications particularly in alloys. It is released into environment by windblown dust, nickel refineries, steel industries, municipal incineration, nickel alloy production and other nickel industries and makes it way to human body through the route of respiratory tract, oral ingestion and contact with skin. Elemental nickel (Ni), nickel oxide (NiO), nickel chloride ($NiCl_2$), nickel sulfate ($NiSO_4$), nickel carbonate ($NiCO_3$), nickel monosulfide (NiS), and nickel subsulfide (Ni_3S_2) are different forms of nickel (Fig. 22.1).

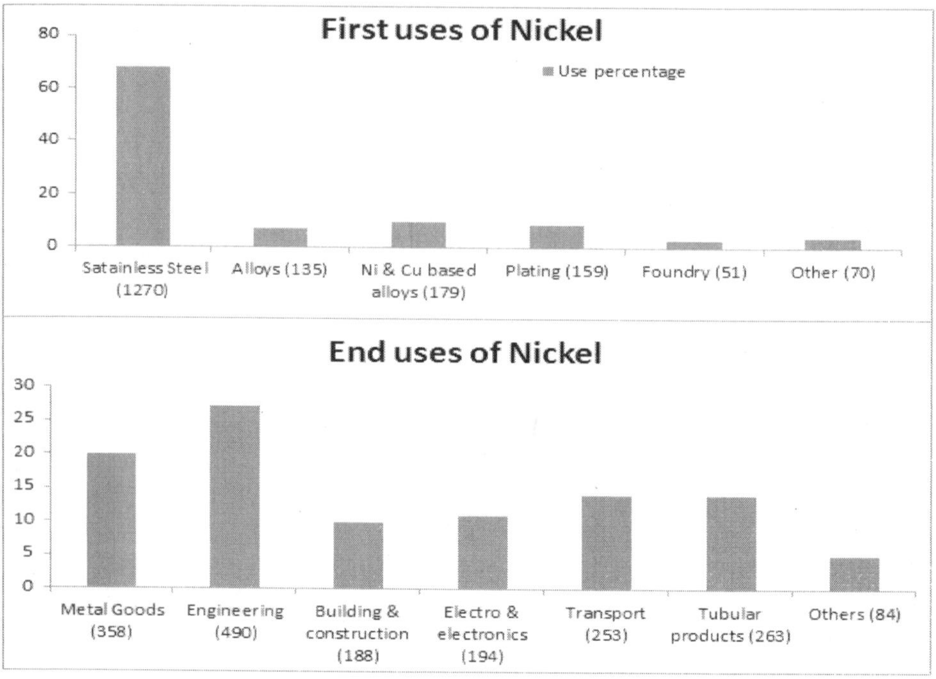

Fig. 22.1: First uses of nickel and End uses of nickel. Figures in parenthesis represent tonnage expressed in thousand tonnes (Adapted from Nickelinstitute.org, 2015).

Nickel and nickel compounds are associated with a multitude of health problems (Morgan *et al.*, 1994) such as contact dermatitis, respiratory diseases (asthma), inflammation, pulmonary fibrosis, and pulmonary edema bronchitis. According to the MAK Value Documentation (2006) various nickel compounds such as metallic nickel, nickel salts, nickel oxides and nickel sulfates, sulfides and carbonates are class 1 category carcinogens.

Nickel refinery dust and nickel sub-sulfide has been classified as group A (carcinogen to humans) and elemental nickel as group 2B (possibly carcinogen to humans) by International agency for research on cancer (IARC), 1990. In the European union directory (Risk Phrase R61 – 'may cause harm to the unborn child') both nickel carbonyl soluble nickel salts (sulphate, chloride, nitrate and carbonate) are classified as toxic to reproduction Category 2 on the basis of their developmental toxicity (Bradberry *et al.*, 1973). It is widely accepted that nickel induced reactive oxygen species (ROS) formation and leads to single strands DNA breaks. Exposure to nickel compounds leads to oxidative stress, DNA strand breaks chromosomal aberrations, and epigenetic changes and interfere with DNA repair processes and thus increases the risk of cancer.

There are many studies which have described the induction of ROS after treatment with $NiCl_2$ in a variety of different cell lines, including normal rat kidney cells and human lymphocytes (Chen *et al.*, 2010). Nickel genotoxicity is linked to the generation of DNA-damaging ROS and inhibition of DNA repair (Kasprzak *et al.*, 2003). Increase in the level of intracellular ROS, lipid peroxidation, hydroxyl radical (OH), and DNA damage have been observed in human lymphocytes after acute exposure to inorganic nickel (Chen *et al.*, 2003). Non occupational population has low risk of nickel exposure due to its low levels in environment but occupational population has higher risk. In IARC report (1990) nickel chloride is observed as an agent that results in DNA breaks, chromosome aberrations, sister chromatid exchange and gene mutations at a light extent in different mammalian cells. Though metallic nickel was able to cause morphological transformation in Syrian hamster embryo cells but didn't cause chromosome aberrations in cultured human lymphocytes. nickel sulfides, subsulfides and nickel oxides also caused morphological transformation in embryo cells of Syrian hamsters. Crystalline forms of nickel sulfides and sub sulfides induced DNA breaks, all types of chromosomal aberrations and gene mutations. Nickel subsulfide was observed to cause mutation in embryo fibroblast line of a transgenic rat (Mayer *et al.*, 1998).

2. Metabolism

Humans absorb about 15-50% of ingested nickel from drinking water after overnight fasting and <15% from ingested food. About 30 % and 20 % of absorbed nickel reaches lungs and circulating blood respectively. A small amount of nickel accumulates in lymph nodes also (Sunderman, 1993). Nickel can be altered and transported as it cannot be destroyed in body. The interference with the physiological processes of manganese, zinc, calcium, and magnesium by nickel is associated with its cytotoxicity (Coogan *et al.*, 1989). Diseases like myocardial infarction, acute stroke and injuries (burn injury) are often associated with altered transport and serum concentrations of nickel. In human serum nickel was found to be associated with albumin, alpha-2 macroglobulin, and ultrafiltrable ligands (eg, histidine) (Nomoto *et al.*, 1988). Studies have found that in non-exposed population the highest nickel concentration is in lungs, thyroid and adrenal glands. Kidneys, heart, brain, liver, spleen and pancreas also contain nickel at lower concentration. Elimination of nickel occurs primarily via the urine and feces. In humans, nickel can also be eliminated via saliva, sweat, milk and tears [International Programme on Chemical Safety (IPCS), 1991; World Health Organization, 2000; Agency for Toxic Cubstances and Disease Registry (ATSDR), 2005]

3. Induction of Oxidative Stress

Nickel enters the nucleus, directly binds to DNA and reacts with H_2O_2 to form reactive nickel oxygen complexes which results in the oxidation of thymine and cytosine residues accompanied by 8-OH-dG formation. Nickel compounds also induce indirect damage through inflammation by stimulating polymorphonuclear leukocytes to produce ROS (Kawanishi *et al.*, 2002).

Use of nickel nanoparticles (NiNPs) in various applications is increasing due to their unique characteristics. Alarifi *et al.* (2013) investigated the cytotoxicity, apoptosis, and DNA damage due to NiNPs in A431 cells. NiNPs induced significant cytotoxicity, generation of reactive oxygen species (ROS) and depletion of glutathione (GSH), significant elevation of lipid peroxidation, catalase, and superoxide dismutase and caspase-3 activity in A431 cells.

Treatment of rat with $NiCl_2$ enhanced the levels of glutathione (GSH), glutathione S-transferases (GST) and glutathione reductase (GR) and decrease in activity of glutathione peroxidase (GPx) (Prasad *et al.*, 2006).

It was observed that nickel (II) induced oxidation of DNA through an oscillatory formation of 8-oxoguanine (8-oxoG) from guanine in DNA (Kelly *et al.*, 2007). This finding is an important step in resolving the mechanism of oxidative DNA damage. 8- oxo- dG is one of the most frequently present DNA damage resulted from DNA oxidation (Souza *et al.*, 2001).

4. Induction of DNA Damage

Nickel compounds can cause a variety of genetic defects in chromatin. To investigate this, different nickel compounds (nickel sulfate, and soluble forms of nickel carbonate hydroxide (NiCH), nickel subsulfide, and nickel oxide) were tested on induction of DNA single-strand breaks (DNA SSBs) in chromosomal and nuclear chromatin of human blood lymphocytes. NiCH and NiO showed significantly higher DNA SSBs in metaphase chromosomes as compared with control samples and nickel sulfate and subsulfides. The genotoxic potency of some nickel compounds studied decreased as follows: NiCH > nickel oxide nickel subsulfide > nickel sulfate in overall study (M'Bemba *et al.*, 2005). Lei *et al.*, 2001 studied nickel transformed BALB/c-3T3 cells to determine the genotoxic lesions of DNA strand breaks and DNA-protein crosslinks. The results obtained indicated that nickel-transformed BALB/c-3T3 cells showed the presence of DNA strand breaks and DNA-protein crosslinks. A high telomerase activity was also found in the NiS and $NiCl_2$-transformed cells. Nickel-refining dusts could induce different degree DNA damage in NIH/3T3 cells in dose and time dependent manner (Zhou *et al.*, 2005). Nickel chloride showed to stimulate L1 retrotransposition. L1 transposons stimulation occurs at the post-transcriptional level, probably during the integration process. The exact mechanism of the L1 stimulation is not determined properly due to the highly complex nature of the effect of nickel on the cell. The stimulation results from a decrease in DNA-repair activities that influence the downstream events of retrotransposition (Farkash *et al.*, 2006). These findings provide a novel mechanism that must be considered when dealing with genomic damage in response to environmental agents.

5. DNA Methylation Changes

Nickel sulfide administration in C57BL/6 mice and p53 heterozygous mice led to development of malignant fibrous histiocytomas and showed very high levels of MAP kinase. It was suggested that hypermethylation of the *p16Ink4a* promoter may be a common feature of nickel-sulfide induced carcinogenesis (Govindarajan *et al.*, 2002). Nickel also damages heterochromatin, thereby silencing the expression of genes located near heterochromatin, which in turn induces an epigenetic loss of histone H4 and H3 acetylation and DNA hypermethylation(Leonard and Jacquet, 1984). Reports show that nickel chloride alters the H3K4 histone methylation by increasing the number of di-methylated H3K9 histones by the inhibition of demethylase enzyme JHDM2 (Chen *et al.*, 2006). Tet-dioxygenases is a class of crucial enzymes for DNA demethylation. Tet proteins catalyze the oxidation of 5-methylcytosine epigenetically. Nickel (II) was found to strikingly decrease the Tet mediated DNA oxidation of 5mC in HEK293T cells and DNA demethylation (Yin *et al.*, 2018).

6. Effect on Gene Expression

According to Costa *et al.*, 2005 nickel compounds also activate hypoxia signaling pathways. Nickel downregulates the tumor suppressor gene p53, activates the proto-oncogene c-myc and induces the AP-1 transcription factor (TF), resulting in increased cellular proliferation and cancer development. The AP-1 TF has been linked to a startling breadth of cellular events including cell transformation, proliferation, differentiation and apoptosis (Henkler *et al.*, 2010).

An investigation on cytotoxic effects of nickel was carried out to check the level of gene expression in nickel exposed L-929 cells (mouse fibroblast cell) with cDNA microarray. It was found that there are 20 upregulated and 19 down regulated genes expressed in the presence of nickel (II). The change in expression levels of genes such as *HsOxip90aa1*, *Ndrg1* and *Saa1* and *Trim23* which are involved in suppression of cell proliferation and differentiation was observed.

Epithelial mesenchymal transition (EMT) is a phenomenon by which epithelial cells gain migratory and invasive characters. EMT associated gene ZEB1 (a key EMT regulatory gene) was found to be irreversibly upregulated epigenetically after Ni exposure in human epithelial BEAS-2B cells. This results in repression of E-cadherin, a cell-cell adhesion molecule in the epithelial cells which leads to cancer and metastasis (Jose *et al.*, 2018).

Exposure to nickel compounds alters histone modification and gene expression globally in peripheral lymphocytes of occupationally exposed population. Eight nickel refinery workers along with ten persons as control group were observed. It was found that nickel exposure caused upregulation of 770 and down regulation of 1986 genes in the PBMC of the workers. Many of these genes were involved in DNA repair, immune and carcinogenetic pathways (Arita *et al.*, 2013).

7. Interference with DNA Repair System

Chen *et al.* (2010) found that nickel inhibits Jumonji-domain-containing histone demethylases demethylase 1A (JMJD1A) and DNA repair enzyme ABH2 by replacing ferrous iron in catalytic centers. There has been accumulating evidence that diverse DNA repair systems including BER and NER are inhibited at low, non-cytotoxic concentrations of Ni (II) (Hartwig *et al.*, 2002) (Fig. 22.2). In presence of H_2O_2 nickel binds to DNA repair enzymes and generate oxygen-free radicals to cause protein degradation in situ. This irreversible damage to the proteins involved in DNA repair, replication, recombination, and transcription could be important for the toxic effects of nickel (Lynn *et al*,. 1997). Nickel exposure in both tumorigenic and non-tumorigenic lung cells leads to transcription down regulation of mismatch repair (MMR) protein MLH1 and homology dependent DNA double strand break repair (HDR) proteins RAD51, BRCA1 and FANCD2 (Scanlon *et al.*, 2017).

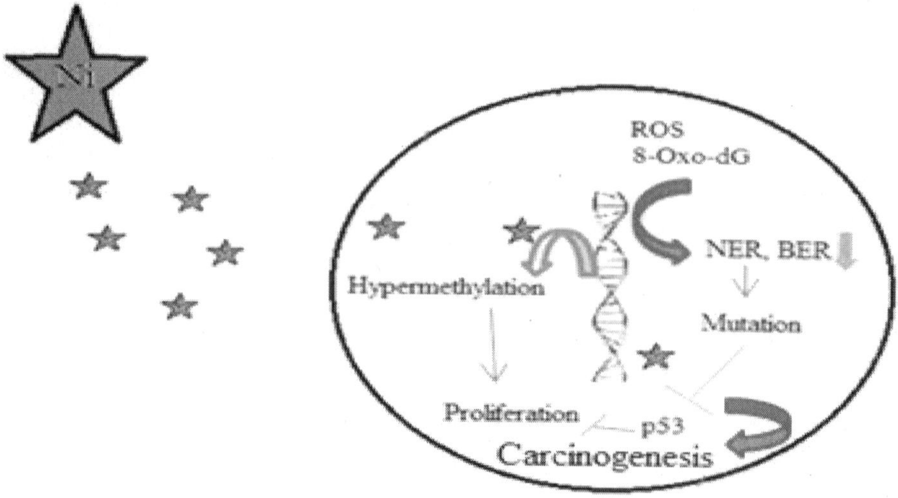

Fig. 22.2: Probable mechanism of nickel induced carcinogenesis. Nickel induced mutations by ROS, 8-oxo-dG generation which furthers downregulates the DNA repair mechanisms [Nucleotide Excision repair (NER) and base excision repair (BER)] that may lead to carcinogenesis.

8. Carcinogenetic Assessment

There is a strong correlation with respiratory cancer and nickel exposure. It has been suggested that suppression of immunological natural killer (NK) cell activity and interferon production may also be involved in cancer development (Shen *et al.*, 1994).

The first confirmation of respiratory cancer in workers exposed to high nickel concentration in environment was reported in Clydach, Wales, 1933. Nickel Refinery Company at Clydach, Wales, U.K., operating since 1903 was using process of nickel refining which formed dust containing nickel oxide (NiO) and nickel arsenide (Ni_5As_2) and other minor components. Before 1923 there was higher incidence of respiratory and nasal cancers (85 nasal mostly ethmoidal and 280 lung cancers mostly squamous carcinomas of bronchi) among the refinery workers. In 1923 nickel refining process was changed which reduced the level of Ni_5As_2 particles to 2.5%. This led to a great decrease in the number of respiratory cancers after 1923 (International Committee on Nickel Carcinogenesis in Man, 1990; Draper, 1997; Heller *et al.*, 2009). Doll and Peto (1981) estimated that occupational exposure of nickel is responsible for 15% and 5% lung cancer in men and women respectively.

In Finland, 1155 workers working in nickel smelter (566 workers), repair shop (239 workers), or refinery (418 workers) for 25 years showed a significant increased risk in nasal and lung carcinogenesis during exposure to nickel compounds (Anttilla *et al.*, 1998). It is concluded that nickel sulfate at a level below 0.5 mg/m^3 along with low levels of other nickel components were responsible for the increased risk of nasal cancers. Grimsrud *et al.* (2003) confirmed a strong dose-related and exposure time on risk in workers in production departments and cumulative exposure to nickel, most clearly seen for water-soluble nickel. Smoking of tobacco and exposure to nickel has an effect on lung cancer in a synergistic way and exposed population also has additional risk of nasal cancer (Anderson *et al.*, 1996). Accumulation of nickel in 60 lung tumor tissues was observed which was significantly higher than the non-cancerous lung tissues. This indicates that increased level of nickel has a close relation with lung cancer (Kuo *et al.*, 2005). In another study on female breast cancer it was found that level of nickel in the cancerous tissue (0.59-4.80 g/g) was higher than the normal tissues (0.37-2.85g/g) (Romanowicz *et al.*, 2011).

Soluble salts of nickel can be utter carcinogens (Diwan *et al.*, 1992) or they can initiate carcinogenesis (Kasprzak *et al.*, 1990) at tissue sites far away from the site of exposure. However due to insufficient exposure data carcinogenetic assessment of soluble nickel compounds is rather difficult.

9. Conclusion

Industrial development has increased a risk of nickel exposure by the production and wide utilization of industrial and commercial products containing nickel. Direct or indirect exposure of nickel induces deregulation of intracellular processes through complex pathway. These processes might be candidates for markers of nickel induced carcinogenesis. Furthermore, complete understanding of these complex mechanisms by pathway analysis will be beneficial to research on the nickel induced cancers and diseases.

References

Agency for Toxic Substances and Disease Registry (ATSDR), 2005. Toxicological profile for nickel. US Department of Health and Human Services. Atlanta, US. https://www.atsdr.cdc.gov/toxprofiles/tp15.pdf.

Alarifi, S., Ali, D., Alakhtani, S., Al Suhaibani, E.S. and Al-Qahtani, A.A. 2014. Reactive oxygen species-mediated DNA damage and apoptosis in human skin epidermal cells after exposure to nickel nanoparticles. *Biological Trace Element Research*, 157: 84-93.

Andersen, A., Berge, S.R., Engeland, A. and Norseth, T. 1996. Exposure to nickel compounds and smoking in relation to incidence of lung and nasal cancer among nickel refinery workers. *Occupational and Environmental Medicine*, 53: 708-713.

Anttila, A., Pukkala, E., Aitio, A., Rantanen, T. and Karjalainen, S. 1998. Update of cancer incidence among workers at a copper/nickel smelter and nickel refinery. *International Archives of Occupational and Environmental Health*, 71: 245-250.

Arita, A., Muñoz, A., Chervona, Y., Niu, J., Qu, Q., Zhao, N., Ruan, Y., Kiok, K., Kluz, T., Sun, H. and Clancy, H.A. 2013. Gene expression profiles in peripheral blood mononuclear cells of Chinese nickel refinery workers with high exposures to nickel and control subjects. *Cancer Epidemiology and Prevention Biomarkers*, 22: 261-269.

Bradberry, S.M., Beer, S.T., and Vale, J.A. 1973. United Kingdom Poison Information Documents (UKPID). Monograph nickel carbonyl. https://www.inchem.org/documents/ukpids/ukpids/ukpid68.htm.

Chen, C.Y., Lin, T.K., Chang, Y.C., Wang, Y.F., Shyu, H.W., Lin, K.H. and Chou, M.C. 2010. Nickel (II)-induced oxidative stress, apoptosis, G2/M arrest, and genotoxicity in normal rat kidney cells. *Journal of Toxicology and Environmental Health, Part A*, 73: 529-539.

Chen, C.Y., Wang, Y.F., Huang, W.R. and Huang, Y.T. 2003. Nickel induces oxidative stress and genotoxicity in human lymphocytes. *Toxicology and Applied Pharmacology*, 189: 153-159.

Chen, H., Giri, N.C., Zhang, R., Yamane, K., Zhang, Y., Maroney, M. and Costa, M. 2010. Nickel ions inhibit histone demethylase JMJD1A and DNA repair enzyme ABH2 by replacing the ferrous iron in the catalytic centers. *Journal of Biological Chemistry*, 285: 7374-7383.

Chen, H., Ke, Q., Kluz, T., Yan, Y. and Costa, M. 2006. Nickel ions increase histone H3 lysine 9 dimethylation and induce transgene silencing. *Molecular and Cellular Biology*, 26: 3728-3737.

Coogan, T.P., Latta, D.M., Snow, E.T., Costa, M. and Lawrence, A. 1989. Toxicity and carcinogenicity of nickel compounds. *Critical Rreviews in Toxicology*, 19: 341-384.

Costa, M., Davidson, T.L., Chen, H., Ke, Q., Zhang, P., Yan, Y., Huang, C. and Kluz, T. 2005. Nickel carcinogenesis: epigenetics and hypoxia signaling. *Mutation Research/Fundamental and Molecular Mechanisms of Mutagenesis*, 592: 79-88.

Diwan, B.A., Kasprzak, K.S. and Rice, J.M. 1992. Transplacental carcinogenic effects of nickel (II) acetate in the renal cortex, renal pelvis and adenohypophysis in F3447/NCr rats. *Carcinogenesis*, 13: 1351-1357.

Doll, R. and Peto, R. 1981. The causes of cancer: quantitative estimates of avoidable risks of cancer in the United States today. *Journal of the National Cancer Institute*, 66: 1192-1308.

Draper, M.H. 1997. A Re-assessment of respiratory cancers at the Clydach nickel refinery: New evidence of causation. *Carcinogenicity of Inorganic Substances: Risks from Occupational Exposure*, p.181.

Farkash, E.A. and Prak, E.T.L. 2006. DNA damage and L1 retrotransposition . Journal of *Biomedicine and Biotechnology*, pp.1- 8. https://doi.org/10.1155/JBB/2006/37285.

Govindarajan, B., Klafter, R., Miller, M.S., Mansur, C., Mizesko, M., Bai, X., LaMontagne Jr, K. and Arbiser, J.L. 2002. Reactive oxygen-induced carcinogenesis causes hypermethylation of p16 (Ink4a) and activation of MAP kinase. *Molecular Medicine*, 8: 1.

Grimsrud, T.K., Berge, S.R., Martinsen, J.I. and Andersen, A. 2003. Lung cancer incidence among Norwegian nickel-refinery workers 1953–2000. *Journal of Environmental Monitoring*, 5: 190-197.

Hartwig, A., Asmuss, M., Ehleben, I., Herzer, U., Kostelac, D., Pelzer, A., Schwerdtle, T. and Bürkle, A. 2002. Interference by toxic metal ions with DNA repair processes and cell cycle control: molecular mechanisms. *Environmental Health Perspectives*, 110(5): 79.

Heller, J.G., Thornhill, P.G. and Conard, B.R. 2009. New views on the hypothesis of respiratory cancer risk from soluble nickel exposure; and reconsideration of this risk's historical sources in nickel refineries. *Journal of Occupational Medicine and Toxicology*, 4(1): 23.

Henkler, F., Brinkmann, J. and Luch, A. 2010. The role of oxidative stress in carcinogenesis induced by metals and xenobiotics. *Cancers*, 2(2): 376-396.

International Agency for Research on Cancer (IARC). 1990. Monographs on the evaluation of carcinogenic risks to humans, 49: 1–648. PMID:2232124.

International Programme on Chemical Safety (IPCS). 1981. Nickel Environmental Health Criteria 108l. http://inchem.org/documents/ehc/ehc/ehc108.htm

Jose, C.C., Jagannathan, L., Tanwar, V.S., Zhang, X., Zang, C. and Cuddapah, S. 2018. Nickel exposure induces persistent mesenchymal phenotype in human lung epithelial cells through epigenetic activation of ZEB1. *Molecular Carcinogenesis*, pp. 1–13. https://doi.org/10.1002/mc.22802

Kasprzak, K.S., Diwan, B.A., Konishi, N., Misra, M. and Rice, J.M. 1990. Initiation by nickel acetate and promotion by sodium barbital of renal cortical epithelial tumors in male F344 rats. *Carcinogenesis*, 11(4): 647-652.

Kasprzak, K.S., Sunderman, F.W. and Salnikow, K. 2003. Nickel carcinogenesis. *Mutation Research/Fundamental and Molecular Mechanisms of Mutagenesis*, 533: 67-97.

Kawanishi, S., Oikawa, S., Inoue, S. and Nishino, K. 2002. Distinct mechanisms of oxidative DNA damage induced by carcinogenic nickel subsulfide and nickel oxides. *Environmental Health Perspectives*, 110(5): 789.

Kelly, M.C., Whitaker, G., White, B. and Smyth, M.R. 2007. Nickel (II)-catalysed oxidative guanine and DNA damage beyond 8-oxoguanine. *Free Radical Biology and Medicine*, 42: 1680-1689.

Kuo, C.Y., Wong, R.H., Lin, J.Y., Lai, J.C. and Lee, H., 2006. Accumulation of chromium and nickel metals in lung tumors from lung cancer patients in Taiwan. *Journal of Toxicology and Environmental Health, Part A*, 69: 1337-1344.

Lei, Y.X., Chen, J.K. and Wu, Z.L. 2001. Detection of DNA strand breaks, DNA-protein crosslinks, and telomerase activity in nickel-transformed BALB/c-3T3 cells. *Teratogenesis, Carcinogenesis, and Mutagenesis*, 21: 463-471.

Leonard, A., Gerber, G.B., Jacquet, P. and Lauwerys, R.R. 1984. Carcinogenicity, mutagenicity, and teratogenicity of industrially used metals. *Mutagenicity, Carcinogenicity, and Teratogenicity of Industrial Pollutants*. pp. 59-126.

Lynn, S., Yew, F.H., Chen, K.S. and Jan, K.Y. 1997. Reactive oxygen species are involved in nickel inhibition of DNA repair. *Environmental and Molecular Mutagenesis*, 29: 208-216.

MAK Value Documentation. 2006. Nickel and its compounds (in the form of inhalable dusts/aerosols). https://doi.org/10.1002/3527600418.mb744002vere0022.

M'Bemba-Meka, P., Lemieux, N. and Chakrabarti, S.K. 2005. Nickel compound-induced DNA single-strand breaks in chromosomal and nuclear chromatin in human blood lymphocytes in vitro: role of oxidative stress and intracellular calcium. *Mutation Research/Genetic Toxicology and Environmental Mutagenesis*, 586: 124-137.

Mayer, C., Klein, R.G., Wesch, H. and Schmezer, P. 1998. Nickel subsulfide is genotoxic in vitro but shows no mutagenic potential in respiratory tract tissues of BigBlue™ rats and Muta™ Mouse mice in vivo after inhalation. *Mutation Research/Genetic Toxicology and Environmental Mutagenesis*, 420: 85-98.

Morgan, L.G. and Usher, V. 1994. Health problems associated with nickel refining and use. *The Annals of Occupational Hygiene*, 38: 189-198.

Nickelinstitute.org, 2015. First uses of nickel https://www.nickelinstitute.org/en/NickelUseInSociety/AboutNickel/FirstAndEndUsesofNickel.aspx.

Nomoto, S. and Sunderman, F.W. 1988. Presence of nickel in alpha-2 macroglobulin isolated from human serum by high performance liquid chromatography. *Annals of Clinical & Laboratory Science*, 18: 78-84.

Prasad, L., Khan, T.H., Jahangir, T. and Sultana, S. 2006. Chemomodulatory effects of *Terminalia chebula* against nickel chloride induced oxidative stress and tumor promotion response in male Wistar rats. *Journal of Trace Elements in Medicine and Biology*, 20: 233-239.

Romanowicz-Makowska, H., Forma, E., Bry , M., Krajewska, W.M. and Smolarz, B. 2011. Concentration of cadmium, nickel and aluminium in female breast cancer. *Polish Journal of Pathology*, 62: 257-261.

Scanlon, S.E., Scanlon, C.D., Hegan, D.C., Sulkowski, P.L. and Glazer, P.M. 2017. Nickel induces transcriptional down-regulation of DNA repair pathways in tumorigenic and non-tumorigenic lung cells. *Carcinogenesis*, 38: 627-637.

Shen, H.M. and Zhang, Q.F. 1994. Risk assessment of nickel carcinogenicity and occupational lung cancer. *Environmental Health Perspectives*, 102: 275.

Souza-Pinto, N.C., Eide, L., Hogue, B.A., Thybo, T., Stevnsner, T., Seeberg, E., Klungland, A. and Bohr, V.A. 2001. Repair of 8-oxodeoxyguanosine lesions in mitochondrial dna depends on the oxoguanine dna glycosylase (OGG1) gene and 8-oxoguanine accumulates in the mitochondrial dna of OGG1-defective mice. *Cancer Research*, 61: 5378-5381.

Sunderman Jr, F.W., 1993. Biological monitoring of nickel in humans. *Scandinavian Journal of Work, Environment & Health*, pp.34-38.

World Health Organization (WHO). 2000. Air quality guidelines for Europe. http://wedocs.unep.org/handle/20.500.11822/8681

Yin, R., Mo, J., Dai, J. and Wang, H. 2018. Nickel (ii) inhibits the oxidation of DNA 5-methylcytosine in mammalian somatic cells and embryonic stem cells. *Metallomics*, 10: 504-512.

Zhou, C.L., WU, Y.H. and Jiao, G.Y. 2005. Application of comet assay to nickel refining fumes-Induced DNA damage. *Industrial Health and Occupational Diseases*, 6: 7.

Chapter 23

Eutrophication: A Causative Factor Behind Avian Species Decline at Regional Wetlands of Ambala, Haryana (India)

Sarita Rana

Department of Zoology, University College, Kurukshetra University Kurukshetra

E-mail: saritarana20@gmail.com

Abstract

Wetlands are important habitats for avian species as they provide wide variability of feeding, roosting and nesting sites for the birds. Also, wetlands are important courtship grounds for variable wetland avian species. Diversity of avian species predicts the quality of wetlands as more heterogeneous habitat inhabits large number of avian species. The study was conducted for the period of three years regarding avian species census of three different wetlands of district Ambala, Haryana i.e. Lord Mahavir park lake, Ambala City; Pond of Village Sultanpur, Ambala City and Rani Ka Talab, Ambala Cantt, Haryana (India). Observations were regularly recorded regarding arrival and departure time of migratory birds. Census regarding resident avian species was also made to predict the variability of avian diversity in three different wetlands. The vegetation structure of three habitats was entirely different from each other. This variation in habitat in turn leads to variation in avian species diversity. Species diversity in Lord Mahavir Lake was found to be maximum in initial years of survey but declined with passage of time due to increased eutrophication in the lake by Eichhornia. Similar scenario was observed in pond of Sultanpur village but it was more affected by Salvinia and Azolla. Least eutrophication was found in Rani Ka Talab but avian diversity in this wetland was low due to homogenous nature of habitat. Infestation of these wetlands by weeds had forced avian fauna to migrate from the habitat and also leads to loss and mortality of large number of avian species. Increased eutrophication in these lakes had changed open water avian habitats to dry scrublands which need to be documented and urgent conservation measures are required to protect the migratory and resident avian species.

Keywords: *Avian diversity, Eutrophication, wetlands, Eichhornia, migratory species*

1. Introduction

Wetlands in India provide a unique habitat to both aquatic flora and fauna and it also provide refuge to numerous migratory species. Out of 310 species of wetland birds found in India, almost half of these are migratory and visit India from cold areas of different part of China, Russia, central Asia, Tibet and from across the entire range of the Himalaya (Gupta *et al.*, 2012). Wetlands are complex and productive ecosystems. Wetlands are important habitats for birds, which use them for feeding, roosting, nesting and rearing young ones. Water birds are only the most prominent group that attract public to wetlands, but also are good bio-indicators and useful models for studying a variety of environmental problems (Junhua *et al.*, 2010). But these days' wetlands are facing tremendous anthropogenic pressure, which can greatly influence the population structure of the bird community. The loss of wetland through human interferences has led to a decline in several water bird populations. Though there are numerous wetlands for birds across Haryana, very few have been systematically surveyed to understand their importance for birds. The wetlands in district Ambala have long been known to support rich diversity of water birds. Three different wetlands of district Ambala harbours large populations of resident and migrant water birds and will be useful for water bird conservation. The diversity, composition and structure of water bird community in wetlands have been poorly documented particularly in these wetlands. Thus, the present survey was conducted to document and assess the avifauna found in wetlands of district Ambala, Haryana.

2. Study Area and Methodology

Study was conducted regarding avian species census of three wetlands of district Ambala , Haryana (India) *i.e.* Lord Mahavir park lake, Ambala City, Pond of Village Sultanpur, Ambala City and Rani Ka Talab, Ambala Cantt. Observations were regularly recorded for the period of two years to know the arrival and departure time of migratory birds as well as stability in census of resident avian species. Observations were made, usually two times in a day between the months of November 2014 to December 2017. Regular surveys were done by observing and counting various bird species found throughout the study area. Observations were made in the morning between 09:00 AM to 10:30 AM and in the afternoon 4:00 PM to 5:00 PM. Surveys were conducted thrice in a week. At each site, birds were counted using Olympus binocular and identified. Extensive photography was done using Canon Powershot SX5000 camera. Study area of all the three sites were divided on the basis habitat variables and classified as heterogeneous and homogenous habitats. Cover and depth of aquatic plant species like *Eichhornia* was calculated using Ellen-Dombois habitat sampling. Multivariate analysis was used to compare the mean population of Purple moorhen which was found in open water, floating vegetation, emergent vegetation and emergent floating vegetation.

3. Results

In the survey sites, 9 orders (Podicipediformes, Pelecaniformes, Ciconiiformes, Anseriformes, Gruiformes, Charadriiformes, Falconiformes, Coraciiformes and

Passeriformes) and a total of 13 families are represented in context of avian diversity in three different wetlands of Ambala. Out of 47 species of wetland birds, 25 species were winter Migratory, 13 were resident and 9 species were summer migratory. Birds like Northern Shoveller, Northern Pintail, Common Teal, Brahminy Shelduck were common between all the visits and hence, these migratory birds stay for longer period but their sightings were rare. However, birds like Mallard and Eurasian Wigeon are found only in December and January and were not observed in any of the wetland of Ambala during this study period.

Floating vegetation mainly consists of *Salvinia* and *Azolla*. The prominent emergent vegetation includes *Oryza* and *Sacciolepis*. The rooted floating vegetation mainly consists of *Nymphaea stellata* and *Utricularia*. The emergent floating vegetation harbors both emergent and floating plants. In this study, habitat characteristics were studied in breeding and non-breeding sites. Density of avian species was observed in three different wetlands and was found to be maximum in Lord Mahavir Park (Fig. 23.1) in initial years of survey but diversity was found to decline drastically due to eutrophication of wetland (Fig. 23.2) infested mainly by *Eichhornia* species (Fig. 23.3). Diversity of avian species was found to be lowest in Rani Ka Talab inspite of low Eutrophication due to homogenous habitat found in this wetland.

Fig. 23.1: Open water wetland (Lord Mahavir Lake, Ambala city)

Fig. 23.2: Eutrophic wetland: (Avian species declined).

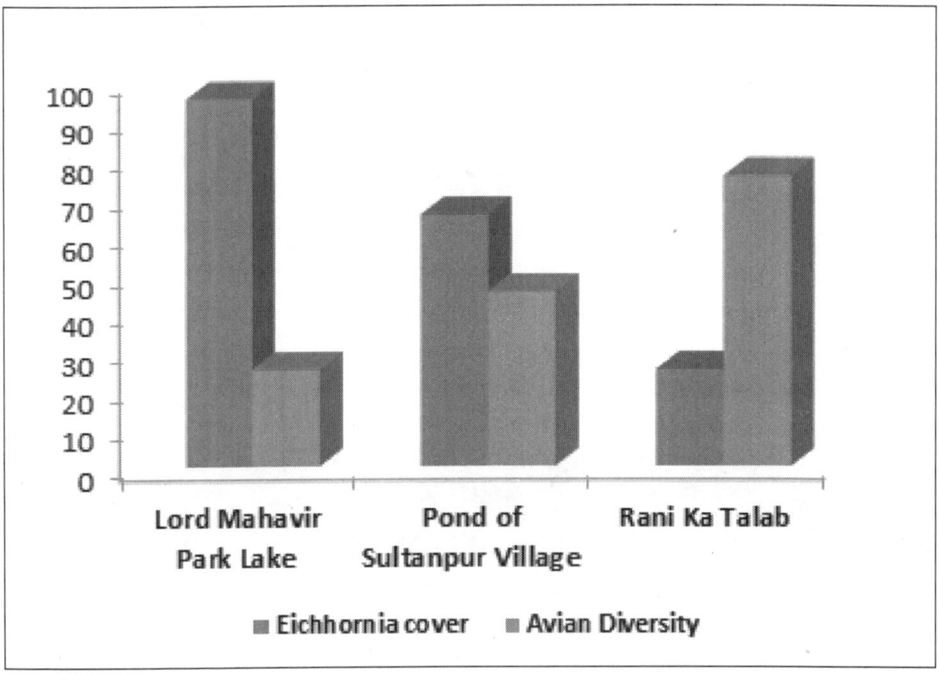

Fig. 23.3: Influence of weed cover upon Avian Diversity.

There was a diversity of resident wetland birds including Pond Herons (*Ardeola grayii*), huge numbers of Black winged Stilts (*Himantopus himantopus*), Common Moorhens (*Gallinula chloropus*), Bronze-winged Jacana (*Metopidius indicus*), Red-wattled Lapwing (*Vanellus indicus*), Yellow wattled Lapwing (*Vanellus malabaricus*), Grey Herons (*Ardea cinerea*), Large Egrets (*Ardea alba*) and Median Egrets (*Mesophoyx intermedia*) (Table 23.1). Spotted Greenshank (*Tringa guttifer*), Eurasian Spoonbill (*Platela leucorodia*), Bar-headed Goose (*Anser indicus*) and Greylag Goose (*Anser anser*) were found in Nighdhu village pond in Karnal but these birds could not be observed from wetland of village Sultanpur in Ambala.

Table 23.1: Avian species sightings variation at three different wetlands of District Ambala.

Sr. No.	Family/Common Name	Scientific Name	Status	Local Abundance
	Family: Podicipedidae	Tachybaptus ruficollis		
1.	Little Grebe		R	Decreased
	Family: Phalacrocoracidae	Phalacrocorax niger		
2.	Little Cormorant		R	Decreased
	Family: Ardeidae			
3.	Little Egret	Egretta garzetta	R	Stable
4.	Grey Heron	Ardea cinerea	R	Decreased
5.	Purple Heron	Ardea purpurea	R	Decreased
6.	Large Egret	Casmerodius albus	WM	Decreased
7.	Median Egret	Mesophoyx intermedia	WM	Decreased
8.	Cattle Egret	Bubulcus ibis	R	Stable
9	Indian Pond- Heron	Ardeola grayii	R	Stable
	Family: Anatidae	Dendrocygna javanica		
10.	Lesser Whistling-Duck		SM	No sightings
11.	Spot- Billed Duck	Anas poecilorhyncha	WM	Decreased
12.	Common Teal	Anas crecca	WM	Decreased
13.	Northern Shoveller	Anas clypeata	WM	Decreased
14.	Red-Crested Pochard	Rhodonessa rufina	WM	No sightings
15.	Common Pochard	Aythya ferina	WM	Decreased
	Family: Rallidae	Amaurornis phoenicurus		Stable
16.	White-Breasted Water Hen		R	
17.	Purple Moorhen	Porphyrio porphyrio	R	Increased
18.	Common Moorhen	Gallinula chloropus	WM	Increased
19.	Common Coot	Fulica atra	WM	Increased
	Family: Scolopaciadae	Tringa totanus	WM	Decreased
20.	Common Redshank			
21.	Common Sandpiper	Actitis hypoleucos	WM	Stable

Sr. No.	Family/Common Name	Scientific Name	Status	Local Abundance
22.	Little Stint	Calidris minuta	WM	Stable
	Family: Recurvirostridae	Himantopus himantopus		Stable
23.	Black-Winged Stilt		R	

The behavior pattern of the species is influenced by the availability and dynamics of the habitat of the species. Species may shift to another habitat depending upon the availability and threats a species has to face in the present habitat. Increased encroachment of wetlands (Fig. 23.4) had reduced foraging and nesting grounds of Purple Moorhen. As a result, species is forced to shift its habitat from one aquatic habitat to another. In relation to change in habitat preference, behavior of the avian species diversity in turn is greatly influenced.

Microhabitat of an organism is its physical and biological factors at finest scale. Changes in the microhabitat can critically influence the survival of an organism. The macrohabitat of the bird is habitat at large scale like open water, floating vegetation, emergent vegetation and emergent floating vegetation which were variously utilized by the bird. The abundance of the birds (Fig. 23.5) in the microhabitats differs in months and seasons mainly depending on availability of vegetation, water and the presence of predators.

Fig. 23.4: Wetland changed to dry scrubland: (Avian diversity nil).

Fig. 23.5: Rich Avian Diversity.

The least number of birds were seen in emergent vegetation and emergent floating vegetation in monsoon because the birds used this vegetation for breeding in monsoon. Purple moorhen so preferred vegetation zones in winter and summer. The availability and requirement of food and water and feeling of safety would be the major factors which influenced the habitat preference of the birds in the various seasons. Nest concealment, or visibility, is a common adaptation for reducing the risk of nest predation. Tall and dense vegetation may confer protection by creating visual barriers, as does increasing the number of available nesting sites and hindering the movement of avian species.

4. Discussion

According to Kaushik and Gupta (2014) birds like Northern Shoveller, Northern Pintail, Common Teal, Bar-headed Goose, Common Pochard, Red-crested Pochard, Tufted Pochard, Pallas Gull and Great Crested Grebe purportedly came to Hathini Kund in Haryana, India from far off destinations like Russia, Siberia, China, and Caspian region. It is pertinent to mention that birds like Northern Shoveller, Northern Pintail, Brahminy shelduck were seen in exceptionally large numbers. Contrary to this, birds like Mallard, Tufted Pochard, and Great-crested Grebe were seen in smaller groups (Robert *et al.*, 2010). However, birds like Mallard were seen in very less number and always seen in pairs.

The winter migratory birds displayed a definite pattern specific to species for arrival at and departure from the Lake (Gabriel *et al.*, 2010). They appeared at the wetland from November and stayed up to March. The peak of winter population of migratory birds was observed during the months of December to February. The Observation throughout the year revealed that Northern Pintail (*Anas acuta*), Common Pochard (*Aythya ferina*), Common Coot, Common Moorhen (*Fulica atra*) and Gadwall (*Aythya ferina)* arrived in late November; Ferruginous Pochard, Garganey, Grey-headed Lapwing arrived in first week of December. As far as the departure time was concerned, Gadwall generally departed in February. Birds like Red-crested Pochard, Common Coot, Common Moorhen, Ferruginous Pochard, Garganey and Northern Pintail departed in March. The summer visitors, namely Cotton Teal (*Nettapus coromandelianus*) were spotted during summer seasons from March to October.

Eichhornia is preferably used microhabitat as it provides appropriate concealment to eggs and chicks in early stages of nesting and incubation for Purple Moorhen but other species do not prefer this plant species as nesting and roosting habitat (Kumar and Balasubramanian, 2010). On the other hand, availability of Oryza is high but preferred less as it does not provide concealment to eggs and chicks during the breeding season. Salvinia is used for foraging and feeding microhabitat as it harbors many invertebrates which are used as food for Purple Moorhen. Purple Moorhen spends most of the time in activities like foraging, care of chicks and resting. Open water bodies are preferred for activities like bathing and swimming. Depending upon the presence and absence of predators, short duration defense activities like attack and escapes are carried out the Purple Moorhen but other species are not able to defend themselves. Other species of birds do not prefer *Eichhornia* and *Salvinia* as nesting and roosting sites and lead to migration of most bird species.

Growth of weeds in these three wetlands is at tremendous rate which will cover whole of the wetland area in few weeks. The weed cover over the water surface prevents invertebrate species from supply of oxygen. As a result, fishes and invertebrate species die out due to asphyxiation (Nicholas and Collias, 1967). Weed cover also reduces food supply as well as foraging site for the avian species. Weed cover destroy the roosting and nesting sites of the avian species. All these factors lead to migration or mortality of the avian species from these wetlands. Continuous growth of weeds in these wetlands had now made avian diversity almost nil in these areas which once use to be an eminent avian spot. There is an urgent need to conserve these sites as loss of species at this alarming rate can lead to an ecological imbalance.

Acknowledgements

I am thankful to staff of City Park, Ambala city, Haryana for their cooperation during the field work. I am thankful to local villagers of Sultanpur for cooperating with us during field work. I thank Department of Zoology, University College, Kurukshetra University for providing me with all the field equipments during my work.

References

Gabriel, L.W., Franchesco, D.F., Anderson, S.B., Josmael, C., Mauro, F.O. and Nilton, C. 2010. Behaviour of the Common Moorhen in Rio Grande do Sul, Brazil. *Acta. Ethologica*, 13(2): 127-135.

Gupta, R.C., Kaushik, T.K. and Gupta, P.K. 2012. Winter migratory wetland birds in Haryana are confronting adverse conditions in rural ponds resulting in reduction in arrival numbers: Case study of village Amin, Kurukshtra. *Indian Journal of Fundamental and Applied Science*, 2(1): 1-7.

Junhua, H.U., Zhigang, J., Daode, Y. and Huijian, H.U. 2010. Nest-site selection by Purple Swamphen. *Chinese Birds*, 1(4): 230–235.

Robert, N.H., David,N. and Phillip, Z.W. 2010. Notes on nesting of the common Moorhen and Purple Gallinule in Southwestern Louisana. *Field Ornithol.*, 58(1): 55-61.

Kumar, S.E. and Balasubramanian, P. 2010. Breeding behaviour and nest treeuse by Indian Grey Hornbill in Eastern Ghats India. *Forktail*, 26: 82-85.

Kaushik,T.K. and Gupta, R.C. 2014. Deteriorating rural ponds: A threat to overseas migrating birds in Kurukshetra, Haryana. *Journal of Applied and Natural Science*, 6(12): 570-577.

Nicholas, E.C. and Collias, E.C. 1967. Quantitative analysis of breeding behaviour in Frican village weaverbird. *The auk*, 84: 396-411.

Chapter 24

Role of Rotifers in Water Quality Assessment

Pooja Devi[1*] and Anita Bhatnagar[2]

[1]*Government P.G. College, Ambala Cantt., Haryana, India*
[2]*Department of Zoology, Kurukshetra University, Kurukshetra 136118, Haryana, India*
**Corresponding Author E-mail: pooja.sangwan@rediffmail.com*

Abstract

Aquatic ecosystems both 'lentic as well as lotic' are natural resources to satisfy agricultural, industrial, domestic, recreational or aesthetic needs of humans. Lentic waters are the vital and vulnerable freshwater systems that are critical for the sustenance of all life. At present, world's water resources are under pressure and are in danger due to various kinds of anthropogenic activities. These activities are responsible for altering the water quality and such change in water quality exerts immense influence on the flora and fauna of an aquatic ecosystem. The water quality assessment is very important for implementation of the monitoring and remediation programs to minimize the risk promoted by hazardous substances in aquatic ecosystems. Biological diversity is mainly governed by plankton distribution of any aquatic ecosystem. The study of planktons is a very fundamental tool for the water quality assessment and productivity of an aquatic body that contributes for the understanding of lentic water bodies since they are strongly affected by environmental conditions and respond quickly to changes in water quality. In freshwater systems, zooplankton organisms form an important group, as most of them eat and incorporate the primary producers, the algae into their bodies and make themselves available to be eaten by higher organisms in food chain. Among zooplanktons, rotifers significantly contribute to the zooplankton richness and also play an important link between the nanoplankton and carnivorous zooplanktons, inhabiting a wide range of habitat in aquatic ecosystem. Role of rotifers is very important in the tropho-dynamics, ecological energetics, cycling of nutrients and aquaculture productivity due to extremely high reproductive rate. So, the study of the rotifer communities of the site can be an important tool in constructing this inventory an important benchmark for evaluating future regional environment. During study of rotifers in religious water bodies of Haryana, appearance of various pollution tolerant taxa viz., Brachionus calyciflorus, Brachionus spp., Filinia sp., Asplanchana sp. and Keratella testudo after the rituals indicated the polluted and eutrophic nature of the water bodies. Although these are pollution tolerant taxa yet their species diversity was affected by the religious activities.

Keywords: *Aquatic system, Monitoring, Rotifer community, Water quality, Zooplankton*

1. Introduction

Rapid industrialization, increasing population pressure, urbanization, modern agricultural activities and other anthropogenic activities affects the water quality (Sinha *et al.*, 2004; Agarwal *et al.*, 2006; Rangarajan *et al.*, 2006; Neamtu *et al.*, 2009; Bhatnagar and Singh, 2010; Kuwabara *et al.*, 2013; Singh and Sharma, 2016). Health of water body is not only determined by physic-chemical parameters but examining aquatic organisms is also essential (Dirican *et al.*, 2009). The abundance of the species composition and condition of the aquatic organisms in natural community are directly related to the water quality. The effect of pollution stress manifests themselves in the aquatic biota *viz.*, changes in the pattern of distribution, elimination of the sensitive species, changes in diversity and subtle morphological and physiological changes (Clarke and Warwick, 1994; Sunkad and Patil, 2004). This forms the basis of biological monitoring of water quality. Biological data of a particular location not only reflects the conditions existing at the time of monitoring but past conditions as well. The physico-chemical methods give information about the type of a substance/pollutant and its concentration, while biological methods indicate their general effect but provide no clue to the nature and quality of a substance. Periodic environmental changes in temperature, water movement, light, soil and other biological factors leads to changes in species composition and food chain of ecosystem (Gray, 1981; Hartley, 1982; Clarke and Warwick, 1994). The phytoplanktons and zooplanktons are ecologically significant as they form the basic link in the food chain of all aquatic animals (Misra *et al.*, 2001). To monitor the water quality of any ecosystem and integrity of water recently plankton has been used as indicators of trophic status of any aquatic system and potentiality of zooplankton as bioindicator is very high as their growth and distribution are dependent on the abiotic (temperature, salinity, turbidity, stratification) and biotic variables (food limitation, predation, competition) (Christou, 1998; Beyst *et al.*, 2001; Escribano and Hidalgo, 2000). In freshwater systems, zooplankton organisms form an important group, as most of them eat and incorporate the primary producers, the algae into their bodies and make themselves available to be eaten by higher organisms in food chain (Michael, 1986). Community size structure of zooplankton can be used as an indicator of trophic status of an aquatic system (Canfield and Jones, 1996). Ramachandra and coworkers (2006) used zooplankton as bioindicator during hydro-biological investigations in selected Bangalore lakes. Sampaio and coworkers (2002) find out the relationship between trophic state of reservoir and diversity of zooplankton during the study of seven reservoirs of Paranapanema River. Among the zooplanktons, rotifers can be considered as the soft-bodied metazoans having very short life cycle playing very significant role in aquatic food chain and acting as a link between the nanoplanktons and carnivorous zooplanktons. Many species of rotifers are primary consumers which feed on phytoplanktons, particulate organic matter and free-swimming algae. The name of rotifers is derived from rotating wheels of cilia, called as corona, used for locomotion and also for sweeping the food particles towards the mouth. Their body is divided into head, trunk and abdomen. These have short developmental period and fast reproductive rate mainly due to parthenogentic production (Herzig, 1983) and can quickly populate in vacant niches converting primary production into a useable form for secondary consumers producing up to 50% of the total plankton biomass (Nogrady *et al.*, 1993). They are very essential ecological group in planktons inhabiting a wide range of habitat in aquatic ecosystem (Dhembare, 2012). Studies on diversity of rotifers have been conducted by various scientists (Duggan *et al.*, 1998, Ferdous and Muktadir, 2009. Dhembare, 2012) but still very little is known to date about the microfauna and zooplankton diversity in India due to general lack of taxonomic expertise. Rotifers can be regarded as the pioneer organisms, appeared first in newly created water body (Paleolog *et al.*, 1997; Sulehria *et al.*, 2009). Raut and Pejaver (2003) studied the rotifer diversity of three macrophytes infested lakes from Thane city, Maharashtra and observed that macrophytes tends to increase the diversity of rotifers and rotifer population varied between mesotrophic and oligotrophic lakes. Virani and Makode (2011) studied the role of rotifer diversity in tropical lentic ecosytem with reference to eutrophication. Occurrence of rotifers and its relation to the water quality during bioremediation process in lake Kacharali, Thane was reported by Somani *et al.* (2012) and revealed mesotrophic and eutrotrophic conditions. Bhatnagar and Devi (2014) observed the biodiversity, distribution and ecology of rotifer communities of religious water bodies of Haryana. In this chapter effort has been made to review the work done on rotifers in aquatic systems of Haryana. The attempt has also been done to highlight how rotifer diversity is affected by religious activities and their role in water quality assessment.

2. Rotifers as Fish Food Organisms

Natural fish food is the chief source of protein and lipids fish diet, but are poor in terms of carbohydrates, containing between 3 and 4.79% of carbohydrates on dry matter (Mitra *et al.*, 2007). The success of fish fingerlings production in the hatchery for stocking in the grow-out production system is mainly dependent on the availability of appropriate live food for feeding fish larvae, fry and fingerlings. Zooplanktons are the microscopic animals that are found floating near the surface of the waters with the help of water currents and constitute a vital food source for fish. Utilization of live zooplankton as a source of food for fish leads to improvement in weight gain and survival rate of fishes (Chakrabarti and Sharma, 1998; Sharma and Chakrabarti, 1999; Singh, 2010; Hassan, 2011). Among zooplankton especially rotifers in the fish culture pond can serve as a rich source of protein and unsaturated fatty acids. Further their small size (mostly ranging from 0.1to 0.5 mm) and high digestibility make them an ideal food for the larval fish. Rotifers are capable to reproduce by both sexual and asexual means so, having high reproduction rate and can be reared in large densities upto 2000 animals/ml. Thus, rotifers can be a great food for silver carps larvae, shrimp larvae and also for marine fin fishes in fish culture ponds. Rotifers act as a basic food for the early stages of the fishes forming several links in food web occupying diversity of trophic level aquatic system (Nikolsky, 1963). *Brachionus plicatilis* and *Brachionus rotundiformis* are the common species of rotifers that can be used to meet the different stages of successful fry production in both marine and freshwater hatcheries.

Three strains of *B. plicatilis* and *B. rotundiformis* can be used in the hatcheries.

1. L type with lorica, size ranged from 100-340μ
2. S type with a size ranged from 100-210μ.
3. SS type with size less than 100μ.

The physicochemical factors in water that should be maintained for mass production of *B. plicatilis* (Santhosh and Anil, 2013) are:

1. Water temperature range should be 20- 28°C.
2. Dissolved oxygen should be above 3 mg L^{-1}.
3. Salinity should be below 35 ppt.
4. pH above 7.5.
5. Ammonia below 1 mg L^{-1}.

In mass cultures of rotifers, there may be some risks of sudden mortality. Thus, stock culture should be maintained separately under aseptic conditions as culture may be contaminated by ciliates and bacteria. The culture tubes and filters being used for stock culture should be properly sterilized. The starter culture of rotifers can be prepared by its collection from wild and then isolation through a series of antibiotic treatments to get pure culture without any contamination. The culture tubes should be gently shaken on a shaker for providing sufficient O_2. Mass culture is usually maintained in large indoor tanks and harvesting is done when rotifer reaches a density of 300-500 nos/ml and rotifer doubles its population daily. A 50μ mesh net can be used for filtering the mass culture during harvest. Algal culture should be pumped in the culture tank on a daily basis and enough is needed to maintain the production (Santhosh and Anil, 2013). Ideal sustainable zooplankton density at the moment of fish larvae stocking ranges from 8.5 to 40 rotifers per ml (Milstein *et al.*, 2006; Valdenberg *et al.*, 2006), or a biomass of 20–40 mg L^{-1} (Li *et al.*, 1996). Nutritional value of rotifers mainly depends on the type of feed used. *Tetraselmis, Nannochloropsis* or *Isochrysis, Chlorella* algae can be used as a rotifer feed. However, the rotifer fed on *Tetraselmis, Nannochloropsis* or *Isochrysis* mixture will be higher in DHA and PUFA than of culture using Chlorella. Formulated diets for rotifers are also available in market e.g. Selco (company producing rotifer feeds). Use of enriched rotifer for feeding larvae is indispensable for better survival rate of larvae.

3. Rotifers as Indicator of Water Quality/Trophic Status/ Bioindicators

Many investigators have shown that the degree of water pollution can be estimated by using some typical organisms. Trophic state of water, other abiotic factors (Bielanska-Grajner and Gladysz, 2010), food composition (esp. algae species), vegetation cover (Viayeh and Spoljar, 2012), and predation (Gannon and Stemberger, 1978) also perform significant roles in determining the abundance and species composition of the rotifer community and thus, preferences of individual rotifer species may differ by region. The short life cycles, quick response to changes in environmental conditions affecting the species composition and abundance of rotifers makes them suitable biological indicators that reflect the changes in water quality. The rotifers not only indicate the level of water pollution, but also serve to determine general tendencies in the changes in environmental conditions over time (Duggan *et al.*, 2001; Tasevska *et al.*, 2010; Ejsmont-Karabin, 2012). The abundance and species dominance of rotifers is increased with increase in nutrient loading. So, rotifer abundance might be more sensitive indicator than species composition for analysis of trophic state (Wen *et al.*, 2011). Rotifers also contributed to the nutrient recycling so they can also play a significant role in cleaning of water by preventing clouds of waste matter.

Asplanchna herricki, Anuraeopsis fissa, Monostyla bulla are regarded as the indicators of fresh and clean waters showing the oligotrophic nature of the waters. *Trichocera pusilla* occurs in mesotrophic lakes, and *Polyurthra dolichoptera* is associated with a low trophic state (Maemets, 1983; Sladecek, 1983; Duggan *et al.*, 2002). Bilgrami and Dutta (1985) reported that *Keratella cochlearis, Brachionus plicatilis, Filinia* sp. as indicators of a eutrophic systems while Saksena (1987) reported *Brachionus, Keratella* sp. as indwellers of moderately clean or mesotrophic waters. Berzins (1949) and Lillieroth (1950) observed *Brachinus angularis, B. quadridendatus, B. calyciflorus, Polyurthera euryptera, Keratella quadrata, Trichocera capucina, T. cyclindrica, Filinia longiseta, F. terminales* as indicators of heavy pollution showing eutrophication of water. *Brachionus, Anuraeopsis fissa, Keratella cochlearis tecta, Keratella quadrata, Filinia longiseta, Pompholyx sulcata, Proales* sp. and *Trichocerca pusilla* indicates the eutrophication of aquatic systems (Karabin, 1985). Virani and Makode (2011) reported *Lecane patella* and *Monostyla closterocera* in the water of Chilai reservoir showing the deterioration of water quality of the reservoir. Some authors estimated lake trophic status on the basis of measuring the individuals per liter as: 500–1000 ind. L^{1}: mesotrophic or mesoeutrophic, 1000–2500 ind. L^{1}: eutrophic and 3000–4000 ind. L^{1}: moderately eutrophic (May and Hare, 2005; Wen *et al.*, 2011). *Monostyla, Lepadella,* and *Brachionus falcatus, M. closterocera, L. patella* are regarded as the pollution indicator rotifers indicating the deterioration of water quality (Virani and Makode, 2011).

The genus *Brachionus* and *Synchaeta* and *Lecane* are considered euryhaline and suggested as mesohaline indicators (Attayde, 1996). According to Kozlowsky Suzuki (1998) and Arcifa *et al.* (1994), these genera are also associated with freshwater input, and they can tolerate high ranges of salinity. *Brachionus* and *Synchaeta* are also associated with low water transparency (Branco, 1998) and according to Berzins and Pejler (1989) the distribution and abundance of rotifers in water with stable oxygen conditions are more conditioned to nutrient levels. They also cited that the genus *Brachionus* is an indicator of eutrophy.

4. Rotifers of Fish Ponds of Haryana

Haryana state with its immense water resources has tremendous scope for augmenting fish production for the growing fish-eating population and also generating rural employment. During last decades, the state has utilized its close water bodies viz. Ponds and tanks through scientific aquaculture. Present day research is being focused to develop sustainable production technology that is highly suited to environmental conditions thereby easing problems of environment degradation. Keeping this into consideration, it was felt that there is a need to generate data on productivity of village ponds, availability of fish food organisms and it was observed that rotifers are integral part of fish food organisms.

A survey of the area along the village ponds of Haryana was made. General survey from the various farmers was made through Performa and four ponds (two managed and two unmanaged) were selected from each of the following four districts of Haryana according to the species they are cultivating and productivity they are obtaining: 1. Kurukshetra, 2. Yamuna Nagar, 3. Nuh (Mewat), 4. Hisar. From Yamuna Nagar only one cultivated pond was selected. Thus, a total of 15 ponds were selected. In fish pond ecosystem, zooplankton organisms form an important group as most of them eat and incorporate the primary producers, into their bodies and make themselves available to be eaten by higher organism in food chain. In the present investigation, the total zooplankton population was observed significantly ($P< 0.05$) high in the unmanaged ponds as compared to the managed ponds in all the four districts. Species diversity index of zooplankton (d_p) ranged from 2.7 to 2.85 in Kurukshetra, 2.92 to 3.00 in Yamuna Nagar, 2.81 to 2.91 in Hisar and 2.82 to 2.96 in Nuh Mewat. Species diversity was higher in managed ponds as compared to the unmanaged ponds in Yamuna Nagar while it was opposite in district Hisar. However, no significant difference in the species diversity and number of species were observed in the managed ponds as compared to the wild ponds in Kurukshetra and Nuh-Mewat. *Branchionus* and *Keratella* genus were the most dominant and abundant among rotifers that were recorded at the village ponds of all the four districts. These are the pollution tolerant taxa showing the deterioration of water quality of ponds from various point or non-point sources. Efforts should be done to improve and conserve the water quality status of these village ponds.

5. Rotifers of Ponds with Religious Activities in Haryana

In present investigation, a total of eight sites from various districts of Haryana were selected for the study of rotifers viz., site1: Kapalmochan (Village- Bilaspur, Yamunanagar), site 2: Kulotaran tirth (Village- Kirmach, Kurukshetra), site3: Banganga tirth (Village- Dayalpur, Kurukshetra), site 4: Brahmsarovar (Kurukshetra), site 5: Jyotisar (Kurukshetra), site 6: Saraswati tirth (Pehowa, Kurukshetra), site 7: Phalgu tirth (Village- Pharal, Kaithal) and site 8: Pandu-Pindara tirth (Village- Pandu-Pindara, Jind) were selected for the study.

Seasonal abundance of rotifer population showed highest no./lit during post monsoon followed by monsoon, summer and winters. The maximum abundance of rotifers was observed at sites 1, 2, 3 and 7 during post monsoon season while at sites 4 and 6 during monsoon season and at site 8 it was during the summer season. The lower number of rotifers was observed during the winter season at all the sites. A total of 44 taxa of rotifers were observed at all the eight religious sites during the study period and these were belonged to the Brachionidae, Asplanachnidae, Trichorceridae, Lecanidae and Filinidae families. Among these Brachionidae was the most dominant one. Among all family members *Brachionus angularis*, *Brachionus forficula, Brachionus calyciflorus, Keratella testudo and Conochilus* sp. were the dominant rotifers recorded during the study period at various sites (Fig. 24.1).

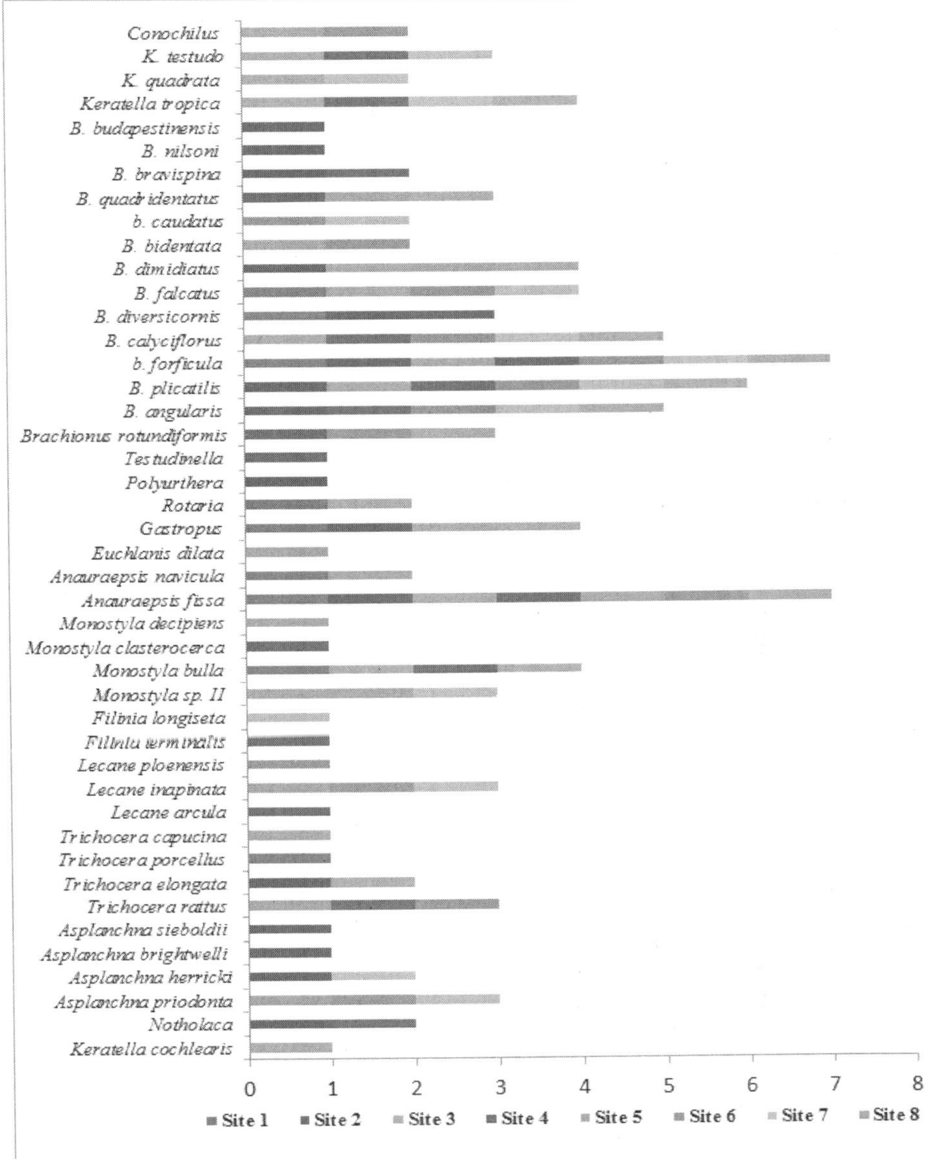

Fig. 24.1: Distribution of rotifers at all the selected religious sites of Haryana.

Presence of *Conochilus* sp. indicates the eutrophic nature of any aquatic system (Gannon and Stemberger, 1978) and it was found in abundance forming rotifers bloom at site 6 during monsoon season indicating eutrophic nature of the water during monsoon season. It may be due to the addition of various organic wastes with the runoff water. *Keratella* sp. indicates the presence of organic pollution in the waters (Goel and Chavan, 1991) and *Keratella testudo* was recorded in dominance at site 7 during the post monsoon season indicating that water was receiving the organic wastes by the activities of the villagers residing near the pond. Dominance of *Keratella tropica* and *Hexarthra* were regarded as the good indicators of high turbidity due to suspended sediments (Ferdous and Muktadir, 2009). Pal *et al.* (2015) suggested *B. calyciflorus* as a considerably pollution tolerant species and can endure high levels of organic nutrients and is an indicator of aquatic eutrophication. Mulani *et al.* (2009) reported presence of *Brachionus* spp. in typical tropical conditions and according to Sampaio *et al.* (2002) is an indicator of eutrophication. In the present study, *Brachionus* spp. were observed in abundance at site 2 and site 6 indicating the eutrophic nature of these water bodies.

Sulehria and Malik (2013) have also reported the dominance of *Brachionus*. Janakiraman *et al.* (2012) found a positive correlation between the density of rotifers and the content of ammonium nitrogen similar to the findings of Bhatnagar and Devi (2014). This illustration further supports the dominance of rotifers in religious ponds after the religious activities as there was significant increase in ammonia level at various sites due to religious activities and mass bathing. The results obtained during the present study clearly depicted that rotifer community was rich in the stagnant water bodies where different anthropogenic or ritual activities are a regular feature.

6. Rotifers of Lotic Water Bodies of Haryana

Rotifers can be regarded as the most important zooplankton faunal component and bio-indicators of freshwater aquatic systems. Nearly 95% species of rotifers have been reported from the freshwaters and remaining 5 % from the marine habitat (Sharma, 1996). The present study was aimed to study the rotifers of Yamuna river region, Haryana. Yamuna is the longest and second largest tributary river of the Ganga in northern part of India. Yamuna river is highly venerated in Hindus and worshipped as goddess. In one of the investigation (Malhotra, 2010), three sites were selected at river Yamuna for the study of distribution and abundance of rotifers. Site one was upstream of the river with no influx of effluent: site two was the point of influx of industrial effluent and site 3 was few kilometers downstream of site 2, which also represented the revival zone of the river. In river Yamuna, Cladocera was dominant at all the stations followed by Rotifera. Out of these three sites, highest percentage of rotifers was observed during the month of November at site 1, in august at site 2 and during April at site 3. The genus *Monostyla, Keratella* and *Brachionus* were observed at all the selected sites indicating the eutrophic nature of the waters.

The river Ghaggar was also investigated for the study of rotifers, an important river originated from outer range of Himalayas. This river of Haryana originates from the Shiwalik Hills of Himachal Pardesh and Haryana and flows from east to west and takes a south westerly course. It is very important in terms of its agricultural purpose and for its spiritual value for the people. At its upstream it receives industrial effluents and domestic sewage from various point and non-point sources, at its downstream it forms a deep region known as 'Ottu head'. Here among zooplanktons, rotifer group was observed as the most dominant group followed by copepod and cladocera. Among rotifers, *Polyarthra, Keratella, Horaella, Brachionus quadridentata, Brachionus angilaris* were the dominant and most abundant taxa that were observed during the study (Bhatnagar and Garg, 1998). The dominance of these species in the river indicates the pollution caused by the industrial effluents and domestic sewage hence deterioration of water quality and eutrophication of the river.

7. Future Prospects in Rotifer Studies

The significant role of rotifers as fish food has been proved by many researchers. Rotifers utilization in the form of food for fish larvae is very crucial, as these are very critical for the future development of first fish stages of fish (Arimoro, 2006). The availability of rotifers as fish food in ponds can be improved by ensuring appropriate environmental conditions. In spite of much nutritional values of rotifers as fish feed a very few studies are available on the production of rotifers in fish culture ponds. In addition to this, studies should be conducted to focus on the balance of nutrients and the different species of rotifers in order to maintain a good water quality, good fish quality in fishponds and increase their conservation value. New techniques should be designed for contamination free mass culture of rotifers in fish culture ponds with high food conversion rate for improving the fish quality. The knowledge regarding the life cycle, peak reproductive period and reproductive means of rotifers can also be very important for mass culture. More research is also needed to know at what population density and which species of rotifers can be helpful to study the trophic state of the aquatic systems. For this, there is need of well taxonomically expertise in this field.

8. Conclusion

The role of rotifers in fish culture ponds has been proved not only for the carp larvae and fry, but also for adults. Application of rotifers for evaluating water quality index is hardly limited by lack of knowledge regarding their taxonomy and form variation acquired so far. The rotifers have versatile capacity to survive in different environments and may be considered as a tolerant group of zooplanktons; however, the stated differences at different sites reflected the micro-environmental differences among different sites. Analysis of rotifer distribution, diversity and ecology can become a very fundamental tool for studying the impacts of external factors on any aquatic ecosystem for evaluating water quality. There is need to designs the techniques for mass culture of rotifers for fish feed and their implementation at the ground level for the benefits of the fish farmers.

References

Agarwal, T., Khillore, P. and Sridhar, V. 2006. PAHs contamination in Bane Sediment of the Yamuna River, Delhi. *Environ. Monit. Assess*, 123(1-3): 151-166.

Arcifa, M.S., Castilho, M.S.M. and Carmouze, J.P. 1994. Composition et évolution du zooplankton dans une lagune tropicale (Brésil) au cours d'une période marquée par une mortalité de poissons. *Rev. Hydrobiol. Trop.*, 27: 251-263.

Arimoro, F. O. 2006. Culture of the freshwater rotifer, *Brachionus calyciflorus*, and its application in fish larviculture technology. *Afr. J. Biotechnol.*, 5: 536–541.

Attayde, J.L. 1996. Limnologia da lagoa de Carapebus e a estrutura de sua comunidade zooplanctônica. Depto. De Ecologia. Universidade Federal do Rio de Janeiro. Master's Thesis. p. 92.

Berzin, B. and Pejler, B. 1989. Rotifer occurrence in relation to temperature. *Hydrobiologia*, 175(3): 223–231.

Berzins, B. 1949. Zur Limnologie der Seen Sudostlettlands. Schweiz. *Z. Hydrol.*, 11: 583-607.

Beyst, B., Buysse, D., Dewicke, A. and Mees, J. 2001. Surf zone hyperbenthos of Belgian sandy beaches: Seasonal patterns. *Estuarine Coastal Shelf Sci.*, 53: 877-895.

Bhatnagar, A. and Devi, P. 2014. Rotifer communities of religious water bodies of Haryana (India): Biodiversity, distribution and ecology. *Journal of Applied and Natural Science*, 6: 729-737.

Bhatnagar, A. and Garg, S.K. 1998. Environment impact assessment in river Ghaggar in Haryana. *J. Natcon.*, 10: 215-224.

Bhatnagar, A. and Singh, G. 2010. Assessment of Culture Fisheries in village ponds: a study in District Hisar, Haryana. *India Int. J. Environ. Res.*, 4: 57-64.

Bielanska-Grajner, I. and Gladysz, A. 2010. Planktonic rotifers in mining lakes in the Silesian Upland: relationship to environmental parameters. *Limnologica*, 40: 67–72.

Bilgrami, K.S. and Dutta Munshi J.S. 1985. A comparative evaluation of some fresh water ecosystems functioning under stress environment. *Proc. 73rd Ind. Sc. Cong.*, Part IV. pp. 58-59.

Branco, C.W.C. 1998. Comunidades zooplanctônicas e aspectos limnológicos de trés lagoas costeiras da região norte Fluminense. Int. De biofísica. Universidade Federal do Rio de Janeiro. 232 p. Doctor's Thesis. p. 145

Canfield, T. J. and Jones, J.R. 1996. Zooplankton abundance, biomass, and size distribution in selected mid-western water bodies and relation with trophic state. *J. Freshwat. Ecol.*, 11: 171–181.

Chakrabarti, R. and Sharma, J. G. 1998. Influence of management protocols on carp growth under nursery conditions: relative importance of food and water quality. *Aquacult. Int.*, 6: 293–301.

Christou, E.D. 1998. Interannual variability of copepods in a Mediterranean coastal area (Saronikos Gulf, Aegean Sea). *J. Mar. System*, 15: 523-532.

Clarke, K.R. and Warwick, M.R. 1994. Changes in marine communities: An approach of stastical analysis and interpretation. *Plymouth Marine Laboratory*, pp. 177.

Dhembare, A. J. 2012. Individual and synchronized correlation between water and Rotifer from Dynaneshwar dam, Ahmednagar, Maharashtra. *European J. of Exp. Biol.*, 2: 122-126.

Dirican, S., Musul, H. and Cilek, S. 2009. Some physicochemical characteristics and rotifers of Camligoze Dam lake, Susehri, Sivas, Turkey. *J. Anim. Vet. Adv.*, 8: 715-719.

Duggan I.C., Green J.D. and Shiel R.J. 2001. Distribution of rotifers in North Island, New Zealand, and their potential use as bioindicators of the lake trophic state. *Hydrobiologia*, 446: 155-164.

Duggan, I.C., Green, J.D. and Shiel, R.J. 2002. Distribution of rotifer assemblages in North Island, New Zealand, lakes: relationships to environmental and historical factors. *Freshwater Biology*, 47: 195–206.

Duggan, I.C., Green, J.D., Thompson, K. and Shiel, R.J. 1998. Rotifers in relation to littoral ecotone structure in lake, Rotomanuka, North Island, New Zealand. *Hydrobiologia*, 113: 77-91.

Ejsmont-Karabin, J. 2012. The usefulness of zooplankton as lake ecosystem indicators: rotifer trophic state index. *Pol. J. Ecol.*, 60: 339–350.

Escribano, R. and Hidalgo, P. 2000. Spatial distribution of copepods in the North of the Humboldt Current region off Chile during coastal upwelling. *J. Mar Biol. Assoc. U.K.*, 80: 283-290.

Ferdous, Z. and Muktadir, A.K.M. 2009. A review: Potentiality of zooplankton as bioindicator. *Amer. J. Appl. Sci.*, 6: 1815-1819.

Gannon, J.E. and Stemberger, R.S. 1978. Zooplankton (especially crustaceans and rotifers) as indicators of water quality. *Trans. Am. Microsc. Soc.*, 97: 16-35.

Goel, P.K. and Chavan, V.R. 1991. Studies on the limnology of a polluted fresh water tank. In: Gopal, B. and Asthana, V. (Eds.), Aquatic Sciences in India. Indian Association Limnol. Oceanogr., New Delhi, pp. 51-64.

Gray, J.S. 1981. Detecting pollution induced changes in communities using the log normal distribution of individuals among species. *J. Mar. Pollut. Bull.*, 12: 173-176.

Hartley, J.P. 1982. Methods for monitoring offshore macrobenthos. *J. Mar. Pollut. Bull.*, 13: 150-154.

Hassan, A.A.E.R. 2011. Zooplankton as natural live food for three different fish species under concrete ponds with mono- and polyculture conditions. *Egypt. J. Aquac.*, 12: 7–41.

Herzig, A. 1983. Comparative studies on the relationship between temperature and duration of embryonic development of rotifers. *Hydrobiologia*, 104: 237-246.

Janakiraman, A., Naveed, M.S. and Altaff, K. 2012. Impact of domestic sewage pollution on rotifer abundance in Adyar estuary. *Int. J. Environ. Sci.*, 3: 689–696.

Karabin, A. 1985. Pelagic zooplankton (Rotatoria + Crustacea): Variation in the process of lake eutrophication. I. Structural and quantitative features. *Ekol. Pol.*, 33: 567–616.

Kozlowsky-Suzuki, B. 1998. O efeito do enriquecimento artificial e das aberturas de barra sobre a dinâmica das variáveis abióticas e sobre a estrutura da comunidade zooplanctônica na lagoa Imboassica, Macaé, Rio de Janeiro. Depto. De Ecologia. Universidade Federal do Rio de Janeiro. Master's Thesis. p. 94

Kuwabara, J.S., Topping, B.R., Moon, G.E., Husby, P., Lincoff, A., Carter, J.L. and Croteau, M.N. 2013. Mercury accumulation by lower trophic-level organisms in lentic systems within the Guadalupe River Watershed, California; (Scientific Investigations Report). USGS: Science for a changing world. pp. 2005-5037

Li, Y.G., Jin, S.D. and Qin, J.G. 1996. Strategies for development of rotifers as larval fish food in ponds. *J. World Aquac. Soc.*, 27: 178–186.

Lillieroth, S. 1950. Uber die Folgen kulturbedingter Wassserstandssenkungen fur Makrophyten and Planktongemeinschaften in seichten Seen des sudschwedischen Oliogotrophiegebiets. *Acta limnol. Lund.*, 3: 188.

Maemets, A. 1983. Rotifers as indicators of lake types in Estonia. *Hydrobiologia,* 104: 357–361

Malhotra, P. 2010. Biomonitoring of river Yamuna for water quality assessment in Yamunanagar (Haryana). PhD. Thesis submitted to Kurukshetra University Kurukshetra

May, L. and Hare, M.O. 2005. Changes in rotifer species composition and abundance along a trophic gradient in Loch Lomond, Scotland, UK. *Hydrobiologia,* 546: 397–404.

Michael, L. 1986. An empirical analysis of zooplankton community size structure across lake trophic gradients. *J. Limnol. Oceanogr.*, 31: 45-55.

Milstein, A., Valdenberg, A. and Harpaz, S. 2006. Fish larvae - zooplankton relationships in microcosm simulations of earthen nursery ponds. *I. Freshwater System. Aquacult. Int.*, 14: 231–246.

Misra, S.M., Pani, S., Bajpai, A. and Bajpai, K.A. 2001. Assessment of trophic status by using Nygaard index with reference to Bhoj wetland. *J. Pollut. Res.*, 20: 147-153.

Mitra, G., Mukhopadhyay, P.K. and Ayyappan, S. 2007. Biochemical composition of zooplankton community grown in freshwater earthen ponds: nutritional implication in nursery rearing of fish larvae and early juveniles. *Aquaculture,* 272: 346–360.

Mulani, S.K., Mule, M.B. and Patil, S.U. 2009. Studies on water quality and zooplankton community of the Panchganga River in Kolhapur city. *J. Environ. Biol.*, 30: 455-459.

Neamtu, M., Ciumasu, I.M., Costica, N., Costica, M., Bobu, M., Nicoara, M.N., Catrinescu, C., VanSlooten, K.B. and DeAlencastro, L.F. 2009. Chemical, biological and ecotoxicological assessment of pesticides and persistent organic pollutants in the Bahlui River, Romania. *Environ. Sci. Pollut. Res. Int.*, 1: 76-85.

Nikolsky, G.V. 1963. The Ecology of Fishes. Academic press, London and New York.

Nogrady T., Wallace, R.L. and Snell, T.W. 1993. Rotifera- Guides to the identification of the macroinvertebtates of the continental waters of the world, 4 SPB. Academic publishing. The Hague. p. 142.

Pal, S., Patra, A.K. and Chakraborty, K. 2015. Prospect of *Brachionus calyciflorus*, a holoplankton, for its potential bio-indicator property: a review. *Int. J. Recent Sci. Res.*, 6(11): 7603-7608.

Paleolog, A., Radwan, S., Kowalik, W., Kowalczyk, C., Stryjecki, R. and Zwolski, W. 1997. Water invertebrates of LasyJanowskie landscape park. *In*: Naural Environment of LasyJanowskie landscape park (Ed. S. Radwan). UMCS Press, Lublin, pp. 83-227 (in Polish).

Ramachandra, T. V., Rishiram, R. and Karthick, B. 2006. Zooplankton as Bioindicators: Hydro-Biological investigations in selected Bangalore Lakes. Pp. 1-115.

Rangarajan, R., Sarin, M. and Swami, K. 2006. Dissolved Uranium and 234U/238U in the Yamuna and the Chambal rivers, India. *J. Aquat. Geochem.*, 12(1): 73-101.

Raut, N.S. and Pejaver, M.K. 2003. Study on biodiversity of some macrophytes infested lakes in Thane city, Maharastra. *Nat. Environ. Pollu. Technol.*, 2(3): 277-281.

Saksena, D. 1987. Rotifers as indicators of water quality. *Acta Hydrochim Hydrobiol.*, 15(5): 481-485.

Sampaio, E.V., Rocha, O., Matsumura, T. and Tundisi, J.G. 2002. Composition and abundance of zooplankton in the limnetic zone of seven reservoirs of the Paranapanema River, Brazil. *Braz. J. Biol.*, 62: 525-545.

Santhosh, B. and Anil, M.K. 2013. Zooplankton for marine fish larval feed. CMFRI Manuel Customized training Book. pp. 107-114.

Sharma, B.K. 1996. Biodiversity of freshwater rotiferan in India- A status report. *Proc. Zool. Soc.*, 49(2): 73-85.

Sharma, J.G. and Chakrabarti, R. 1999. Larval rearing of common carp *Cyprinus carpio*: a comparison between natural and artificial diets under three stocking densities. *J. World Aquac. Soc.*, 30: 490–495.

Singh, B.M. and Sharma, R.C. 2016. Studies of physicochemical and bacteriological characteristics of Jaisamand Lake, Alwar (RAJ.). *J. Environ. Sci. Toxicol. Food Technol.*, 10(7): 1-9.

Singh, G. 2010. Fish Food Organisms and Intestinal Enzymes of Indian Major Carps from Fish Culture Ponds in Haryana. PhD. Thesis submitted to Kurukshetra University Kurukshetra.

Sinha, P.C., Guliani, P., Jena, G.K., Rao, A.D., Dube, S.K., Chatterjee, A.K. and Murty, T. 2004. A breadth averaged numerical model for suspended sediment transport in Hooghly estuary, East Coast of India. *Natural Hazards*, 32: 239-255.

Sladecek, V. 1983. Rotifers as indicators of water quality. *Hydrobiologia*, 100: 169–201.

Somani, V., Quadros, G. and Pejaver, M.K. 2012. Occurrence of rotifers and its relation to the water quality during the bioremediation process in lake Kacharali, Thane, MS, India. *ISCA J. Biol. Sci.*, 1(3): 54-58.

Sulehria, A.Q.K. and Malik, M.A. 2013. Diversity indices of pelagic rotifers in Camp Balloki water park, Lahore, Pakistan. *Turk. J. Zool.*, 37: 699-705.

Sulehria, A.Q.K., Qamar, M.F., Anjum, R.F., Ejaz, M. and Hussain, A. 2009. Seasonal fluctuations of rotifers in a fish pond at District Bahawalnagar, Pakistan. *Biologia (Pakistan)*, 55(1&2): 21–28.

Sunkad, B.N. and Patil, H.S. 2004. Water quality assessment of phytoplankton density in relation to environmental variables in Gopalswamy pond at Chitradurga, Karnataka. *J. Environ. Biol.*, 25(3): 113-116.

Tasevska, O., Kostoski, G. and Guseska, D. 2010. Rotifers based assessment of the Lake Dojran water quality, BALWOIS 2010, Ohrid, Republic of Macedonia, 25–29 May, 2010.

Valdenberg, A., Milstein, A. and Harpaz, S. 2006. Effects of timing of common carp larvae stocking on zooplankton succession in earthen nursery ponds: a microcosm simulation. *J. World Aquac. Soc.*, 37: 378–387.

Viayeh, R.M. and Spoljar, M. 2012. Structure of rotifer assemblages in shallow water bodies of semi-arid northwest Iran differing in salinity and vegetation cover. *Hydrobiologia*, 686: 73–89.

Virani, R.S. and Makode, P.M. 2011. Role of rotifer diversity in tropical lentic ecosystem with reference to eutrophication. *Biosci. Biotechnol. Res. Commun.*, 4: 55-64.

Wen, X., Xi, Y., Qian, F., Zhang, G. and Xian, X. 2011. Comparative analysis of rotifer community structure in five subtropical shallow lakes in East China: role of physical and chemical conditions. *Hydrobiologia*, 661: 303–316.

Chapter 25

Biodiversity, Classification and Taxonomy of Arbuscular Mycorrhizal Fungi Associated with Some Spices of Haryana

Esha Jangra[1*], Ashok Aggarwal[1], Vipin Prakash[2] and Alpa Yadav[3]

[1]*Department of Botany, Kurukshetra University, Kurukshetra, Haryana, India- 136119.*

[2]*Forest Pathology Division, Forest Research Institute (Deemed to be) University, Dehradun*

[3]*Department of Botany, Indira Gandhi University, Meerpur, Rewari (Haryana) - 122502 India*

**Corresponding Author E-mail: ashokbotanykuk@gmail.com; eshajangra13@gmail.com*

Abstract

The diversity of Arbuscular Mycorrhizal Fungi (AMF) has significant ecological and biological consequences because each AM species vary in their potential to promote growth of the plant and adaptation to various biotic and abiotic factors. Thus, the dynamics and diversity of AMF population have a marked impact on the structure and diversity of the associated plant species (Van-der Heijden et al., 1999). An important prerequisite to the analysis of AMF is the isolation and correct identification of individual isolates. Studies conducted at Haryana have shown that the taxonomic composition of AMF and the dominant AM fungal taxa colonizing plant roots may differ between habitat type, host plant species and location. It is concluded that more studies are required in this aspect for a more comprehensive knowledge of AMF diversity.

Keywords:

1. Introduction

According to Redecker *et al.* (2003), the fungal phylum-Glomeromycota that comprises of Arbuscular Mycorrhizal Fungi (AMF) have co-evolved with their plants since plants acquired the terrestrial ecosystem more than 430 million years ago. In 1995, Taylor and his associates reported a fossil specimen based on the morphology of the structures similar or resembling spores and mycelium of AM fungi. Later Stubblefield *et al.* (1987) observed mycelial hyphae, vesicles and AM spores in the preserved roots from the Triassic period. Finally, molecular data authenticated these findings showing the origin of AMF at a time between the Ordovician and Devonian period. According to Dotzler *et al.* (2009), mycorrhizal (Greek mykes= Fungi+ rhiza= roots) colonization appeared before plants had true roots.

Considering the long evolutionary history of AMF and their non-specificity with the host, it is not surprising to say that mycorrhizal association is found more than 80 percent of vesicular angiosperms (Wang and Qiu, 2006). Although AM fungi are obligate biotrophs as they are unable to grow and complete their life cycle without a living host (Smith and Read, 2010) yet AM spore of one AM fungus was reported to produce mycelium and spore when grown with *Paenibacillus validus* without living host (Hilderbrandt *et al.*, 2006). For establishing mycorrhizal association with the living host roots, the somatic structure of AM fungi discrete arbuscules, vesicles, auxillary cells, intraradical and extra radical mycelium along with asexual spores used to group disparate geographical populations into species (Morton, 1990).

2. Taxonomy of Arbuscular Mycorrhizal Fungi

Take any soil in which plants have grown and there are good chances that it will have the AM spores but it requires a careful eye to spot them. AM fungi have always created great taxonomic problems mainly because of difficulties in isolation from host plant and to maintain the AM fungi in living cultures. Thaxter (1992) in his monograph of the family Endogonaceae described four AM genera which produce spores in the sporocarp. The four genera were *Endogone, Glaziella, Sclerocystis* and *Sphaerocreas*. Today, a large number of data regarding the taxonomy of AM fungi have been generated by different workers belonging to family Endogonaceae from their studies in pot cultures. The first taxonomic key for the identification of AM spores has been proposed by Mosse and Bowen (1968), which was later revised by Gerdmann and Trappe (1974).

Morton and Benny (1990) transferred AM fungi forming arbuscules to a new order Glomales having two suborders Glomineae and Gigasporineae. The suborder Glomineae is having one family Glomeraceae with the genera *Glomus* and *Sclerocystis* and the second family Acaulosporaceae is containing the genera *Acaulospora* and *Entrophospora*. The suborder Gigasporineae is having one family Gigasporaceae with known genera *Gigaspora* and *Scutellospora*. Morton and Redecker (2001) errected two new families Arhaeosporaceae and Paraglomaceae with order Glomales (now Glomerales) on the basis of molecular, morphological and biochemical parameters. The family Archeosporaceae contained one genus, *Archaeospora* and Paraglomaceae with *Paraglomus* as type genus.

The classification of AM fungi presented here is basically of Gerdmann and Trappe (1974) with later amended by Morton and Benny (1990), Schüßler *et al.* (2001), Schüßler and Walker (2010) and Oehl *et al.* (2011). (Table 25.1)

Table 25.1: Classification of Glomeromycotan fungi (Stürmer, 2012).

Phylum	Class	Order	Family	Genera
According to Gerdemann and Trappe (1974)				
Zygomycota	Zygomycetes	Endogonales	Endogonaceae	*Glomus*
				Gigaspora
				Acaulospora
				Sclerocystis
According to Morton and Benny (1990)				
Zygomycota	Zygomycetes	Glomerales	Glomeraceae	*Glomus*
				Sclerocystis
			Gigasporaceae	*Gigaspora*
				Scutellospora
			Acaulosporaceae	*Acaulospora*
				Entrophospora
According to Schüβler *et al.* (2001)				
Glomeromycota	Glomeromycetes	Glomerales	Glomeraceae	*Glomus*
		Diversisporales	Gigasporaceae	*Gigaspora*
				Scutellospora
			Diversisporaceae	*Diversispora*
			Acaulosporaceae	*Acaulospora*
				Entrophospora
		Archaeosporales	Geosiphonaceae	*Geosiphon*
			Archaeosporaceae	*Archeospora*
		Paraglomerales	Paraglomeraceae	*Paraglomus*
Schüβler and Walker (2010)				
Glomeromycota	Glomeromycetes	Glomerales	Glomeraceae	*Glomus*
				Fenneliformis
				Rhizophagus
				Sclerocystis
			Claroideoglomeraceae	*Clariodeoglomus*
		Diversisporales	Gigasporaceae	*Gigaspora*
				Scutellospora
				Racocetra
			Acaulosporaceae	*Acaulospora*
			Enterophosphoraceae	*Entrophospora*
			Diversisporaceae	*Diversispora*
				Otospora
			Pacisporaceae	*Pacispora*
		Archaeosporales	Archaeosporaceae	*Archaeospora*
			Geosiphonaceae	*Geosiphon*
			Ambisporaceae	*Ambispora*
According to Oehl *et al.* (2011)				
Glomeromycota	Glomeromycetes	Glomerales	Glomeraceae	Glomus
				Septaglomus
				Fenneliformis

Phylum	Class	Order	Family	Genera
				Simiglomus
			Claroideoglomeraceae	*Claroideoglomus*
				Viscospora
		Diversisporales	Diversisporaceae	*Diversispora*
				Otospora
				Redeckera
			Entrophosporaceae	*Entrophospora*
			Pacisporaceae	*Pacispora*
			Acaulosporaceae	*Acaulospora*
				Kuklospora
		Gigasporales	Gigasporaceae	*Racocetra* *Cetraspora*
			Racocetraceae	*Scutellospora* *Orbispora*
			Scutellosporaceae	*Scutellospora* *Orbispora*
			Dentiscutataceae	*Dentiscutata* *Fuscutata* *Quatunica*
	Archaeosporomycetes	Archaeosporales	Archaeosporaceae	*Archaeospora*
				Intraspora
			Geosiphonaceae	*Geosiphon*
			Ambisporaceae	*Ambispora*
	Paraglomeromycetes	Paraglomerales	Paraglomeraceae	*Paraglomus*

3. Morphological Diversity of AM Fungi

There are many morphological characters of AM fungi that are stable and discrete are used to identify and classify. The different morphological characters are given below:

3.1 Hyphal Characters

According to Morton *et al.* (1995) AM fungal hyphae are filamentous, which tend to form various shapes and arrangements such as H-shaped, parallel connections in *Glomus*, constricted hyphae in *Acaulospora* and *Entrophospora* and irregular knobs, swollen and coiled in *Gigaspora* and *Scutellospora*.

3.2 Vesicles

Vesicles are sac like globular or oval terminal swellings of the hyphae which develop after the formation of arbuscules. In Glomaceae, vesicles are generally sub-globose to elliptical whereas in Acaulosporaceae they are knobly (Abbott and Robson, 1991).

3.3 Arbuscules

Arbuscules are tree like hyphae present in cortical cells. Abrupt narrowing off branch hyphae with reduction in width form arbuscules in *Glomus* and *Gigaspora* (Brundrett and Kendrick, 1990).

3.4 Auxillary cells

They are clusters of thin walled cells having different shape, size and ornamentation and are found in Suborder Gigasporineae (Morton, 1990).

3.5 Sporocarp(s)

They are formed by the aggregation of fungal hyphae and spores and are covered by wall layer called peridium. Most of the AM fungi are identified on the basis of morphology of their sporocarps (Gerdemann and Trappe, 1974).

3.6 Zygospores/ Azygospores

These spores may be formed singly or in groups, colour varies from hyaline to black and texture ranges from smooth walled to ornamented and size also varies from 10 µm (*Glomus*) to 1000 µm in *Gigaspora* (Aggarwal *et al.*, 2012).

3.7 Chlamydospores

They are formed terminally, laterally or intercalary and are thin walled but later develops some ornamentation (Mehrotra and Baijal, 1994).

4. Biodiversity of AM fungi

The biodiversity of AM fungi has important role in ecosystem because each individual species isolated differ in their effectiveness to enhance plant growth and adaptation under varying abiotic and biotic factors. Thus, the dynamics and composition of AMF populations have significant impact on the diversity and structure of the associated host plant both in natural and agricultural ecosystems (Van-der Heijden *et al.*, 1998). All ecological studies indicate that dynamics of AMF species decreases with land use intensification (Tchabi *et al.*, 2008). The decrease is attributed to the agricultural practices such as ploughing, fungicide application and fertilization (Roldan *et al.*, 2007). Apart from agro ecosystems, plant species, soil type, previous crops and weeds also have effect on the population and dynamics of AM fungi (Vestberg *et al.*, 2005).

In nature, most of the spices are known to harbour AM fungi which directly or indirectly effect the growth and production (Garampalli *et al.*, 2012; Dutta and Neog, 2015). Garampalli *et al.* (2012) studied mycorrhizal status of *Ocimum* sp., *Piper nigrum* and *Mentha piperata* and reported AM species richness in Western Ghats of Karnataka region.

Dutta and Neog (2015) while studying biodiversity of AM fungi with *Curcuma longa* reported three species of *Glomus*, one each of *Gigaspora* and *Acaulospora*. AM fungal diversity also depends on the time of sampling, age, season and growth stage of the test plant which generally increases from vegetative state to completely grown stage (Javaid and Riaz, 2008). According to D'Souza and Redrigues (2013), minimum AM spore species richness can be observed during monsoon season in India while maximum during pre-monsoon or winter season, which could be due to absence of organic matter and oxygen during rainy season (Verma *et al.*, 2010). The mycorrhizal colonization and spore number was found minimum in *Piper nigrum* which could be due to presence of fungitoxic compounds present in root cortical tissue (Garampalli *et al.*, 2012) whereas in *Mentha piperata* 98% root colonization was observed.

Therefore, there is an urgent need to study the AM fungi diversity in natural condition especially on spices where little work has been done. Keeping this in view, the present study was aimed to carry out (i) to determine the status of AM fungi associated with spices (ii) pattern and percentage root colonization with various spices.

5. Materials and Methods

5.1 Study Site

Seven districts of Haryana *i.e.* Ambala, Karnal, Kurukshetra, Yamunanagar, Rohtak, Kaithal and Jind have been selected for the present investigation as cultivation of different spices are done in these districts. During survey soil samples and plant roots were collected during 2015-2016.

5.2 Sample Collection

Fine plant roots, rhizome and rhizospheric soil samples (8-15 cm depth) were collected at random from the fields. Root pieces of different spices were carefully washed to clear the adhered soil particles and stored in FAA (Formalin, Acetic acid and Alcohol) solution for further analysis of AM root colonization. One part of rhizospheric soil sample was stored at 4^0C for further analysis and other part was air dried in shade and sieved through 2mm for AM spore counting.

5.3 Isolation of AM Spores

AM spores were isolated by "Wet sieving and Decanting Technique" of Gerdemann and Nicolson (1963).

5.4 Quantification of AM Spores

Quantification of isolated AM spores was done by "Grid Line Intersect Method" given by Adholeya and Gaur (1994).

5.5 Occurrence of AM Colonization

The root colonization was studied by "Rapid Clearing and Staining Technique" given by Phillips and Hayman (1970). The percentage AM root colonization was calculated by the formula:

$$\text{Percentage AM root colonization} = \frac{\text{Number of root segment colonized}}{\text{Total number of root segments studied}} \times$$

5.6 Identification of AM Spores

Identification of AM spores was done by using different keys of Walker (1986; 1992), Trappe (1982), Schenck and Perez (1990), Morton and Benny (1990), Mukerji (1996), Morton and Redecker (2001). These spores were photographed using Nikon Coolpix 54000 camera with trinocular microscope.

6. Results and Discussion

6.1 AM Morphological Types

The extent of AM fungal colonization, percentage of AM colonization and various types of mycelium, structure varied significantly in different spices. Total root AM colonization was variable, ranging maximum in *Allium sativum* and minimum in *Brassica* sp. The majority of investigated spices (11) had intracellular hyphal coils (Paris type) whereas 8 spices had linear (Arum) type of intracellular hyphae. Besides that, other type of mycelium such as Y shaped, H shaped, X shaped and T shaped was also observed in the roots (Fig. 25.1).

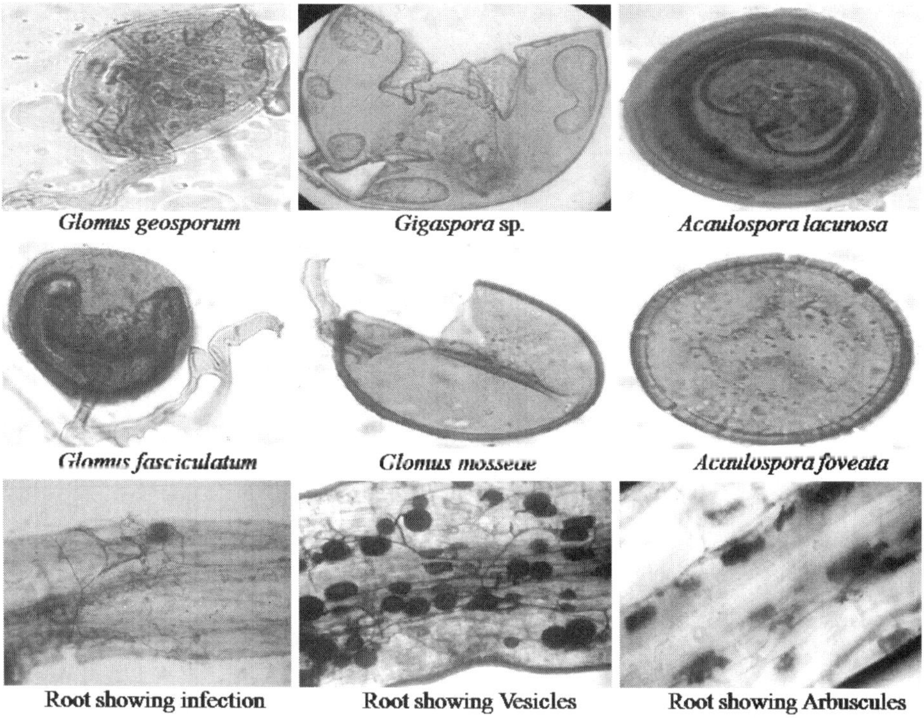

Fig. 25.1: AM spore diversity associated with rhizosperic soil of different spices.

6.2 AM Spore Density

Analysis of soil samples revealed that the AM spore number varied greatly among plants which ranged from 15.1±0.01 to 400.3±1.04. AM spore were higher in *Allium cepa* followed by *Murraya, Capsicum annum, Trigonella* sp. and *Allium sativum*. AM spore number less than 100 were observed in *Brassica, Trachyspermum* sp., *Cuminum cyminum* and *Medicago* sp. respectively. It was found that soil samples collected from Kaithal and Karnal having neutral pH exhibited more spore density in comparison to alkaline and acidic soils.

6.3 Pattern of AM Colonization

Mycorrhizal colonization was mainly observed as intracellular arbuscular, intercellular hyphae, vesicles and hyphal coils. Vesicles of various shapes like round, oval, pear shaped, elliptical were observed in majority of spices except in *Brassica* where vesicles are absent and only small fragments of intracellular hyphae was observed in *Brassica* sp.

6.4 AM Spore Diversity

The details of AM spore diversity associated with various spices are furnished (Table 25.2) and it was found that AM spores are well distributed throughout the investigated area but the species diversity varies. A total of 37 species of different genera of AM fungi i.e. *Glomus, Acaulospora, Gigaspora* and *Sclerocystis* were isolated and then identified on the basis of morphological characteristics. Out of these genera *Glomus spp.* and *Acaulospora* spp. were found to be dominant AM genera followed by *Gigaspora* sp. and *Sclerocystis* sp. (Table 25.2; Fig. 25.2). Twelve species of *Acaulospora,* twenty two species of *Glomus,* two species of *Gigaspora* spp. and one of *Sclerocystis* were found in rhizospheric soil of different spices. The most dominant species was *Glomus mosseae* occurring in all the studied samples. Similarly in *Acaulospora, A. laevis* was most frequently observed species seen in almost all the plants. In *Glomus,* out of 22 species, eleven species were identified as *Glomus mosseae, G. fasiculatum, G. etunicatum, G. aggregatum, G. luteum, G. geosporum, G. macrocarpus, G. lamellosum, G. deserticola, G. monosporum, G. clarum* and eleven species were designated as unidentified (Table 25.3). Similarly, in *Acaulospora,* five species identified were as *Acaulospora laevis, A. lacunosa, A. bireticulata, A. foveata,* and *A. scrobiculata* while 7 could not be identified.

Maximum number of spores were isolated from Kurukshetra, Ambala and Yamunanagar districts while the soils of other area were found to be comparatively poor in AM propagules. Considering the high production cost of various spices in India, it seems that the application of the dominant AM fungal inocula may help in lowering the input cost of spices. Before that knowing the AM fungal diversity, presence of dominant AM spore and extent of colonization is important for the selection of the efficient and dominant AM strain. Therefore, isolation and proper identification of AM spores was undertaken in the present study to evaluate the AM fungal diversity, host preference and extent of root colonization in spices of some districts of Haryana.

Table 25.2: Diversity and Distribution of arbuscular mycorrhizal fungi in some spices of Haryana.

S. No.	Crop	District	Pattern of Mycorrhization			AM spore density in 50 g of soil	AM Root Colonization (%age)
			Mycelium	Vesicles	Arbuscules		
1.	*Foeniculum vulgare*	Kurukshetra	+(A)	+	+	135.6±13.9	70.43±2.35
2.	*Coriandrum sativum*	Rohtak	+(A)	+	+	250.7±10.3	75.13±1.44
3.	*Alium cepa*	Karnal	+(P)	+	+	400.3±1.04	90.14±2.04
4.	*Alium sativum*	Rohtak	+(A)	+	-	320.3±2.45	91.13±1.40
5.	*Brassica nigra*	Jind	+(P)	-	-	15.1±1.01	3.1±0.01
6.	*Trigonella foenum-graecum*	Kaithal	+(P)	+	+	350.2±2.04	85.4±2.13
7.	*Capsicum annum*	Karnal	+(A)	+	+	360.1±2.15	85.3±2.13
8.	*Curcuma longa*	Kurukshetra	+(P)	+	-	125.3±1.13	55.1±1.16
9.	*Elettaria cardamomum*	Kurukshetra	+(P)	+	-	150.4±4.13	70.1±2.13
10.	*Mentha*	Jind	+(A)	+	-	135.3±1.45	65.1±3.10
11.	*Ocimum basilicum*	Kurukshetra	+(P)	+	-	150.3±2.15	70.3±1.45
12.	*Murraya*	Rohtak	+(A)	+	+	370.4±2.17	88.2±1.11
13.	*Punica granatum*	Jind	+(P)	+	-	135.4±1.45	69.3±2.11
14.	*Tamarindus indica*	Rohtak	+(P)	+	-	112.3±1.35	59.4±1.49
15.	*Cinamomum tamala*	Kurukshetra	+(P)	+	-	110.4±2.14	69.3±2.01
16.	*Trachyspermum*	Kurukshetra	+(A)	+	-	90.3±1.54	55.3±2.11
17.	*Cuminum cyminum*	Kurukshetra	+(P)	+	-	130.1±1.45	70.4±1.45
18.	*Medicago*	Yamuna Nagar	+(A)	+	-	60.3±2.15	55.4±2.15
19.	*Ocimum sanctum*	Kurukshetra Ambala	+(P)	+	++	200.1±1.3	78.3±1.01

Each value is a mean of five replicates,± = Standard deviation,A= Arbuscules,M= Mycelium,V= Vesicles, += Present, = Absent, A= Arum type hyphae, P= Paris type hyphae

Table 25.3: Biodiversity of AM fungal species on the various spices of Haryana.

S. No.	Name of the crop	Species richness	Diversity of AM fungal species*
1.	*Foeniculum vulgare*	17	1,2,3,4,5,6,9,13,14,16,19,21,22,23,31,32,35
2.	*Coriandrum sativum*	13	1,2,3,7,8,9,11,13,14,15,21,36
3.	*Allium cepa*	19	1,2,3,4,5,10,11,12,13,14,15,17,18,20,21,23 ,27,29,31
4.	*Alium sativum*	13	1,3,4,6,13,14,16,19,20,23,24,32,34
5.	*Brassica nigra*	2	12,29
6.	*Trigonella foenum-graecum*	15	1,3,4,10,11,13,14,17,18,20,21,26,27,28,34
7.	*Capsicum annum*	9	1,13,14,16,5,9,10,31,32
8.	*Curcuma longa*	4	13,14,18,36
9.	*Elettaria cardamomum*	4	1,2,13,14
10.	*Mentha*	13	2,3,6,11,12,13,14,16,18,19,22,24,25
11.	*Ocimum basilicum*	13	1,3,7,10,13,14,18,19,21,23,27,30,31
12.	*Murraya*	10	1,2,4,5,11,13,14,15,32,33
13.	*Punica granatum*	8	9,10,12,13,14,17,30,33
14.	*Tamarindus indica*	6	13,15,16,23,31,34
15.	*Cinamomum tamala*	8	5,6,9,12,14,13,17,36
16.	*Trachyspermum ammi*	9	4,6,10,12,14,15,17,20,21
17.	*Cuminum cyminum*	13	1,3,4,5,7,9,10,15,14,16,19,33,34
18.	*Medicago*	14	1,2,3,4,5,7,9,10,15,14,16,19,33,34
19.	*Ocimum sanctum*	15	1,2,5,6,7,9,13,14,16,18,19,21,30,31,33

*1- *Acaulospora laevis*, 2- A. *lacunosa*, 3- *A. scrobiculata*, 4- A. *bireticulata*, 5- A. *foveata*, 6- *Acaulospora* sp.1, 7- *Acaulospora* sp.2, 8- *Acaulospora* sp.3, 9- *Acaulospora* sp.4, 10- *Acaulospora* sp.5, 11- *Acaulospora* sp.6, 12- *Acaulospora* sp.7, 13- *Glomus mosseae*, 14- G. *fasiculatum*, 15- G. *etunicatum*, 16- *G. aggregatum*, 17- G. *luteum*, 18- G. *geosporum*. 19- G. *macrocarpum*, 20- G. *lamellosum*, 21- G. *deserticola*, 22- G. *monosporum*, 23- G. *clarum*, 24- *Glomus* sp.1, 25- *Glomus* sp.2, 26- *Glomus* sp.3, 27- *Glomus* sp.4, 28- *Glomus* sp.5, 29- *Glomus* sp.6, 30- *Glomus* sp.7, 31- *Glomus* sp.8, 32- *Glomus* sp.9, 33- *Glomus* sp.10, 34- *Gigaspora* sp.1, 35- *Gigaspora* sp.2, 37- *Sclerocystis*

Arbuscular mycorrhizas are multifunctional both in natural system and agroecosytems and hence the quality of soil (Cardoso and Kuyper, 2006). The occurrence of AM fungal association in all the spices examined from the different districts of Haryana confirms the widespread distribution of AM fungal association. Some of the plants examined under present investigation such as *Allium cepa, Allium sativum, Curcuma longa, Ocimum sanctum* and *Mentha* sp. were reported to be mycorrhizal in earlier work from different places (Garampalli *et al.*, 2012; Dutta and Neog, 2015; Muthu Kumar and Sathya, 2017). Our findings of the presence of AM spores and AM colonization in roots of *curcuma longa* stands in accordance with the observation of Dutta and Neog (2015) while working at Assam.

AM fungi are non host specific but are to some extend host preferential (Helgason *et al.*, 2002).

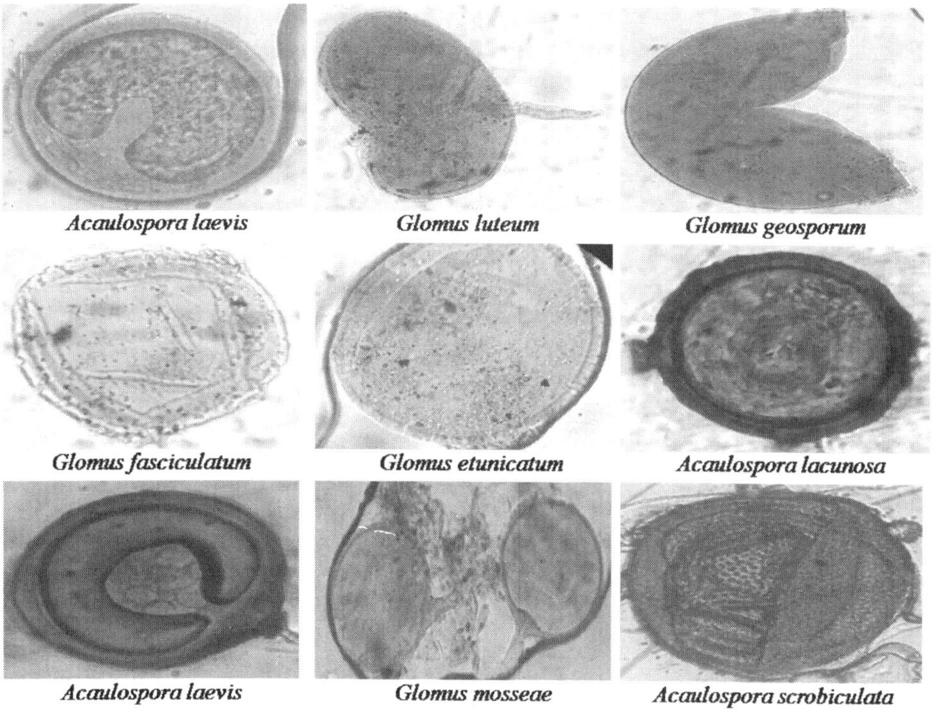

Fig. 25.2: AM spore diversity associated with rhizosperic soil of different spices.

Although the spices studied showed the presence of AM association showing that the AM fungi are widely distributed in this region yet lot of variation was seen among members of same family. This shows host preference as the main factor determining AM symbiosis in soil. Highest AM spore density was seen in *Allium cepa* followed by *Murraya*, *Capsicum annum* and *Trigonella* sp. Likewise, Sinegani and Sharifi (2007); Jangra *et al.* (2017) also reported AM spore abundance in *A. cepa* and *C. frutescens* which is in accordance with our findings. The environmental conditions and edaphic factors also influence AM fungal spore density and root colonization as observed in the present study.

In the present investigation, only 5 plants showed 75% root colonization and none of the plants showed 100% colonization. The higher root colonization is mainly due to the high dependency of spices on AM fungi (Smith and Read, 2010). The variation of root colonization could be accredited to the exudation of certain metabolites from the roots that attract the efficient and dominant fungi in disparate colonization pattern amidst different plants. It was observed in the present investigation that low level of spore sporulation is related to reduce AM root colonization while in some plants high spore number corresponds to increased AM root colonization. The results corroborate with the finding of Jangra *et al.* (2017)

mentioning correlation between spores number and root colonization. Presence of low AM spores (15.1±1.01) in *Brassica*-a non-mycorrhizal plant is in agreement with the work of Jaison *et al.* (2012).

The low AM spore number and less AM root colonization in the present study could be due to the input of chemical fertilizers by the farmers over the years might be another reason for reduced level of root colonization and low AM spore count (Dutta and Neog, 2015). The variation in AM root colonization and AM spore density may be due to changes in soil edaphic factors (Hayman, 1982) or it could be due to suitability of AM fungal spores to a particular plant of particular plant of particular geographic regions.

The moderate level of AM fungal colonization and AM spore number observed in the present study is consistent with the findings of Muthukumar and Udaiyan (2000); Fracchia *et al.* (2009) in Western Ghats of India and Argentina respectively. However, high root colonization and AM spore number we observed may be due to winter weather, when plants have good photosynthetic activity, resulting in more allocation of carbon to roots and AM fungi, resulting in higher AM colonization (Shamim *et al.*, 1994).

Soil EC is an excellent indicator of nutrient availability and it was low in our soil samples, do not negatively influence AM fungal colonization (Halder *et al.*, 2015; Wang and Jiang, 2015). Considering economic importance and production of spices in India, it seems that application of AM fungi as biofertilizer may represent a major contribution to lower the use of chemical fertilizers. But to fulfill the requirement of biofertilizers knowing the fungal diversity of AM is very important for the selection of efficient and dominant strain. Therefore, identification of AM was done to evaluate the AM fungal diversity and host preference of various species of Haryana.

According to Chaudhary *et al.* (2013), abundance and distribution of AM fungi are dependent on abiotic factors (e.g. climate) and biotic factors (type of vegetation and interspecific interactions) and inherent properties of AM fungal species. Apart from this occurrence of some AM fungal species has also been related to phosphorus status of soil, pH, salinity and soil disturbance (Abbott and Robson, 1991). In the present study, the species of *Glomus* such as *Glomus mosseae, G. fasiculatum* are widely distributed AM fungal species in many locations. The review of literature also shows that these genera have been reported from tropical as well as temperate zone, dry and wet climate, sandy and clay soils, low pH and higher pH. This indicates that these AM fungal species have the ability to invade most existing ecosystems and establish a permanent presence (Oehl *et al.*, 2003).

Acaulospora spp. are comparatively less common in the present study and are generally less in disturbed agroecosytems (Siqueria *et al.*, 1989). The result of present study is in accordance with the finding of Oehl *et al.* (2004) who have reported that there is a pattern of succession of AM fungal sporulation in natural ecosystem, *Glomus* species are the first to sporulate followed by *Acaulospora, Gigaspora, Sclerocystis* and *Scutellospora*. In the present study, some species were recorded in low numbers as one species of *Sclerocystis* reflect their narrow adaptability in contrast to several species of *Glomus* which have proved more adaptable. All the information currently available shows that *Glomus* spp. is for more frequently distributed than any of the other genera of AM fungi (Castillo *et al.*,2016). Schenck and Kinloch (1980) also reported the abundance of *Glomus* species in the soils of their wide adaptability to different plants and environmental conditions citing that *G. intraradices* was found in 9 out of 18 agroecosystems. Further Douds and Schenck (1990) attributed that *G. intraradices* has been found to be less sensitive to nitrogen and phosphorus fertilization mentioning that the intracortical sporulation of *G. intraradices* may reduce the parts of carbon flux and could contribute to the relative insensitivity of this species to chemical fertilization. Similar results were also found by different workers when compared with other studies (Alguacil *et al.*, 2014; Pagano *et al.*, 2013).

7. Conclusion

The present paper has been envasigate with an aim to provide better inside into the diversity and taxonomic characteristics of AMF associated with some spices of Haryana. It is clear from the text that the taxonomic position of different genera and species are difficult to identify on the basis of morphological characters and molecular approach is perhaps more authentic in identifying the exact taxonomic position of different genera and species of AMF. The genetic variation among AMF species and between same species may influence their comparability with plants and determine their beneficial effects on host plants. Knowledge of these things will help in better understanding to elucidate the role of AMF in different agro-ecosystems.

References

Abbott, L.K. and Robson, A.D. 1991. Factors influencing the occurrence of vesicular-arbuscular mycorrhizas. *Agric. Ecosyst. Environ.*, 35(2-3): 121-150.

Adholeya, A. and Gaur, A. 1994. Estimation of VAM fungal spores in soil. Mycorrhiza News, 6(1):10-11.

Alguacil, M.M., Torrecillas, E., García-Orenes, F. and Roldán, A. 2014. Changes in the composition and diversity of AMF communities mediated by management practices in a Mediterranean soil are related with increases in soil biological activity. *Soil Biol. Biochem.*, 76: 34-44.

Brundrett, M.C. and Kendrick, B. 1990. The roots and mycorrhizas of herbaceous woodland plants. II. Structural aspects of morphology. *New Phytol.*, 114: 469–479.

Cardoso, I.M. and Kuyper, T.W. 2006. Mycorrhizas and tropical soil fertility. *Agric. Ecosys. Environ.*, 116(1-2): 72-84.

Castillo, C.G., Borie, F., Oehl, F. and Sieverding, E. 2016. Arbuscular mycorrhizal fungi biodiversity: prospecting in Southern-Central zone of Chile. *J. Soil Sci. Plant Nutr.*, 16(2): 400-422.

Chaudhary, M.S., Rahman, S.U., Ismaiel, M.S., Sarwar, G. and Saeed, B. 2009. Coexistence of arbuscular mycorrhizae and dark septate endophytic fungi in an undisturbed and a disturbed site of an arid ecosystem. *Symbiosis*, 49(1): 19-28.

Dotzler, N., Walker, C., Krings, M., Hass, H., Kerp, H., Taylor, T.N. and Agerer, R. 2009. Acaulosporoid glomeromycotan spores with a germination shield from the 400-millionyear-old Rhynie chert. Mycol. Prog., 8(1): 9-18.

D'Souza, J. and Rodrigues, B.F. 2013. Biodiversity of Arbuscular Mycorrhizal (AM) fungi in mangroves of Goa in West India. *J. For. Res.*, 24(3): 515-523.

Douds, D.D. and Schenck, N.C. 1990. Relationship of colonization and sporulation by VA mycorrhizal fungi to plant nutrient and carbohydrate contents. *New Phytol.* 116: 621-627.

Dutta, S.C. and Neog, B. 2015. Arbuscular Mycorrhizal status in Turmeric (*Curcuma longa*) from Assam, India. *Mycorrh. News.*, 26(4): 2-5.

Fracchia, S., Aranda, A., Gopar, A., Silvani, V., Fernandez, L. and Godeas, A. 2009. Mycorrhizal status of plant species in the Chaco Serrano Woodland from Central Argentina. *Mycorrhiza*, 19(3): 205-214.

Garampalli, R.H., Seema, H.S. and Kumar, C.P. 2012. Diversity of Arbuscular mycorrhizal fungi associated with some medicinal plants of Western Ghats of Karnataka Region, India. *World J. Sci. Tech.*, 2(1): 13-20.

Gerdemann, J.W. and Nicolson, T.H. 1963. Spores of mycorrhizal Endogone species extracted from soil by wet sieving and decanting. *Trans. Br. Mycol. Soc.*, 46(2): 235-244.

Gerdemann, J.W. and Trappe, J.M. 1974. The Endogonaceae in the Pacific Northwest. *Mycol. Mem.*, 5: 1–76.

Halder, M., Dhar, P.P., Nandi, N.C. and Akhter, S. 2015. Arbuscular Mycorrhizal Colonization in some plant species and relationship with the soil properties in the BCSIR campus of Chittagong, Bangladesh. *Res. J. Biotechnol.*, 10: 51-57.

Hayman D S 1982 Practical aspects of Vesicular-arbuscular mycorrhiza. In: Advances in Agricultural Microbiology (ed) N S Subba Rao, Oxford: IBH Pub. Co., pp. 325-373.

Hayman, D.S. 1982. Influence of soils and fertility on activity and survival of VAM Fungi. *Phytomorphol.*, 72: 1119-1125.

Helgason, T., Merryweather, J.W., Denison, J., Wilson, P., Young, J.P.W. and Fitter, A.H. 2002. Selectivity and functional diversity in arbuscular mycorrhizas of cooccurring fungi and plants from a temperate deciduous woodland. *J. Ecol.*, 90(2): 371-384.

Hildebrandt, U., Ouziad, F., Marner, F.J. and Bothe, H. 2006. The bacterium *Paenibacillus validus* stimulates growth of the arbuscular mycorrhizal fungus *Glomus intraradices* up to the formation of fertile spores. *FEMS Microbiol. Letters*, 254(2): 258-267.

Jaison, S., Rajeshwari, K. and Muthukumar, T. 2012. Patterns of Endomycorrhizal associations in fruit crops of Southern India. *J. Plant Nutr. Soil Sci.*, 175: 572-581.

Jangra, E., Yadav, K. and Aggarwal, A. 2018. Potential Role of Arbuscular Mycorrhizal Fungi and Pseudomonas fluorescens on growth, physiological parameters, and yield of Capsicum frutescens L. Int. J. Veg. Sci., 24(2): 137-145.

Javaid, A. and Riaz, T. 2008. Mycorrhizal colonization in different varieties of *Gladiolus* and its relation with plant vegetative and reproductive growth. *Int. J. Agric. Biol*, 10: 278-282.

Mehrotra, V.S. and Baijal, U., 1994. Advances in the taxonomy of vesicular arbuscular mycorrhizal fungi. In: Biotechnology in India (Eds.) B.K. Dwivedi, and G. Pandey, Bioved. Res. Soc., Allahabad, India, p. 227-286.

Morton, J.B. 1990. Evolutionary relationships among arbuscular mycorrhizal fungi in the Endogonaceae. *Mycologia*, 82: 192– 207

Morton, J.B. and Benny, G.L. 1990. Revised classification of arbuscular mycorrhizal fungi (Zygomycetes). A new order, Glomales, two new suborders, Glomineae and Gigasporineae, and two new families, Acaulosporaceae and Gigasporaceae, with an emendation of Glomaceae. *Mycotaxon.*, 37: 471–491

Morton, J.B., Bentivenga, S.P. and Beve, J.D. 1995. Discovery, measurement, and interpretation of diversity in arbuscular endomycorrhizal fungi (Glomales, Zygomycetes). *Can. J. Bot.*, 73: 25-32

Morton, J.B. and Redecker, D. 2001. Two new families of Glomales, Archaeosporaceae and Paraglomaceae, with two new genera *Archaeospora* and *Paraglomus*, based on concordant molecular and morphological characters. *Mycologia*, 93: 181–195

Mosse, B. and Bowen, G.D. 1968. A key to the recognition of some Endogone spore types. *Trans. Brit. Mycol. Soc.*, 51(3-4): 469-483.

Mukerji, K.G. 1996. Taxonomy of endomycorrhizal fungi. In: Mukerji, K.G., Mathur,B., Chamola, B.P. and Chitralekha, P. (Eds.), Advances in Botany. APH Pub. Corp. New Delhi, pp. 211-221.

Muthukumar, T. and Sathya, R. 2017. Endomycorrhizal Fungal Association and Colonization Patterns in Solanaceae. *Pol. Bot. J.*, 62(2): 287-299.

Muthukumar, T. and Udaiyan, K. 2000. Influence of organic manures on arbuscular mycorrhizal fungi associated with *Vigna unguiculata* (L.) Walp. in relation to tissue nutrients and soluble carbohydrate in roots under field conditions. *Biol. Fert. Soil.*, 31(2): 114-120.

Oehl, F., Sieverding, E., Ineichen, K., Mäder, P., Boller, T. and Wiemken, A. 2003. Impact of land use intensity on the species diversity of arbuscular mycorrhizal fungi in agroecosystems of Central Europe. *App. Environ. Microbiol.*, 69(5): 2816-2824.

Oehl, F., Sieverding, E., Mäder, P., Dubois, D., Ineichen, K., Boller, T. and Wiemken, A. 2004. Impact of long-term conventional and organic farming on the diversity of arbuscular mycorrhizal fungi. *Oecologia*, 138(4): 574-583.

Oehl, F., Silva, G.A., Goto, B.T and Sieverding, E. 2011. Glomeromycota: three new genera and glomoid species reorganized. Mycotaxon, 116: 75-120

Pagano, M.C., Zandavalli, R.B. and Araújo, F.S. 2013. Biodiversity of arbuscular mycorrhizas in three vegetational types from the semiarid of Ceará State, Brazil. *App. Soil Ecol.*, 67: 37-46.

Phillips, J.M. and Hayman, D.S. 1970. Improved procedures for clearing roots and staining parasitic and vesicular-arbuscular mycorrhizal fungi for rapid assessment of infection. Trans. Br. Mycol. Soc., 55(1): 158-161.

Redecker, D., Hijri, I. and Wiemken, A. 2003. Molecular identification of arbuscular mycorrhizal fungi in roots: perspectives and problems. *Folia Geobot.*, 38(2): 113-124.

Roldan, A., Salinas-Garcia, J.R., Alguacil, M.M. and Caravaca, F. 2007. Soil sustainability indicators following conservation tillage practices under subtropical maize and bean crops. *Soil Tillage Res.*, 93(2): 273-282.

Schenck, N.C. and Kinloch, R.A. 1980. Incidence of mycorrhizal fungi on six field crops in monoculture on a newly cleared woodland site. *Mycologia*, 72: 445–456.

Schenck, N.C. and Perez, Y. 1990. Manual for the identification of VA mycorrhiza(VAM) fungi. University of Florida. (Vol. 286). Synergistic Publications, Florida, USA,241.

Schüßler, A., Schwarzott, D. and Walker, C. 2001. A new fungal phylum, the Glomeromycota: phylogeny and evolution. *Mycol. Res.*, 105: 1413–1421

Schüβler, A. and Walker, C. 2010. The Glomeromycota. A species list with new families and new genera. *Gloucester.*, England, pp. 1-58.

Shamim, D., Ahmed, T. and Ayub, N. 1994. Influence of seasonal variation on VAM infection in perennial plants. *Pak. J. Plant Pathol.*, 6: 77-80.

Sinegani, A.S. and Sharifi, Z. 2007. Changes of available phosphorus and phosphatase activity in the rhizosphere of some field and vegetation crops in the fast growth stage. *J. App. Sci. Environ. Manag.*, 11: 113-118.

Siqueira, J.O., Carneiro, M.A.C., Curi, N., da Silva Rosado, S.C. and Davide, A.C. 1998. Mycorrhizal colonization and mycotrophic growth of native woody species as related to successional groups in Southeastern Brazil. *For. Ecol. Manag.*, 107(1-3): 241-252.

Smith, S.E. and Read, D.J. 2010. Mycorrhizal symbiosis. Academic Press, London,769.

Stubblefield, S.P., Taylor, T.N. and Trappe, J. M. 1987. Vesicular-arbuscular mycorrhizae from the Triassic of Antarctica. *Am. J. Bot.*, 74: 1904-1911.

Sturmer,S.L. 2012. A history of the taxonomy and systematics of arbuscular mycorrhizal fungi belonging to the phylum Glomeromycota. Mycorrhiza, 22(4): 247-258

S korová, Z., Ineichen, K., Wiemken, A. and Redecker, D. 2007. The cultivation bias: different communities of arbuscular mycorrhizal fungi detected in roots from the field, from bait plants transplanted to the field, and from a greenhouse trap experiment. *Mycorrhiza*, 18(1): 1-14.

Taylor, T.N., Remy, W., Hass, H. and Kerp, H. 1995. Fossil arbuscular mycorrhizae from the Early Devonian. *Mycologia*, 87: 560-573.

Tchabi, A., Coyne, D., Hountondji, F., Lawouin, L., Wiemken, A. and Oehl, F. 2008. Arbuscular mycorrhizal fungal communities in sub-Saharan Savannas of Benin, West Africa, as affected by agricultural land use intensity and ecological zone. *Mycorrhiza*, 18(4): 181-195.

Thaxter, R. 1992. A revision of the Endogonaceae. *Proc. Am. Acad. Arts Sci.* 57: 291-351.

Trape, J.M. 1982. Synaptic keys to the genera and species of zygomycetous mycorrhizal fungi. Phytopathology. 72: 1102- 1108.

Trappe, J.M. and Gerdemann, J.W. 1974. The endogonaceae in the Pacific Northwest. *Mycologia Memoir*, 5: 1-76.

Van der Heijden, M.G., Boller, T., Wiemken, A. and Sanders, I.R. 1998. Different arbuscular mycorrhizal fungal species are potential determinants of plant community structure. *Ecology*, 79(6): 2082-2091.

Verma, N., Tarafadar, J. C. and Shrivastava, K.K. 2010. Periodic changes in *Prosopis cineraria* associated AM population at different soil depth and its relationship with organic carbon and soil moisture. *Afr. J. Microbiol.*, 4: 115-121.

Vestberg, M., Saari, K., Kukkonen, S. and Hurme, T. 2005. Mycotrophy of crops in rotation and soil amendment with peat influence the abundance and effectiveness of indigenous arbuscular mycorrhizal fungi in field soil. *Mycorrhiza, 15*(6):447-458.

Wang, B. and Qiu, Y. L. 2006. Phylogenetic distribution and evolution of mycorrhizas in land plants. *Mycorrhiza, 16*(5):299-363.

Wang, M. and Jiang, P. 2015. Colonization and diversity of AM fungi by morphological analysis on medicinal plants in Southeast China. *Sci. World J.*, 2015.

Walker, C. 1986. Taxonomic concepts in the Endogonaceae. II. A fifth morphological wall type in Endogonaceous spores. *Mycotaxon* 25: 95–99.

Walker, C. 1992. Systematics and taxonomy of the arbuscular endomycorrhizal fungi (Glomales)-a possible way forward. *Agronomie*, 12: 887–897.

Chapter 26

Impact of Seed Priming with Sulfosalicylic Acid on Seed Germination and Seedlings Growth of Chickpea under different Salinity Conditions

Savita and Somveer Jakhar*

Plant Physiology & Biochemistry Lab, Department of Botany, Kurukshetra University, Kurukshetra 136119, Haryana, India

**Corresponding Author E-mail: somveerjakhar@hotmail.com*

Abstract

Salinity is one of the limiting environmental factors for crop production. Seed priming is an efficient method for increasing seed vigour and integration of germination, as well as the growth of seedlings of many crops under stressful conditions. In this study, seed germination and seedling growth of chickpea was assessed using three replicates of 10 seeds in each petriplate under laboratory conditions. Four levels of salinity (0, 50, 100 and 150 mM) were given to the petriplates having seeds primed with four levels of sulfosalicylic acid (0, 10^{-4} M, 10^{-5} M and 10^{-6} M). Salinity reduced the plumule length, radical length, seedling length, plumule fresh weight, radical fresh weight, seedling fresh weight, seed vigour index and seed germination percent and priming the seeds with sulfosalicylic acid alleviated the inhibitory effects of salt stress by enhancing the tested parameters. Our results showed that the maximum response was recorded by 10^{-5} M sulfosalicylic acid priming treatment. Seedling length was significantly increased followed by plumule length in all priming treatments in comparison with control.

Keywords: *Chickpea, salinity, seed germination, seed priming and sulfosalicylic acid.*

1. Introduction

Chickpea (*Cicer arietinum* L.) belongs to family leguminoseae, is an annual vegetarian food plays an important role in human food and domestic animal feed. Emergence and rapid germination are important determinants of successful growth and establishment of seedling. However, salinity stress is considered as important problem for sowing of chickpea. The impact of growth regulators in manipulating physiological processes in crop production include seed germination, vigour, nutrient uptake etc. Priming of seeds with sulfosalicylic acid has been reported to be a simple and safe technique for increasing the capacity of seed to osmotic adjustment and enhancing seed germination (Afghani and Taheri, 2012; Ghasemi and Hosseini, 2012). Seed priming with sulfosalicylic acid induced salinity tolerance. Following are the objectives of the present study:

- To study the effect of seed priming with sulfosalicylic acid on seed germination percentage and seed vigour index under salt stress.
- To study the effect of seed priming with sulfosalicylic acid on radicle length, plumule length, seedling length, radicle fresh weight, plumule fresh weight and seedling fresh weight under different salt stress.

2. Materials and Methods

The experimental treatments included salinity at four levels [0, 50 (S_1), 100(S_2) and 150(S_3) mM] and sulfosalicylic acid (SSA) at four levels (0, 10^{-4}, 10^{-5} and 10^{-6} M). The seeds were sterilized by 10% hypochlorite solution and then, were rinsed with distilled water. The seeds were soaked in 10^{-4}, 10^{-5} and 10^{-6} M SSA at 25 C for 4 hours. Distilled water was used as control treatment. Afterwards, the primed seeds were rinsed with distilled water and dried between two layers of paper (23±2 C with relative humidity of 60%). Ten pre-treated seeds were placed in each petri dish and given the same required amount of salinity solutions and distilled water (as control). After 24 hours, the seeds were daily checked for 7 days and the number of germinated seeds was recorded. A seed was considered germinated upon emergence of radical about 2 mm in length. Seed germination percentage (SGP) was determined at the end of the test. After 7 days, to determine the radical length (RL), plumule length (PL), seedling length (SL), radicle fresh weight (RFW), plumule fresh weight (PFW), seedling fresh weight (SFW), normal plumules and radicals produced in each petri dish were separated from the seeds, and their lengths and fresh weight were measured. Seedling vigour index (SVI) was calculated by multiplying the germination percentage by the seedling length and then dividing the product by 100 (Abdul-Baki and Anderson, 1970). Analysis of Variance was calculated using ANOVA test using software SAS. Differences among the means of the treatments were estimated using the Duncan's multiple range tests at the 5% probability level.

3. Results and Discussion

3.1 Germination Percentage

Salinity decreased the germination percentage as compared to non-salinity condition. Salinity treatment lowered the germination percentage by 12, 33 and 46 percent at S_1, S_2 and S_3 levels of salinity respectively as compared to controls (Fig. 26.1). Under salinity conditions, germination percentage was significantly increased by SSA treatment as compared to the seedlings without SSA. The most effective concentration of SSA was 10^{-5} M which increased percentage germination by about 14, 12 and 10 percent at different levels of salinity (50, 100 and 150 mM) respectively corresponding to their controls. Similar trends was observed in earlier studies by Shakirova *et al.* (2003) and El-Tayeb *et al.* (2005) where they observed that emergence percentage was increased by SA treatment under salinity conditions. Seed germination and early seedling growth are the most sensitive stages to salinity stress in many crop species (Akazawa *et al.*, 1990). Seeds primed with different concentrations of SSA proved to be effective in inducing salt tolerance at the germination stage (Afzal *et al.*, 2006). It is quite clear from present investigation.

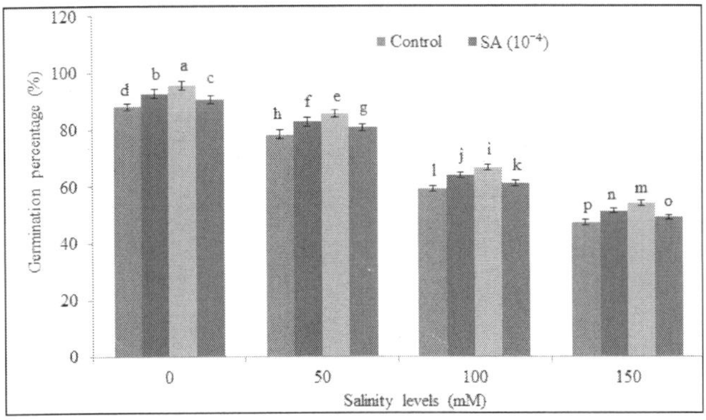

Fig. 26.1: Effect of priming by sulfosalicylic acid on germination percentage in chickpea (*Cicer arietinum*) under different salinity levels.

3.2 Seed Vigour Index (SVI)

SVI is the ability of emergence of seed during stress and it decreased with increasing salinity stress in the present study (Fig. 26.2) indicated that SSA at 10^{-5} M concentration increased SVI by 54, 47 and 40 percent respectively at different levels of salinity (50, 100 and 150 mM) as compared to control. Makawi *et al.* (1999) reported that survival ability has reduced when seedling cannot make a powerful root system and therefore, root growth can be considered as a screening scale for seed vigour.

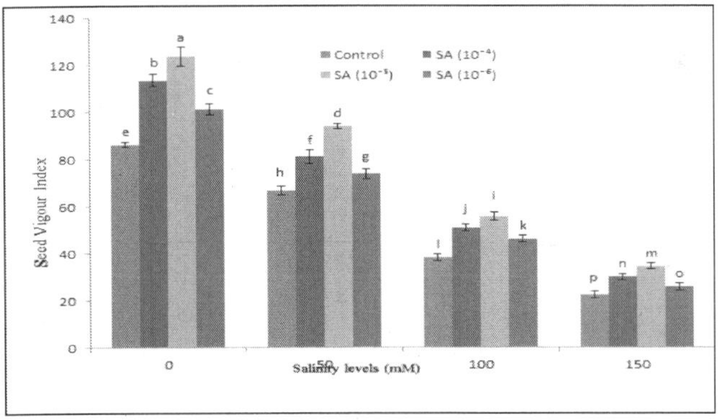

Fig. 26.2: Effect of seed priming by sulfosalicylic acid on seed vigour index in chickpea (*Cicer arietinum*) under different salinity levels.

3.3 Radicle Length

In the current research, radicle length decreased progressively with increasing salinity. It may be due to toxic effects of the unbalanced nutrient uptake by the seedlings due to presence of high concentration of NaCl. SSA at 10^{-5} M concentration enhanced radicle length by 62, 57 and 49 percent respectively at S_1, S_2 and S_3 levels of salinity as compared to their controls. SSA at 10^{-6} M also increased the radicle length but the magnitude was less as compared to 10^{-5} M concentration. Root elongation was inhibited by high salinity due to slowing down the uptake of water by the plants (Werner *et al.*, 1995). The seeds primed with 10^{-5} M SSA showed maximum radicle length (6.93 cm). The seeds treated with SSA produced larger radicle length in all salinity levels than untreated seeds (Fig. 26.3). Similarly Jaleel *et al.* (2007) on *Catharanthus roseus* reported that root length was affected by different salinity levels.

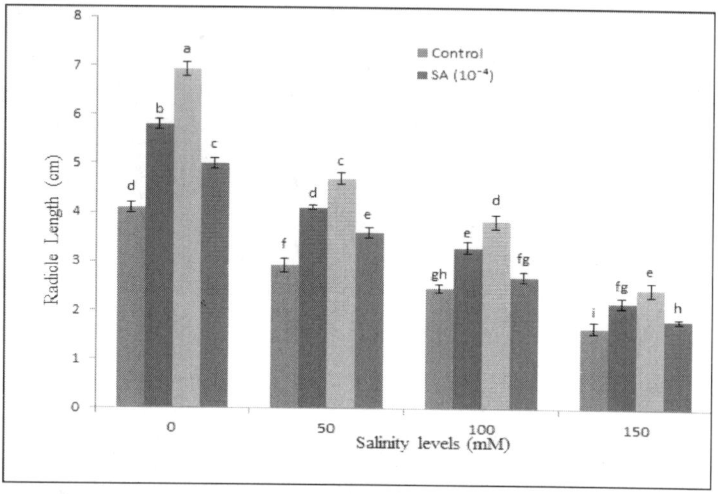

Fig. 26.3: Effect of seed priming by sulfosalicylic acid on radicle length in chickpea (*Cicer arietinum*) under different salinity levels.

3.4 Plumule Length

Plumule length decreased significantly with increase in salinity levels. The shortest plumule length was found in the highest salinity level (150 mM). Sulfosalicylic acid at 10^{-5} M caused significant alleviation in length of plumule at different salinity levels (Fig. 26.4). SSA at 10^{-5} M mitigated the effects of salinity by increasing plumule length by 78, 63 and 45 percent respectively at S_1, S_2 and S_3 levels corresponding to their controls. Decrease in length may be due to decrease in cell elongation and decrease in turgor. Similar results have been observed in salinity treatment of two sensitive and resistant types of rice (Kang *et al.*, 2002) and cowpea (Dantas *et al.*, 2005).

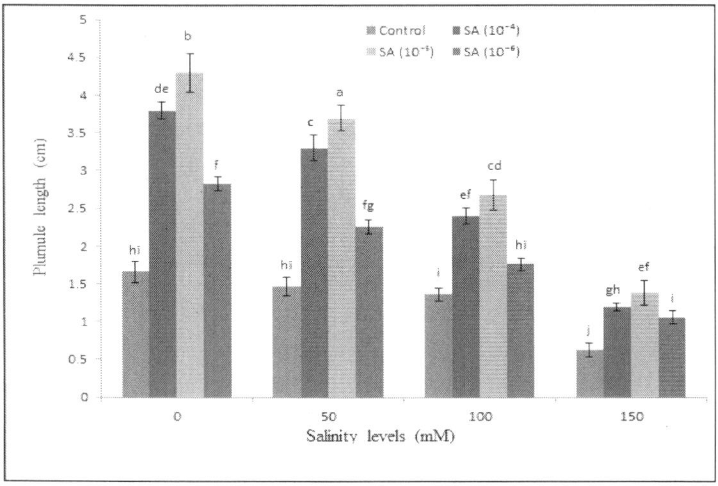

Fig. 26.4: Effect of priming by sulfosalicylic acid on plumule length in chickpea (*Cicer arietinum*) under different salinity levels.

3.5 Seedling Length

Results showed the effect of seed pretreatment by different concentrations of sulfosalicylic acid on seedling length during stress (Fig. 26.5). Seeds pre treatment with sulfosalicylic acid alleviates the seedling length in both stressed and non-stressed conditions. 10^{-5} M concentration of SSA increased the salt stressed seedling length at S_1, S_2 and S_3 levels by about 129, 89 and 56 percent corresponding to their controls.

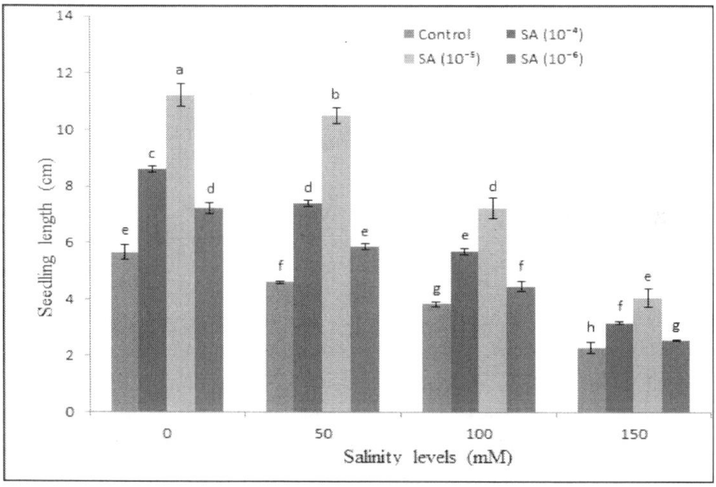

Fig. 26.5: Effect of priming by sulfosalicylic acid on seedling length in chickpea (*Cicer arietinum*) under different salinity levels.

3.6 Radicle Fresh Weight

Radicle fresh weight decreased under salinity treatment by about 14 %, 35 % and 52 % at S_1, S_2 and S_3 levels as compared with control (Fig. 26.6). Previous studies have shown that NaCl treatment decreased the some growth parameters such as fresh weight of root of cucumber plants (Yildirim *et al.*, 2008). In this study, we found that the presence of NaCl reduced the radical fresh weight of chickpea compared to the control as a consequence of salt osmotic effects, which reduced water availability. During imbibition, water enters through aquaporins, which have reduced expression in the presence of salt (Boursiac *et al.*, 2005). Seed priming with different concentrations of sulfosalicylic acid alleviated the radical fresh weight of salt stressed chickpea seeds but 10^{-5} M concentration was the most effective which increased radical fresh weight by 48, 39 and 24 percent respectively at S_1, S_2 and S_3 levels corresponding to their controls.

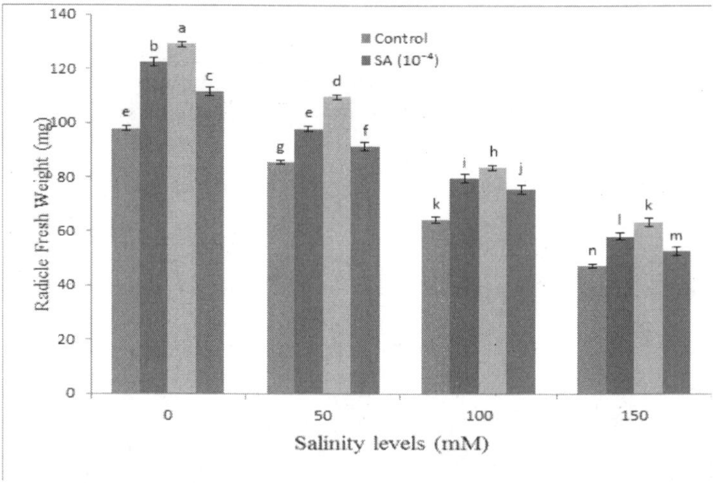

Fig. 26.6: Effect of priming by sulfosalicylic acid on radicle fresh weight in chickpea (*Cicer arietinum*) under different salinity levels.

3.7 Plumule Fresh Weight

Salinity reduced the fresh weight of plumule of chickpea compared to the control by about 28, 32 and 62 % at S_1, S_2 and S_3 levels respectively (Fig. 26.7). Priming of seeds by different concentrations of salicylic acid mitigated the negative effect of salt stress. 10^{-5} M SSA was the most effective and alleviated the salt stressed plumule fresh weight of chickpea by about 45, 29 and 16 percent respectively at S_1, S_2 and S_3 levels corresponding to their controls. Khodary (2004) also reported that SA increased fresh weight of shoot of salt stressed maize plants. Similarly, Mori *et al.* (2011) reported that salinity reduced the plumule fresh weight of plants.

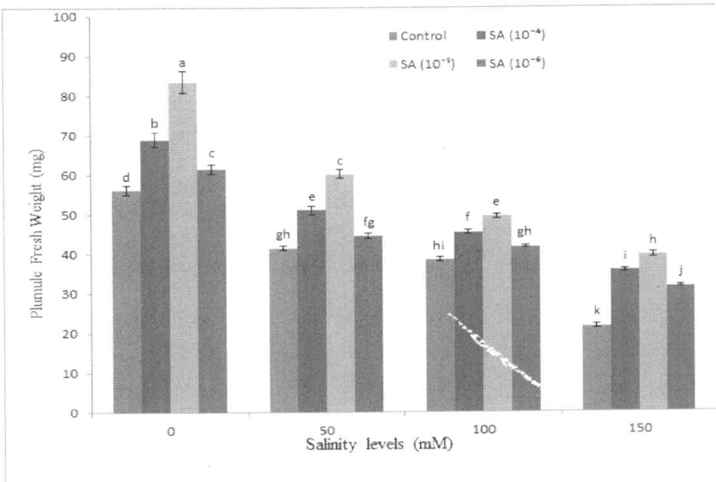

Fig. 26.7: Effect of priming by sulfosalicylic acid on plumule fresh weight in chickpea (*Cicer arietinum*) during salt stress.

3.8 Seedling Fresh Weight

(Fig. 26.8) Indicated the effect of SSA on seedling fresh weight on chickpea seedling during salinity stress. Three different salinity levels (50, 100 and 150 mM) decreased seedling fresh weight by about 18, 34 and 55 percent respectively. Priming with different concentration of SSA mitigated the negative effect of salt stress on seedling fresh weight. 10^{-5} M SSA concentration was the most effective and increased the seedling fresh weight by 36, 29 and 22 % at S_1, S_2 and S_3 salinity levels respectively. The results are in agreement with the findings of Yagmur and Kaydan (2008) while working on triticale.

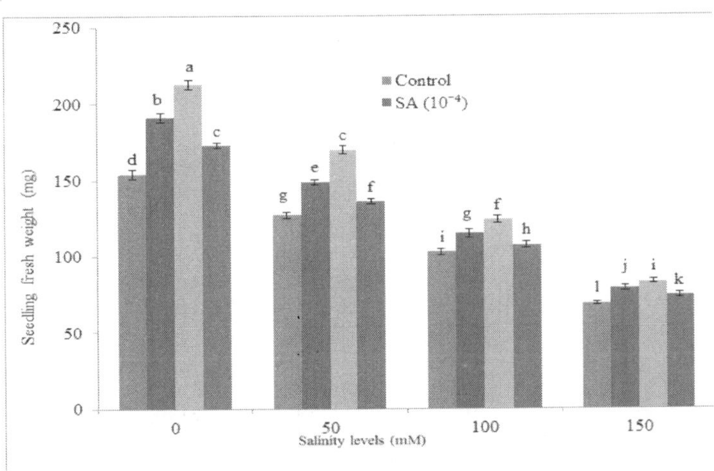

Fig. 26.8: Effect of priming by sulfosalicylic acid on seedling fresh weight in chickpea (*Cicer arietinum*) under different salinity levels.

4. Conclusions

Priming increases seed and seedling quality so it is very useful for chickpea seedlings. Primed seeds germinate faster and this will increase their competition power against the weeds. The priming by sulfosalicylic acid was the most effective on plumule length. The cost of sulfosalicylic acid is nominal and affordable. SSA is required in minute quantity therefore making it an excellent tool for enhancing productivity.

References

Abdul–Baki, A.A. and Anderson, J.D. 1970. Viability and leaching of sugars from germinating barley. *Crop Sci.* 10: 31-34.

Afghani, A. and Taheri, G. 2012. Survey the effect of seed priming on germination and physiological indices of cotton khordad cultivar. *Annals Biol. Res.* 3(2): 1003-1009.

Afzal, I., Basra, S.M.A., Faroog, M. and Nawaz, A. 2006. Alleviation of salinity stress, in spring wheat by hormonal priming with ABA, Salicylic Acid and Ascorbic acid. *Int. J. Agr. Biol.* 1: 23-28.

Akazawa, T., Yamaguchi, J. and Hayashi, M. 1990. Rice-Amylase and gibberellin action a person view. In gibberellin (Eds), Takahashi BN, Phineey BO and Mac Millan J. Springer-Varlag. 114-124.

Boursiac, Y., Chen, S., Luu, D.T., Sorivul,M., Drias, N. and Maural, C. 2005. Early effects of salinity on water transport Arabidopsis roots: Molecular and cellular features of aquaporin expression. *Plant Physiol.* 139: 790-805.

Dantas, B.F., Saribeiro, L. and Aragao, C.A. 2005. Physiological response of cowpea seeds to salinity stress. *Rev. Bras. Sementes.* 27 (1): 144-148.

El-Tayeb, M. A. 2005. Response of barley grains to the interactive effect of salinity and salicylic acid. *Plant Growth Reg.* 45: 215-224.

Ghasemi Lemrasky, M. and Hosseini, S.Z. 2012. Effect of seed priming on germination behaviour of wheat. *Int. J. Agric. Crop Sci.* 4(9): 564-567.

Jaleel, C.A., Gopi, R., Sankar, B., Manivannan, P., Kishorekumar, A., Sridharan, R. and Pannneerselvan, R. 2007. Studies on germination, seedling vigor, lipid peroxidation and proline metabolism in *Catharanthus roseus* seedlings under salt stress. *South Africa J. Bot.* 73: 190-195.

Kang, H.M. and Saltveit, M.E. 2002. Chilling tolerance of maize, cucumber and rice seedlings leaves and roots are differently affected by salicylic acid. *Physiol. Plantarum.* 115: 571-576.

Khodary, S.E.A. 2004. Effect of salicylic acid on the growth, photosynthesis and carbohydrate metabolism in salt-stressed maize plants. *Int. J. Agric. Biol.* 6: 5-8.

Makawi, M., Balla, M.E.L., Bishaw, Z. and Van-Gastel, A.J.G. 1999. The relationship between seed vigor test and field emergence in lentil (*Lens culinaris* L.). *Seed Sci. Technol.* 26: 657-667.

Mori, M., Di-Mola, I. and Quaglietta-Chiaranda, F. 2011. Salt stress and transplant time in snap bean: growth and productive behavior. *Int. J. Plant Prod.* 5: 49-63.

Shakirova, F.M., Sakhabutdinova, A.R., Bezrukova, M.V., Fatkhutdinova, R.A. and Fatkhutdinova, D.R. 2003. Changes in the hormonal status of wheat seedlings induced by salicylic acid and salinity. *Plant Sci.* 164: 317-22.

Werner, J.E. and Finkelstein, R.R. 1995. Arabidopsis on growth of roots and shoots in soybean. Plant mutants with reduced response to NaCl and osmotic stress. *Physiol. Plant.* 93: 659-666.

Yagmur, M. and Kaydan, D. 2008. Alleviation of osmotic stress of water and salt in germination and seedling growth of triticale with seed priming treatments. *African J. Biotechnol.* 7(13): 2156-2162.

Yildirim, E., Turan, M. and Guvenc, I. 2008. Effect of foliar salicylic acid applications on growth, chlorophyll and mineral content of cucumber grown under salt stress. *J. Plant Nut.* 31: 593-612.

Chapter 27

Standardization of Herbal Drugs: Challenges and Resolutions

Prerna Sarup[1,*], Vipin Saini[2] and Manisha Bhatia[1]

[1]MM College of Pharmacy, Maharishi Markandeshwar Deemed University, Mullana-Ambala

[2]Maharishi Markandeshwar Deemed University, Solan, H.P.

**Corresponding Author E-mail: psarup_pu@hotmail.com*

Abstract

The popularity of herbal medicines has risen worldwide. India is sitting on a gold mine of well-recorded and well-practiced knowledge of traditional herbal medicine. This increase in usage renders safety issues important. Herbal drug development is possible only through the development of standardized herbal products. Many undesirable effects of herbal medicines can be attributed to the substandard quality of the raw materials or the finished products. Quality issues of herbal medicines can be external or internal. Present review includes contamination with pesticides residues and microbes, adulteration and misidentification. Intricacy and non-consistency of the ingredients in herbal medicines are the internal issues which affects their quality. The thorough execution of Good Agricultural and Collection Practices (GACP) and Good Manufacturing Practices (GMP) would without doubt reduce the risk of external issues. We assume that drug manufactured in accordance with principles of Ayurveda, Siddha and Unani will reach new horizons and make them the best in the world if the quality of the herbal drugs is maintained, efficacy would itself be maintained and then there is nothing to stop them from competing with the modern medicine with added advantages of fewer side effects and lower costs.

Keywords: *Standardization, herbal drugs/medicines, adulteration, Ayurveda*

1. Introduction

Plants contain several hundred constituents and some of them are present at very low concentrations. In spite of the modern chemical analytical procedures available, only rarely do phytochemical investigations succeed in isolating and characterizing all secondary metabolites present in the plant extract. According to guidelines as per

European Agency for the Evaluation of Medicinal Products, the source and quality of raw materials play a pivotal role in guaranteeing the quality and stability of herbal preparations. Other factors like use of fresh plants, temperature conditions, exposure to light, availability of water and nutrients, season & time of collection, method of collection & drying, packing, storage and transportation of raw material etc., can greatly affect the quality and the therapeutic value of herbal medicines (Anonymous, 1998). Some plant constituents are heat labile and the plants containing them need to be dried at low temperatures.

Also, other active principles are destroyed by enzymatic processes that continue for long periods of time after plant collection. This explains why frequently the composition of plant based drugs is quite variable. Thus, proper standardization and quality control of raw material and the herbal preparations themselves should be permanently carried out (Anonymous, 1998c). In the cases where the active principles are unknown, marker substance(s) should be established for analytical purposes. However, in most cases these markers have never been tested to see whether they really account for the therapeutic action reported for the herbal drugs. Apart from these variable factors, others such as the method of extraction and contamination with microorganisms, heavy metals, pesticides, etc., can also interfere with the quality, safety and efficacy of herbal drugs (Calixto, 2000). For these reasons, pharmaceutical companies prefer using cultivated plants instead of wild-harvested plants because they show smaller variation in their constituents (Grunwald, 1995). Furthermore, and certainly more relevant, when medicinal plants are produced by cultivation, the main secondary metabolites can be monitored and this permits definition of the best period for harvesting.

Compared with well-defined synthetic drugs, standardization of herbal medicines exhibit some marked differences, namely:

- The active principles are frequently unknown;
- Standardization, stability and quality control are feasible but not easy;
- The availability and quality of raw materials are frequently problematic;
- Well-controlled double-blind clinical and toxicological studies to prove their efficacy and safety are rare;
- Empirical use in folk medicine is a very important characteristic;
- They have a wide range of therapeutic use and are suitable for chronic treatments;
- The occurrence of undesirable side effects though less frequent but they do exist (Farnsworth, 1993)
- They usually cost less than synthetic drugs.

2. Standardization/Evaluation Parameters

In order to assure a consistent and acceptable quality herbal product, care should be taken right from the identification and authentication of herbal raw materials to the verification process of final product. Standardization is a system to ensure that every unit of a medicine has the correct substances in the correct amount and produces the desired therapeutic effect (Yadav and Dixit, 2008).

Standardization and quality assurance has three attributes *viz.* authenticity, purity and assay. Authenticity relates to proving that the material is true i.e. it corresponds to the right identity. It in itself involves many parameters including morphology, microscopy and chemical analysis. Purity pertains to evaluating that there are no adulterants present in the plant material. It can be evaluated by pharmacognostic evaluations especially microscopic characters (Patel *et al.*, 2006).

In assay, chemical and biological profiling is established, which assures the curative value of the drug. Safety for use could also be assessed through this parameter. In biological assay, the drug activity is evaluated through pharmacological models. However, biological testing is highly subjective in nature due to lot of variations and moreover there are strict regulatory controls on animal testing to limit its applicability in plant standardization. Chemical profiling on the other hand is a versatile technique. Fingerprints can be developed by using techniques of HPTLC, HPLC, GC and marker compounds can be quantified. The following are the parameters to be considered in standardization of herbal drugs as well as finished products (Bhutani, 2000).

The following parameters are recommended for the standardization of herbal drugs as follows:

2.1 Authentication

It needs to be assured that the plant material used in the formulation has been collected from an appropriate geographical region at an appropriate stage of its growth and the botanical identity has been correctly established (Handa, 1995).

2.2 Macroscopic Evaluation

This refers to drug evaluation by means of our sense organs and includes other sensory organs like odour, colour, taste and texture. It includes the study of organoleptic and morphological characters.

Microscopical identification of plant materials is based on shape, size, colour, odour, taste, texture, fracture and appearance of the cut surface. Since these characters are judged instinctively and substitutes or adulterants may closely resemble the genuine material, it is often necessary to substantiate the findings by microscopy and/ or by physicochemical tests as per WHO guidelines given under quality control methods for medicinal plant materials (Anonymous, 1998.

2.3 Microscopic Evaluation

Microscopic evaluation of plant drugs is essential to study the adulterants and is indispensable in establishing the correct identity (Handa, 1995). Qualitative microscopy like stomatal number, stomatal index and palisade ratio is of help in differentiation of closely allied species of leafy drugs (Evans, 1997).

2.4 Physical Evaluation

2.4.1 Foreign Matter

The WHO describes foreign matter as a material consisting of any or all of the following: (Anonymous, 1998).

- ☆ Parts of the medicinal plant or materials other than those named with specified limits;
- ☆ Any organism, part or product of an organism, other than that named in the definition and description;
- ☆ Mineral admixtures not adhering to the medicinal plant materials, such as soil, stones, sand and dust.

2.4.2 Ash Value

When vegetable drugs are incinerated, they leave an inorganic ash which in case of many drugs varies within narrow limits and these values are of significance for the purpose of plant drug evaluation (Handa, 1995). The presence of ash in the medicinal plant materials is determined as total ash, acid-insoluble and sulphated ash. **Ash values** helps in determining the quality and purity of powdered crude drugs. The objective of ash formation is to remove all traces of organic matter, which can interfere in an analytical determination.

2.4.3 Extractive Values

The determination of extractable matter refers to the number of constituents present in a given amount of medicinal plant material that is extracted in a given volume of a solvent. Such extractive values provide an indication of the extent of polar, medium-polar and non-polar components present in the medicinal plant material. It is yet again an important parameter for standardization of crude drugs (Anonymous, 1998).

2.4.4 Pesticide Residues

Pesticide residues finds an entry to the plant materials from agricultural practices, such as spraying, fumigation, soil treatment during cultivation and storage. Since many medicinal preparations of plant origin are taken over long periods of time, limits for pesticide residues should be established following the recommendations of the Food and Agriculture Organization (FAO) and the World Health Organisation (WHO). The recommended guidelines include the analytical methodology for the assessment of specific pesticide residues (Anonymous, 1998).

2.4.5 Microbial Contamination

Medicinal plant materials normally carry a great number of bacteria and moulds, often originating in soil. A large number of bacteria and fungi form the naturally occurring microflora of herbs. Aerobic spore forming bacteria frequently predominate this microflora. Current practices of harvesting, handling and production may cause additional contamination and microbial growth. The determination of *E. coli* and moulds may indicate the quality of production and harvesting practices. In addition, the presence of aflatoxins in plant material can cause health hazards, even if present in very small amounts. Therefore, they should be determined using a suitable procedure (Anonymous, 1998).

2.4.6 Heavy Metals

Environmental pollution and traces of pesticides that comes from agricultural practices such as spraying and fumigation are the major causes that are responsible for the contamination of medicinal plant materials with arsenic, lead, mercury and other heavy metals. The limits (parts per million) of such heavy metals in medicinal plants should remain within specifications. Different countries have enforced different limits of heavy metals which normally range between 0.1 to 20 ppm for lead, 0.5 to 5.0 ppm for mercury, 0.3 ppm for cadmium, traces to 5 ppm for arsenic and 150 ppm for copper (Anonymous, 1998).

2.4.7 Radioactive Contamination

Irradiation is used as a procedure of microbial decontamination and sterilization of plant materials during packaging and storage. Accidental contamination of an area and of plant materials growing in the area can result from industrial accidents. Contamination occurs when material that contains radioactive atoms is deposited on materials, skin, clothing, or any place where it is not desired. It is noteworthy that radiation does not spread any contamination rather it is radioactive contamination that can be spread (Anonymous, 1998).

2.4.8 Chromatography

One of the best methods of standardizing herbs and herbal formulations based on the modern scientific tools is chromatography (Dobriyal and Naryana, 1998). Chromatography is a universally adopted analytical technique in pharmaceutical analysis of herbal drugs. The technique is used by the chemists to separate and determine species in variety of organic, inorganic, and biological materials. Basically, chromatography is based on the differences in the rate at which the components of a mixture move through a stationary phase under the influence of a solvent or gas called mobile phase. Thin layer chromatography studies are among the key identity tests in most Pharmacopoeial monographs. Pharmacopoeial standards are typically used by industry as a basis for meeting quality control requirements (Dobriyal and Naryana, 1998).

High-performance thin layer chromatography (HPTLC) is an extension of TLC and is robust, rapid and efficient tool in quantitative analysis of compounds. It uses the same principles as in thin layer chromatography and column chromatography. HPTLC confirms as well as establish identity of herbal drugs. Further, it is an ideal screening tool for adulteration and is highly appropriate for evaluation and monitoring of cultivation, harvesting, and extraction processes and testing of stability.

The chromatographic method of separation in general involves the following steps:(Chatwal and Anand, 2003).

- Adsorption or retention of the substances on the stationary phase
- Separation of the adsorbed substances by the mobile phase

- ☆ Elution of the adsorbed substances by a continuous flow of the mobile phase
- ☆ Qualitative and quantitative analysis of the eluted substances.

A chromatographic fingerprint of a herbal medicine is a chromatographic pattern of the extract showing chemical components of pharmacologically interest and / or chemical interest (Wagner and Bladt, 1984). This chromatographic profile should feature the fundamental attributes of "integrity" and "fuzziness" (or "sameness" and "differences") so as to chemically represent the plant material under investigation (Poole, 2003). It is suggested that with the help of chromatographic fingerprints, the authentication and identification of herbal medicines can be accurately conducted even if the amount and concentration of the characteristic chemical constituents are not exactly the same for different samples of the herbal medicine. Thus, it is now widely agreed to consider multiple constituents in the herbal medicines for evaluating the quality of the medicinal plant extracts. In any plant drug or its extract, there are hundreds of unknown components and many of them are present in extremely low amounts. To obtain reliable chromatographic fingerprints that represent pharmacologically active and chemically characteristic components is not an easy work. Surprisingly, chromatography offers very powerful separation ability such that the complex chemical components in plant products can be separated into many relatively simple sub-fractions (Shrikumar and Ravi, 2007).

Chromatographic fingerprinting has been in use for a long time for the single chemical entity drug substances. The use of chromatographic fingerprinting for herbal drugs tends to focus on identification and assessment of the stability of the chemical constituents observed by chromatography. Chemical and chromatographic techniques may also be used to aid in identification of a herbal material or extract. Fingerprinting of the herbals is done by HPLC, HPTLC, TLC and LC- MS etc (Yadav and Dixit, 2008).

i. Thin Layer Chromatography

This is a separation technique wherein a stationary phase consists of a suitable material spread in a uniform thin layer on a glass, metal or plastic support. The mobile phase moves through the stationary phase by capillary action, sometimes assisted by gravity or pressure. The numerous sorbents have been used like silica gel, cellulose, aluminium oxide, polyamide and chemically bonded silica gel. TLC separations take place in 'open' layer with each component having the same migration time but covering different distances (Chatwal and Anand, 2003).

TLC has become widely adopted for the rapid and positive analysis of plant drugs. Out of many chromatographic methods, presently available, since the time required in this technique for the demonstration of most of the characteristic constituents is very short. In addition to qualitative detection, it also furnishes semi-quantitative information about the major constituents of the drug which enables assessment of drug quality easier (Kokate *et al.*, 2002).

It has the advantage of many fold possibilities of detection in analyzing herbal medicines. In addition, it is rather simple and can be employed for multiple sample analysis.

ii. Effective use of Thin Layer Chromatography

The attractive features of TLC include parallel sample processing for high sample throughout; accessibility of the sample for post-chromatographic evaluation; free of time constraints; detection in the presence of the stationary phase independent of mobile phase properties; and normally the single use of stationary phase (Wagner and Bladt, 1984). It is effective for the low-cost analysis of samples requiring minimal clean up or using fewer steps in sample preparation. It is also preferred for the analysis of substances with poor detection characteristics requiring post-chromatographic treatment for detection. Since all sample components are located in the chromatogram, it is the most suitable technique for surveying sample properties (Wagner and Bladt, 1984).

Many of these advantages are realized in large scale surveillance programs. Examples include the identification of drugs of abuse and toxic substances in biological fluids, the identification of unacceptable residue levels of drugs used to prevent disease or promote growth in farm animals, maintaining a safe water supply by monitoring natural and drinking water resources for crop protecting agents in modern agriculture and confirmation of label claims for content of pharmaceutical products. For these applications, it is used as a component of a pyramid strategy. It requires an efficient screening step to identify suspect samples and a confirmation step to establish the contaminant level in suspect samples. The benefits of this strategy are lower costs and an increase in the number of samples processed. TLC is selected for the screening step for the following advantages:

- ✰ Single use of the stationary phase minimizes sample preparation;
- ✰ Parallel separations enhancing productivity; and
- ✰ Easy post chromatographic derivatization which improves method selectivity and specificity (Evans, 1997).

In the pharmaceutical industry, TLC is used for the analysis of complex and dirty samples with poor detection characteristics and for testing stability and content uniformity. It continues to be widely used for the standardization of plant materials used in traditional medicines. Combined with digital autoradiography, it is frequently selected as the method of choice to study the metabolism and fate of radio-labelled compounds in the body and environment. In addition, it retains a historic link with the characterization of dyes and inks and the control of impurities in industrial chemicals (Wagner and Bladt, 1984).

iii. High Performance Thin Layer Chromatography (HPTLC)

HPTLC is an advancement of TLC principle requiring shorter time and giving better resolution over shorter runs. The basic difference between conventional TLC and HPTLC is in the particle, pore size and uniformity of the sorbent. The adsorbent particles are smaller in size and more uniform in HPTLC plates. HPTLC is applied to obtain "fingerprint" pattern of herbal formulations, quantification of active ingredients and also in detection of adulterants.

iv. Gas Chromatography

It is very important for the analysis of volatile compounds of herbal medicines. GC of a volatile oil gives a fingerprint which can be used to identify the source plant. Gas chromatography is a similar method to that of HPLC. It provides a quick and easy way of determining alkaloids in a mixture. The only requirement is some degree of stability at the temperature necessary to maintain the substance in the gas state (Stauffer *et al*, 2008).

v. High Performance Liquid Chromatography

It is a popular method for the analysis of herbal medicines for its reliability, coupled with numerous options of analysis and its use is not limited by the volatility or stability of the sample. It can be used to analyze almost all kinds of compounds in the herbal medicines. It has received the most extensive application in the analysis of herbal medicines and most advancements have taken place in this technique (Chatwal and Anand, 2003).

2.5 Chemical Evaluation

The preliminary photochemical, screening was carried out using the extracts for different types of chemical constants. The qualitative chemical tests gave the general idea regarding the nature of chemical constituents of wide drugs. The extracts were subject to preliminary phytochemical investigation for detection of alkaloids, carbohydrates, glycosides, phenolic compounds, flavonoids, proteins & amino acids, saponins, steroids and acidic compounds (Saxena *et al.*, 2014). List of important tests to detect various phytoconstituents have been summarized below in Table 27.1.

Table 27.1: Qualitative analysis methods for various phytoconstituents (Saxena *et al.*, 2014).

S. No.	Phytoconstituent	Chemical test
1.	Alkaloids	Mayer's test
		Hager's test
		Dragendroff's test
		Wagner's test
2.	Carbohydrates & reducing sugars	Molisch's test
		Fehling's test
		Benedict's test
3.	Cardiac glycosides	Killer-Killiani's test
4.	Fixed oils & Fats	Spot test
		Saponification test
5.	Flavanoids	Lead Acetate test
6.	Gums & Mucilages	Ruthenium Red Solution test
7.	Phytosteroids & Terpenoids	Salkowski's test
		Liebermann Burchard's test

S. No.	Phytoconstituent	Chemical test
8.	Tannins & Phenols	Ferric Chloride test
		Lead Acetate test
9.	Proteins and Amino acids	Millon's test
		Biuret's test
		Ninhydrin's test
10.	Saponins	Foam test

2.6 Biological Evaluation

Estimation of biological efficacy is considered to be the most effective method for evaluation of herbal drugs. The specific protocols have been given to evaluate various types of biological activities. For all these protocols, there are a sequential animal model for testing and controlling requirements, for the use and evaluation of the results. Biological evaluation is generally preferred when the drug cannot be evaluated satisfactorily by chemical and physical methods. The response produced by the test drug on a living system is compared with that of the standard preparation. The biological activity is represented in units as International Units (I.U) (Barre- Sinoussi, 2015).

We share approximately 99% of our DNA with mice (1), and moreover, we can use "knockout" mice to work out what effect individual human genes have in our body. We do this by "turning off" one of the genes in a mouse, common to a human, and seeing what effect this has on the mouse. By recreating human genetic diseases in this way we can begin to look for treatments. Animal models allow closer approximation to a human response (Anonymous, 2010).

3. Conclusions

Herbal medicines are prepared from materials of plant origin which are prone to contamination, deterioration and variation in composition. This gives rise to inferior quality of herbal products with reduced or no therapeutic efficacy. The first step is the collection of raw material wherein there are chances of mistaken identity of crude drugs. The next steps involving storage and processing further adds to variation in crude drug.

The relative proportion of active constituents can vary from plant to plant of the same species due to the climatic and geographical factors. So, there is a need to lay standards for selection of authentic material for development of formulation. Thus, standardization of herbal medicines is necessary to evaluate the quality of crude drug in terms of content of active constituents present in the drug.

India can play a major role in production of standardized and therapeutically effective herbal formulations. It can be achieved only if herbal drugs are analyzed using sophisticated modern techniques of standardization such as HPLC, HPTLC, GC-MS, spectrofluorimetric and other methods. It is strongly recommended that various government agencies should strictly adopt WHO guidelines for developing monographs for crude drugs. This will reinforce the regulatory process and minimize violation of use of spurious or below quality herbal drugs.

References

Anonymous. 1998. Quality control of methods for medicinal plant materials. World Health Organization. Geneva.

Anonymous. 1998. Quality of Herbal Medicinal Products. Guidelines European Agency for the Evaluation of Medicinal Products EMEA, London.

Anonymous. 1998c. Basic Tests for Drugs, Pharmaceutical Substances, Medicinal Plant Materials and Dosage Forms. World Health Organization, Geneva.

Anonymous. 2010. Biological similarity of humans and other animals. https://speakingofresearch.com/facts/the-animal-model.

Barre-Sinoussi, F., and Montagutelli, X. 2015. Animal models are essential to biological research: issues and perspectives. *Future Sci. OA*, 1(4): FSO63.

Bhutani, K.K. 2000. Fingerprinting of Ayurvedic Drugs. *East. Pharm.*, 43: 21-26.

Calixto, J.B. 2000. Efficacy, safety, quality control, marketing and regulatory guidelines for herbal medicines (phytotherapeutic agents). *Braz. J. Med. Biol. Res.*, 33(2): 179-189.

Chatwal, R.G. and Anand, K.S. 2003. Instrumental methods of chemical analysis. 5th edition. Mumbai: Himalaya Publishers, pp: 567.

Dobriyal, R.M. and Naryana, D.B.A. 1998. Ayurvedic herbal raw material. *East. Pharm.*, 45: 31-35.

Evans, W.C. 1997. Trease and Evan's Pharmacognosy, 14th edition. India. Harcourt Brace and Company Asia Pvt. Ltd. Pp: 576.

Farnsworth, N.R. 1993. Relative safety of herbal medicines. *Herbal Gram*, 29 (Special Suppl.): 36A-36H.

Grunwald, J. 1995. The European phytomedicines market: figures, trends, analysis. *Herbal Gram*, 34: 60-65.

Handa, S.S. 1995. Herbal raw materials and traditional remedies. *East. Pharm.* 8: 23-27.

Kokate, C.K., Purohit, A.P. and Gokhale, S.B. 2002. *Pharmacognosy*, 20th edition. India: Nirali Prakashan, pp: 99.

Patel, P.M., Patel, N.M. and Goyal, R.K. 2006. Quality control of herbal products. *Indian Pharm.*, 5(45): 26-30.

Poole, F.C. 2003. Thin layer chromatography challenges and opportunities. *J. Chromatogr. A*, 1000: 963-984.

Saxena, R., Sharma, R., Chandra Nandy, B., Jasuja, N. D. 2014. Qualitative and quantitative estimation of bioactive compounds in *Mimosa hamata*. *Int. J. Pharm. Pharm. Sci.*, 6(6): 72-75.

Shrikumar, S. and Ravi, T.K. 2007. Approaches towards development and promotion of herbal drugs. *Pharmacog. Rev.*, 1: 180-184.

Stauffer, E., Dolan, J. A., Newman, R. 2008. Gas Chromatography and Gas Chromatography Mass Spectrometry in Fire Debris Analysis. pp: 235-293.

Wagner, H. and Bladt, S.Z. 1996. Plant drug analysis. A thin layer chromatography atlas. 2nd Eds. Springer- verlag, Berlin.

Yadav, N.P. and Dixit, V.K., 2008. Recent approaches in herbal drug standardization. *Int. J. Integr. Biol.*, 2: 195-203.

Chapter 28

Bioinformatics Tools in Drug Discovery

Sangeeta Singh[1,*], Shilpi Verma[1], Swechha Mishra[1], Shikha Jaggi[2] and Shruti Gupta[1]

[1]*Department of Applied Science, Indian Institute of Information Technology, Allahabad-211015, India.*

[2]*Department of Zoology, Gandhi Memorial National College, Ambala Cantt.-133005, Haryana, India*

**Corresponding Author E-mail: swechha.iiita@gmail.com*

Abstract

In this article, the drug discovery pipeline and various bioinformatics tools for designing a rational drug have been discussed. Rational way of developing a drug is too expensive and lengthy, leading people to opt for more prominent and effective methodology. Popularity of bioinformatics in pharmaceutical field lies on the fact that it is both time saving as well as cost effective. Pharmaceutical industry has now adopted bioinformatics for finding drug targets and for drug discovery. In this process, first and foremost is requirement of a target. Target is screened out thorough literature search. Later, 3D-structure of these targets was downloaded from RCSB database and various modeling methods are used when 3D-structure is not available. Overall various prospects of drug discovery like functional prediction, mapping information, gene expression data are taken care of by various bioinformatics tools which gather the aforementioned information from the respective databases. This is beneficial not only in terms of cost but also saves time of many researchers.

Keywords: *Drug discovery, target identification, homology modeling, virtual screening, target validation, lead discovery, docking, lead optimization, preclinical test.*

1. Introduction

The discovery of a drug begins when there are no medical products available for the treatment of a particular disease. The first step towards drug designing is to make a presumption about a particular biological entity whose activation or inactivation will be curative. The drug molecules can modulate the activity of the target biomolecule by showing higher binding affinities than the natural ligands and thereby inactivating it or by activating the target which has been down-regulated in the diseased state like cancer. After the identification of the target biological entity, its validation needs to be carried out. The next step in the drug discovery process involves lead discovery where a vigorous search is made for finding the drug molecule that effectively binds to the target. After lead discovery and optimization, preclinical and clinical development of the drug ensues followed by marketing of the drug (Hughes *et al.*, 2011). The schematic showing the drug discovery pipeline is shown in (Fig. 28.1).

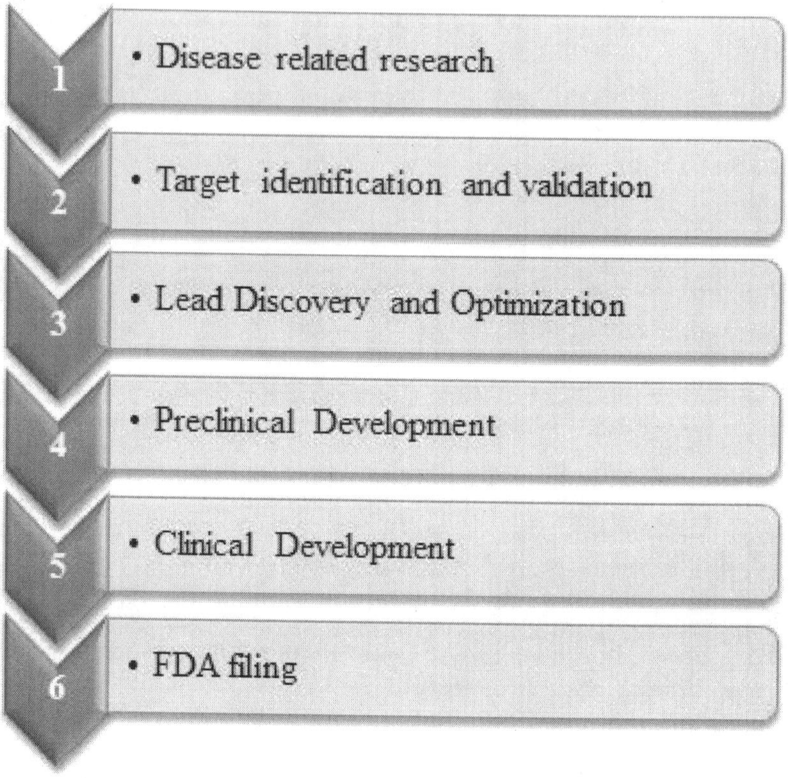

Fig. 28.1: Work flow of drug discovery process.

The innovation process of a drug is a very tedious and a time-consuming task. Shortcuts are required to find the structures of the target biological entities and the drug molecules that appropriately bind to them to bring about a therapeutic response. Computational tools have brought pharmaceutical industry closer to engineering-based disciplines by simulating every aspect of drug discovery and development. Bioinformatics tools are used in many areas of computer aided drug designing research like virtual high throughput screening, drug bioactivity and bioavailability prediction, sequence inspection, homology modeling, finding similarities, physiochemical modeling, drug lead improvement etc.

One of the main bioinformatics tool used widely for drug designing is Schrödinge. Schrödinger is software that develops the chemical simulation for use in pharmaceutical, biotechnology, and material science research. It provides molecular modeling programs of chemical simulation and series of drug design software. It also includes ligand and structure based methods. This software helps in generating automatically the structural analogs of the compound. It can be viewed, analyzed and filtered by using the different tools to reduce the hundreds or thousands of analogs to a manageable number for synthesis. Maestro is the portal to all of Schrödinger computational technology, which helps the researchers to organize and analyze data. For the application in Schrödinger, it requires dedicated licenses (https://www.schrodinger.com/suites/biologics-suite) (Table 28.1).

Table 28.1: Some of the applications and its functions for the drug designing in Schrödinger.

Application	Function
Canvas	Cheminformatics analysis
CombiGlide	Combinatorial Library Design
Epik	pKa calculation in biological conditions
Glide	Ligand - receptor docking
Induced -Fit Docking	Ligand-receptor docking with target dynamics
Jaguar	ab initio quantum mechanics
Liaison	Binding affinity calculations
LigPrep	2D to 3D conversion program to prepare ligand libraries
Macro Model	Classical molecular modeling application
Phase	Ligand-based pharmacophore modeling
Prime	Protein structure prediction package
QikProp	ADME p properties prediction of drug candidates
QSite	QM/MM application
SiteMap	Identification and characterization of active sites
Strike	Scientific data analysis program

2. Target Recognition

A drug target can encompass various biological entities present in our body like genes, RNA and proteins. The target needs to be 'druggable' that is, the drug molecule should be able to bind to the target and should be able to bring about a biological effect which is determinable and quantifiable both *in vitro* and *in vivo*. Proteins are considered as good targets for drug designing. If the structure of the target protein is already known, it can be extracted from the respective database like Protein Data Bank, Dynameomics, PDBWiki, PDB Lite, OCA and the drug designing step can be taken to the next level. However, if the amino acid sequence is known but the structure is not known then homology modeling is used. In homology modeling, a 3D representation of the objective protein is created from an obtainable amino acid sequence and a 3D-structure of a protein comparable to the target protein serves as a template. BLAST (Basic Local Alignment Search Tool) is the most popular server that gives the structures of all the proteins that align with the sequence of the target protein by more than 30% by searching various databases (Blundell *et al.*, 2006). Sometimes homologues of the target protein may not be available in the databases. Another template-based target protein structure prediction approach called threading can be used in such cases. In threading, a 3D-structure of the target protein can be generated from a template protein having the same folds as the target protein (Blundell *et al.*, 2006).

3. Protein Modeling (Table 28.2)

Table 28.2: Protein Modeling Softwares.

Name	Description	URL
Modeller	Automated Webserver & Downloadable Program	https://salilab.org/modeller/download_installation.html
Swiss Model	Automated Web server	https://swissmodel.expasy.org/
CPHModel	Automated Webserver	http://www.cbs.dtu.dk/services/CPHmodels/
Robetta	Automated Webserver	http://www.robetta.org/submit.jsp
RaptorX	Automated Webserver & Downloadable Program	http://www.robetta.org/submit.jsp

3.1 Modeller

MODELLER is used for the homology modeling of a particular protein having three-dimensional structure. It is a computer program originally written and maintained by Andrej Sali at the University of California, San Francisco. The target sequence and the related protein structure need to be provided by the user and MODELLER generates a model containing all non-hydrogen atoms. In this, the user provides an alignment of a sequence that has to be modeled and MODELLER will automatically generate and calculates the model having all non-hydrogen atoms. It gives a comparative protein structure modeling by giving the full information about the spatial restraints, and it includes *de novo* modeling of loops in protein structures, optimization of protein structure, clustering, searching of sequence databases etc. It develops a model by using a method that tries to overcome any spatial constraints (Vyas *et al.*, 2012). It can carry out numerous tasks like loop modeling in protein structures from the scratch, modifying the protein structure to get high accuracy, manifold protein sequence alignments, comparison of protein structures etc. It can run on Unix, Linux, OS X, and Windows operating system (Šali and Blundell, 1993).

3.2 Swiss Model

It is a structure-based bioinformatics web-server for modeling of protein 3D structure. Swiss Model is based on homology-modelling for constructing the protein architecture. Its access can be gained by ExPASy and DeepView (Swiss PDB-Viewer). It takes a target protein sequence as input and uses BLAST to find the protein structures to which the target protein sequence aligns to in the sequence database of proteins in the PDB (Protein Data Bank). The SWISS-MODEL works on the basis of its integrated components: 1) SWISS-MODEL pipeline -for protein structure modeling software tools and database; 2) SWISS-MODEL Workspace- web based graphical user; 3) The SWISS-MODEL Repository- for continuously updating of the database of homology model.

For building up a protein structure, the SWISS-MODEL pipeline first identifies the structural template, then alignment of target sequence and template structure occurs. When model is build up, its energy minimization occurs. It implements the rigid fragment assembly for modeling. Finally, through QMEAN the model is evaluated (Schwede, 2003).

3.3 CPH Model

CPHmodels-3.0 known to be a web-server used for predicting protein having 3D-structure by using the single template homology modeling. This server provides a hybrid of the scoring function from the CPHmodels-2.0 and a novel remote homology-modeling algorithm. By using the CPHmodels -2.0, the query sequence is first exerted and then modeled. The template recognition depends on the profile-profile alignment guided by the secondary structure and the exposure predictions. One of the very important features of this server is its speed, and the response time of the server is less than 20 minutes. In CPHmodels 3.2 Server, the sequence which is used to be pasted should be in FASTA format. The restriction in this server is that the only one sequence per submission with not less than 15 and not more than 4,000 amino acids can be processed (Nielsen *et al.*, 2010).

3.4 Robetta

This server provides an automated tool for prediction of protein structure and its analysis. The sequence is submitted to the server for the structure prediction and it is resolved into presumptive domains and the structural model is generated by using the comparative modeling of the structure or by *de novo* prediction of structure methods. The server gives the model for an entire protein sequence whether the sequence of homology protein of known structure is present or not. The results are given as in the form of molecular coordinates of models spanning the full-length query. The sequence used to be submitted in this server must be in one letter amino acid format. The sequence is either pasted into the submission form or can be uploaded from a file. The user can submit the sequence for domain identification or for the structure prediction. The user has to mention the PDB id or chain for the comparative modeling (Kim *et al.*, 2004).

3.5 RaptorX

It is software as well as a web server which is associated with the protein structure and function prediction. It is free to use and available for the non-commercial purposes. It able to produce a definite protein models with a large number of sequence homologs by manipulating the structure information. Through this web portal, various servers are running. Firstly, RaptorX Structure Prediction predicts the structure of protein sequence (tertiary) by using the templates. RaptoX Property Prediction helps in prediction of the structure properties of a protein by using the given templates. RaptorX contact Prediction without using the template helps in predicting the contact map of the protein sequence and tertiary structure (Sutcliffe *et al.*, 1987).

4. Protein Threading (Table 28.3)

Table 28.3: Protein Threading Softwares.

Name	Description	URL
I-TASSER	Structure & Function Prediction	http://zhanglab.ccmb.med.umich.edu/I-TASSER/download/download.php
LOOPP	Webserver	http://clsb.ices.utexas.edu/loopp/web/
Phyre & Phyre2	Webserver, updated fold library,genome searching	http://www.sbg.bio.ic.ac.uk/phyre/html/
SPARKS-X	Webserver	http://sparks-lab.org/yueyang/server/SPARKS-X/
MUSTER	Webserver	http://zhanglab.ccmb.med.umich.edu/MUSTER/

4.1 I-TASSER

I-TASSER is a bioinformatics tool which is used to predict 3-D structure model of a protein by using the amino acid sequences. It helps in detecting the structure template from the PDB by fold recognition or threading. Rearrangement of the structural fragments from threading templates constructs the full-length structures models by using replica exchange, Monte Carlo simulations. It provides annotations on ligand binding site and gene ontology by matching the structure of the structural models of the target protein to the already known proteins function database (Yang *et al.*, 1987).

4.2 LOOPP

LOOPP is a tool for fold recognition and threading which is based on the collection of various signals, and combining them into a single score and then generating the atomic coordinates depending upon the alignment into a homologous template structure. Straightforward sequence alignment, sequence profile, threading, and exposed surface area prediction signals are used to be included. The signals are individually merged locally to produce a combined models and to provide overall scores such as Z score calculation. The atomic model are produced by using an alignment generated by the scoring scheme, the final atomic structure which is evaluated by adding the other energy scores, and the energies used is determined by the mathematical programming algorithm (Teodorescu *et al.*, 2004).

4.3 Phyre and Phyre 2

Phyre and Phyre2 are a protein Homology/Analogy recognition Engine and a web-based services for protein structure prediction. This tool is able to generate a highly reliable protein models in comparison to other tools. In this tool, the protein sequence which is submitted by the user is firstly scanned against a large sequence database by using the PSI-BLAST. Then the profile produced by this is processed by the neural network secondary structure prediction program known as PsiPred and a protein disorder predictor program Disopred. The graphical representation shows all the presence of alpha-helices, beta -strands and the disordered regions with a color-coded confidence bar. Phyre2 which is an advanced version used as a homology detection method to build 3D-models and predicting the ligand binding sites and analyze the effect of the amino acid variants for a user protein sequence (Kelley *et al.*, 2015).

4.4 Sparks-X

Sparks-X is a fold recognition server and it uses a single-method fold recognition technique known as SPARKS by manipulating the alignment scoring function and integrate with the SPINE-X techniques which leads to the improved prediction of secondary structure along with the torsion angle and solvent accessible surface area. It helps in the recognizing the structurally similar folds even in the absence of the sequence similarity (Yang *et al.*, 2011).

4.5 MUSTER

MUSTER is a newly developed threading algorithm which combines different sequence and structure information into a single body form which can be easily used in the dynamic programming search. The results of this server include the top five template proteins and the query-template alignments along with the full-length built by MODELLER (Wu and Zhang, 2008).

5. Ab Initio Structure Prediction (Table 28.4)

Table 28.4: Ab initio Protein Structure Prediction.

Name	Description	URL
EVFold	Webserver	http://evfold.org/evfold-web/evfold.do
QUARK	on-line server for protein modeling	http://zhanglab.ccmb.med.umich.edu/QUARK/
FALCON	Webserver	http://protein.ict.ac.cn/TreeThreader/
PEP-Fold	On-line server for peptide structure prediction	http://mobyle.rpbs.univ-paris-diderot.fr/cgi-bin/portal.py#forms::PEP-FOLD
CABS-Fold	Webserver	http://biocomp.chem.uw.edu.pl/CABSfold/

5.1 EVfold

The EVfold or EVcoupling calculates the evolutionary coupling of a protein. It shows the network of interactions between residue pairs which is based on the evolutionary information of the protein family. The interaction between the residue pairs can be easily visualized without having the 3D-structure, but if the 3D-structure is available than the interaction between the residue-residue can be viewed. It is used as distance constraints on unfolded proteins, which is then folded using the distance geometry and simulated annealing. It works on both the globular and transmembrane proteins and optimized for the individual domains (Hopf *et al.*, 2014) (Fig. 28.4).

5.2 QUARK

QUARK is a computer algorithm for prediction of ab initio protein folding and its structure, and leads to construct the 3D-protein model from the given amino acid sequence. This tool stands from the small fragments around 1-20 residues long by replica-exchange Monte Carlo simulation using the atomic level knowledge-based force field. In the QUARK, it starts from the collection of the fragments (1-20 residues) from PDB structure libraries (Xu and Zhang, 2012). Then a distance profile is obtained which contains long-range interactions in a two-step method: 1) for each pair of the residues, a histogram of a certain distance is calculated only from the pair of top 200 fragments; 2) Histograms convert the distance profile: if there is a histogram peak in the middle range of the distances. Now finally, the fragments are assembled into full-length models by using the replica-exchange Monte Carlo simulations which is guided by some knowledge-based potential that include hydrogen-bonding, van der Waals, solvation, coulomb forces and other secondary structure element (Zhang *et al.*, 2016).

5.3 FALCON

FALCON is a protein structure prediction server which is used for focusing on the homologue identification. The form of the model depends on the homologous proteins which consist of the conserved region associated with the highly variable regions and it leads to the nebulous alignments in the threading approaches. It first retrieves the conserved regions from the template and then aligns it in a query protein with the conserved region, and this helps in the neglecting of nebulous alignment in highly variable regions which improves the remote homologue recognition (Wang *et al.*, 2015).

5.4 PEP-Fold

PEP-Fold is a *de novo* approach which leads to the prediction of peptide structure from the amino acid sequences. The method used in this is depends on the structural alphabet SA letters and it describes the conformations of the four consecutive residues and the SA letters is predicated by the greedy algorithm and a coarse-grained force field. The first step in this server is to start with the single amino acid sequence and the advanced version of this runs up to 200 simulations and generate the models, according to the details about the clusters and the best conformation. The 5 best clusters are then selected (Maupetit *et al.*, 2009).

5.5 CABS-Fold

CABS-Fold is a web server which provides tools for the prediction of protein structure from the sequence only and by using the alternative templates. The web server is based on the critical assessment of techniques for protein structure prediction. The different kinds of qualitatively approaches are used for structure modeling such as de novo methods, which include molecular simulation of protein folding processes or the fragment assembly (Blaszczyk *et al.*, 2013).

6. Target Validation

Target validation is a very time consuming and costly process that determines whether the target chosen in the drug discovery process is of significance or not. As a general rule, the target is not really validated until late stage clinical trials are finished and the therapeutic effects are observed. Target validation is additionally used to portray the early phases of the drug discovery procedure that pave the way for moving to the next step in the drug discovery pipeline. Some of the protein validation softwares (Table 28.5) are:

Table 28.5: Target validation Softwares.

Name	Description	URL
HARNONY	Protein Structure Validation Web server	http://caps.ncbs.res.in/harmony/
RCSB PDB	Web server	http://www.rcsb.org/pdb/home/home.do
RESPROX	Web server	http://www.resprox.ca/
PSVS	Protein Structure Validation Software Suite	http://www.resprox.ca/
PROSESS	Web server	http://www.prosess.ca/

6.1 Harmony

This web-based unique tool probes into the resemblance among the sequence and structure of the target protein by accrediting values to amino acid residues and their different arrangements in the wake of considering their local environment. Local Environment is depicted by the access to solvent, conformation of the backbone, and the arrangement of hydrogen bonds. The user can upload the documents related to the protein structure and the details of its homologue if present and scores are accredited by the tool (Pugalenthi *et al.*, 2006).

6.2 RCSB PDB

The RCSB PDB (Research Collaboratory for Structural Bioinformatics Protein Data Bank) is a worldwide accessible resource that provides assistance in not only research and important inventions but also in achieving a deep understanding of the biological system and its processes at the molecular stage. It provides tools which, with the help of data stored in the Protein Data Bank (PDB), are able to procure and envision data. It is a representative of the worldwide PDB (wwPDB) which is a result of the joint effort by various scientists to make a globally accessible resource for providing important structural information about biological macromolecules that have been experimentally reported (http://www.rcsb.org/pdb/home/home.do).

6.3 RESPROX

RESPROX is an internet web server which takes PDB coordinate data as an input and predicts the atomic resolution of NMR (Nuclear Magnetic Resonance) protein structures. To accurately determine the atomic resolution of a protein structure. RESPROX uses the machine learning techniques that uses 25 measurable features that can be inferred from its atomic coordinates. RESPROX offers X-ray crystallographers and NMR spectroscopists the chance to effectively evaluate the accuracy and caliber of their 3D-protein structures because atomic resolution is a straightforward and near-universal measure of structure quality (Berjanskii *et al.*, 2012).

6.4 PSVS

PSVS (Protein structure validation software suite) is appropriate when a protein structure needs to be evaluated, whose structure has been determined by X-Ray Crystallography, NMR or homology modeling techniques. It helps in estimating the goodness of a structure by using various structure authenticating tools like PROCHECK, Verify3D, Prosall, PDB software etc. for structure approval. It assigns scores based on how well the query structure matches with the experimentally available data and its homologues (Bhattacharya *et al.*, 2006).

6.5 PROSESS

Protein Structure Evaluation Suite & Server (PROCESS) is a web server which is designed for the validation of the protein structure from the X-ray crystallography or NMR spectroscopy. It evaluates various factors such as covalent and geometric quality, torsion angle quality, and chemical shift quality. PROSESS gives a detailed table, explaining the structural images and graph which gives a outline of the results and compare the values of the observed in the high-quality or high-resolution protein structures (Berjanskii *et al.*, 2010).

7. Lead Discovery

Before the advent of advanced technologies for finding suitable drug for the target, the drug target interactions were analyzed through *in vivo* studies. Since doing such studies was time consuming and expensive, *in silico* approaches were developed for finding the candidate drugs. Ligand docking tools help in finding the most appropriate orientation of the compound to bind with the active site of the protein. To short list a large number of drug molecules based on their affinities for the protein's active site, virtual screening is adopted which makes docking more efficient. In virtual screening, an extensive search is made in the databases of small molecules for finding an optimum ligand for the protein. Web based docking servers accommodate a multitude of *in silico* chemistry software that help in the estimation of constants and specifications which are required at various stages of the docking process. It performs definite optimization of ligand configuration, minimization of energy, charge prediction and estimation, and the protein structure modeling.

Some protein-ligand docking softwares (Table 28.6 and Fig. 28.8) are:

Table 28.6: Lead Discovery Softwares.

Name	Description	URL
Autodock	Downloadable software	http://autodock.scripps.edu/
		MGL tools :- http://mgltools.scripps.edu/downloads
		Cygwin:- http://www.cygwin.com/install.html
		Discovery Studio Visualizer:- http://accelrys.com/products/discovery-studio/visualization-download.php
		Binary Files:- http://autodock.scripps.edu/downloads/autodock-registration/autodock-4-0-1-and-autogrid-4-0.0
		Java :- http:/www.java.com/en/download/index.jsp

Name	Description	URL
Dock	Downloadable software work on the basis of Geometric	http://dock.compbio.ucsf.edu/
Swiss Dock	Web server	http://www.swissdock.ch/
idTarget	Web server	http://idtarget.rcas.sinica.edu.tw/
MEDock	Webserver	http://medock.csbb.ntu.edu.tw/

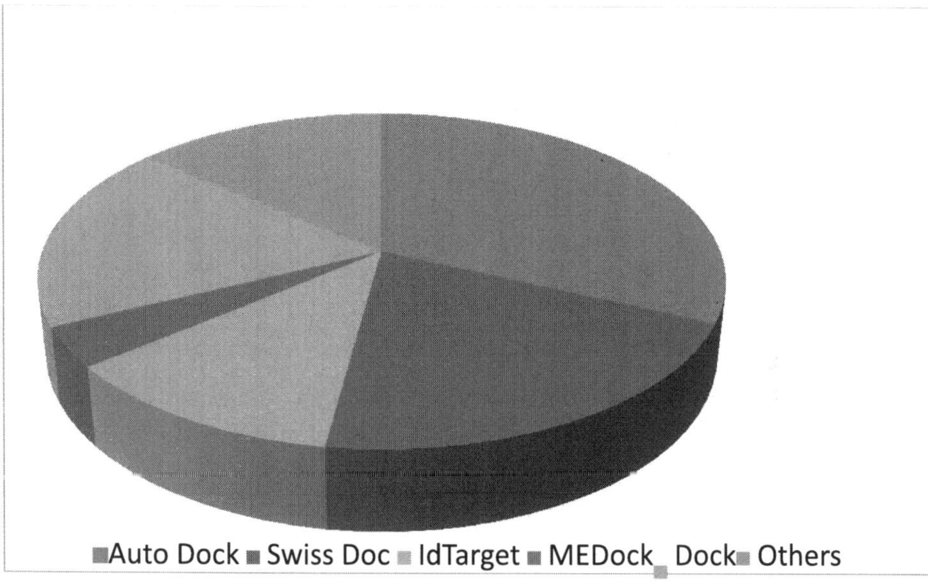

Fig. 28.8: The pie chart showing the docking softwares which are used frequently.

7.1 Auto Dock

AutoDock is software that helps in simulating the interaction and intercommunication between the ligand and the protein. It comprises of automated docking tools. The Scripps Research Institute and Olson Laboratory currently own this unique docking software. It performs structure-based drug designing, lead optimization, high throughput screening (HTS), conjunctional designing of library, protein-protein docking and imparts knowledge of various chemical processes (Morris *et al.*, 2008). It has two generations of software: Auto Dock 4 and Auto Dock Vina. The Auto dock 4 consists of two main programs: (i) Autodock for placing the ligand in a framework that represents the protein of interest; (ii) Autogrid for predetermining the framework. In AutoDock 4 calculations are performed in various steps:

- ☆ Coordinate file preparation: The protein and ligand models used can have a polar hydrogen bond but it should be free from carbon atom. The PDBQT format is used for the preparation of coordinate files, which include the partial charge on the atom and its type.
- ☆ Autogrid calculation: The protein is deep-seated in a three-dimensional framework and an inquiring atom is placed at each point in the framework. This grid helps in the estimation of compatibility of each type of atom present in the ligand with that of the target protein.
- ☆ Docking using AutoDock: Various techniques like Lamarckian Genetic Algorithm, simulated annealing are used for doing the task of docking. The AutoDock executes multiple times and gives many configurations produced by docking the ligand on the target protein. The best result is selected by analyzing the predicted energies of all the docked conformations and the consistency of the results (http://autodock.scripps.edu/). Some freely available tools are required for running AutoDock in Windows XP or Windows 7 which are MGL tools, Cygwin, Discovery Studio Visualizer, Binary files and Java.

7.2 Dock

Dock brings about the placement of the ligand in a suitable site on the target by using a geometric matching algorithm. The ligand is allowed to overlap with the negative image of the binding site. The algorithm's efficiency is improved by using various other features like the ability to find the site where binding involves the least energy, using functions based on force fields, maximization of efficiency, and algorithms for flexible ligand docking (Baldi, 2010).

7.3 Swiss Dock

It is a web service which is used to predict the molecular interactions that occur between the targeted protein and a small molecule. Swiss dock proposes a binding mode for ligand, produces a complex for subsequent calculations, creates graphs and structures for the molecule and design inhibitors for the target. It is a docking software based on the EADock DSS algorithm which goes according to the steps such as firstly generation of binding modes in all the target cavities that is blind docking. Secondly, CHARMM energies are estimated on a grid. Then the favorable energies binding modes are evaluated with FACTS, and clustered. Lastly, the most favorable clusters can be observed online and can be downloaded (Grosdidier *et al.*, 2011).

7.4 IdTarget

It is a web server for identifying the bimolecular target for the chemical molecules with vital scoring functions and approach of docking. It predicts the possible binding sites of a molecule by the method of divide-and -conquer docking approach. The affinity profile of the protein target molecule used for the prediction of confidence levels. Through the method of divide-and -conquer docking approach, it helps in achieving the better docking efficiency (Wang *et al.*, 2012).

7.5 MEDock

Maximum Entropy-Based Docking (MEDock) is a web server used for the prediction of the ligand binding sites. It works on the predefined binding sites of the target for the design of site-directed mutagenesis and to find the structural features that effect the selection of the binding for minimizing the drug's adverse effect. This software is incorporated as global search strategy that accomplishes the maximum entropy property of the Gaussian probability distribution (Chang *et al.*, 2005).

8. Lead Optimization

Once the drug molecule has been discovered to target the biological entity (protein of interest), its properties can be improved by overcoming any deficiencies in its structure (Table 28.7).

Table 28.7: Lead Optimization Softwares.

Name	Description	URL
CODESSA	Downloadable software	http://www.semichem.com/codessa/
3DQSAR	Downloadable software	http://open3dqsar.sourceforge.net/?Downloads
E-DRAGON	Web server	http://www.vcclab.org/lab/edragon/
Pentacle	Downloadable software	https://www.moldiscovery.com/software/pentacle/
Mol inspiration	Web server	http://www.molinspiration.com/

8.1 CODESSA

CODESSA (Comprehensive Descriptors for Structural and Statistical Analysis) is ingenious software that aids in the establishment of quantitative structure/activity relationship (QSAR). CODESSA's graphical interface guides through the process of the correlation, pairing property values with data files and determining which descriptors are to be computed. It computes the constitutional (number of atoms of a specific element, number of different type of bonds, rings, molecular weight), topological (atomic connectivity of the molecule), geometric (optimized 3D structure, volume and surface area of the molecule of interest), electrostatic (largest and the smallest charge, dipole moment and ability to become polarized by the molecule), charged partial surface areas (positive and negative charge, total surface area), quantum chemical (quantum mechanical energy), fragment (user defined fragments), and thermodynamic descriptors (vibration, rotational and translational energies, heat of formation, entropy) (Lewis, 2005).

8.2 3DQSAR

3DQSAR is a web server tool which is directed at pharmacophore exploration by the high throughput chemometric analysis of the molecular interaction field. It produces the steric potential, electron density and import the binary files generated by the GRID. It leads to achieve the high computational performance through the implementation of the algorithm for molecular interaction field generation, validation and variable selection. It is controlled by the command line interface, by installing the PyMOL and the Open3DQSAR will operate interactively. The output is in ASCII text. The files are generated to store and retrieve data. The results are computed and the 3D maps are exported for the visualization (Mignani *et al.*, 2016).

8.3 E-DRAGON

E-Dragon is software used for the calculation of molecular descriptors and these descriptors are helpful in evaluating the molecular structure-activity or structure property relation. It also helpful for the similarity resolution and high throughput screening of the molecule databases. It can analyze maximum up to 149 molecules and 150 atoms per molecule. It needs molecular structure files to run DRAGON which are retrieved by the specific molecular modeling software. It accepts the molecular structure files such as SMILES or MOL2 (Sybyl) files (Tetko *et al.*, 2005).

8.4 Pentacle

It is brilliant software that helps in predicting the structure activity relationship which is not dependent on alignment. A set of structures are used which take the help of the energy computed as a result of interactions between the ligand and the target protein to construct three dimensional maps. Pentacle includes the MIF sampling and encoding algorithm for improving the Almond GRIND 3D QSAR program which was previously implemented. The aim of the tool is to estimate the descriptors, build and optimize the model within minutes (Pastor *et al.*, 2000).

8.5 Mol Inspiration

It supports a broad range of cheminformatics software tools for molecule manipulation and processing, which include SMILES and SDF (Standard Data File) conversion, generation of the tautomer, fragmentation of the molecule, calculation of molecular factors needed in QSAR. It supports the fragment-based virtual screening, bioactivity prediction of the molecule and helps in data visualization. Mol inspiration works on JAVA platform (http://www.rdchemicals.com/chemistry-software/molinspiration.html).

9. Preclinical Tests

After the discovery and optimization of the lead molecule, the ADMET (Absorption, Distribution, Metabolism, Excretion and Toxicity) characteristics of the drug need to be evaluated. Nowadays, ADMET properties of the drug are tested in the initial phases of drug development to filter out molecules with poor ADMET properties. Before advancing to the clinical phases, various drug molecules are screened based on the aforementioned properties which save time and developmental costs (Fig. 28.3).

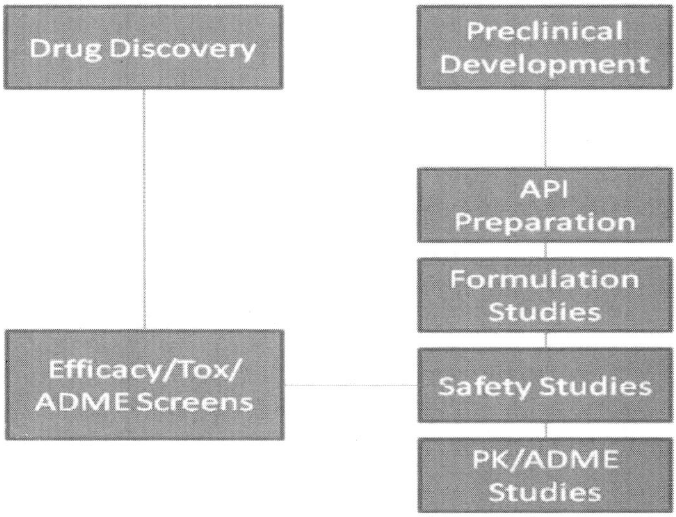

Fig. 28.3: Flowchart for drug designing and ADMET property prediction.

ADMET property is used for analyzing the creditability of the lead molecule as a potent drug molecule (Table 28.8).

Table 28.8: Few ADMET tools (Preclinical Softwares) are:

Name	Description	URL
DSSTox	Web server	https://www.epa.gov/chemical-research/distributed-structure-searchable-toxicity-dsstox-database
Virtual ToxLab	Web server	http://www.biograf.ch/index.php?id=projects&subid=virtualtoxlab
OCHEM	Webserver	https://www.ochem.eu/home/show.do
TOPKAT	Web server	http://accelrys.com/products/discovery-studio/toxicology/
LAZAR	Web server	https://lazar.in-silico.ch/predict

9.1 DSSTox

DSSTox (Distributed Structure-Searchable Toxicity Database) helps in predicting toxicology of the input sample by mapping the bioassay and physiochemical properties of the data linked with chemical substances with their corresponding chemical structures. DSSTox database comprises of cheminformatics workflows and chemical infrastructures for promoting safer chemical research by EPA (US Environmental Protection Agency). Features like chemical searching, SDF (Standard Data File) downloads and web services are yet to be incorporated in it (Richard and Williams, 2002).

9.2 Virtual ToxLab

Virtual ToxLab is software developed by the Biographics Laboratory $3R^2$. It can predict toxicity of a drug towards 16 different receptors present in the human body. These 16 receptors comprise of 10 receptors present in the nucleus, 4 representatives of cytochrome P450 clan and a transcriptional factor present in the cytosol, hERG (Fitzpatrick, 2008). It follows a fully automated protocol and makes it easier for the people who do not have much knowledge about molecular modeling. One of the major drawbacks of this tool is its inability to take into consideration the molecular stability of ligand-protein complex (http://www.biograf.ch/index.php?id=home).

9.3 OCHEM

OCHEM is a software package that helps chemists and toxicologists to predict the ADMET properties of a drug. It comprises of a database of molecules along with their ADMET properties. It acts like a repository for data and its resources, enables syncing data in many formats, helps in automatic transformation of quantifiable entities, stores essential knowledge related to experiments, corrects one or more molecules at the same time, provides restricted and public access to data, permits visualization of any modifications done to the data, helps in stockpiling essential physiochemical and biological characteristics of molecules and commissions benefits to its operator. It also gives opportunity to the user to spell out new units, characteristics and conditions for experiments. A directory of designs and models constructed by various linear and non-linear schemes is also present in OCHEM. Various biological models help in estimating ADMET properties, namely imbibitions in the alimentary canal, relative rate of skin penetration and traversal, seepage through the blood brain barrier to name a few (http://www.eadmet.com/en/ochem.php).

9.4 TOPKAT

TOPKAT (Toxicity Prediction by computer Assisted Technology) uses the Optimum Prediction Space (OPS) technology to find whether the query compound fits in the criteria of the training set. It is also possible to predict the toxicity of the compound that is similar to the query molecule and evaluate the performance of DS-TOPKAT. It is discovered by Health Inc. and now maintained by Accelrys in its Discovery Suite. Its older versions used structural, electronic, topological and electro-topological descriptors but the current version uses molecular fingerprint, Bayesian networks for classification and partial least squares for making regression models (Cronin and Madden, 2010).

9.5 LAZAR

LAZER (LAZy structure-Activity Relationships) software was developed by Professor Christoph Helma of the University of Freiburg, Germany. It follows a modified K-nearest neighbor algorithm by searching the query compound against a database having structural and experimental data related to the compounds that closely resemble the query compound. It also generates a confidence index depending upon the query compound being applicable to the training database (Kapetanovic, 2008).

10. Conclusion

The use of the bioinformatics tools in the drug discovery process cannot be appreciated more. It has enabled researchers and scientists to discover and design drugs in a cost and time effective manner. Apart from the tools discussed for carrying out target determination and its confirmation, lead discovery and maximization of its accuracy to bind to the target biomolecule, the tools for testing ADMET properties of the drug compounds have made the *in silico* drug discovery process more efficient by lowering the probability of rejection of the of the drug in the clinical phases. These tools have assisted in providing deep rooted understanding about the intercommunication between various proteins and its ligands (http://autodock.scripps.edu/). The drug research has been set around years ago. The chemistry field reached the maturity level that set the principles and methods itself and by the beginning of pharmacology it is recognized as a scientific discipline. Although the algorithms and approaches adopted by various bioinformatics tools in discovering drugs are not fully accurate in predicting the results and require wet labs for validation of the results, nevertheless the future of bioinformatics tools in the area of drug discovery is bright (Esposito *et al.*, 2014).

In series of Bioinformatics technologies, Next Generation Sequencing (NGS) is a most recent one, which offers the determination of precise order of nucleotides. Identification of nucleic acid sequences has become an indispensible tool in every area of biological sequencing. Correspondingly, the field of bioinformatics is central to the interpretation and application of this biological data. Tools like illumina have made it much easier to interpret the correct sequences. These new generation tools use various programmatic language, mathematical formulas and statistical methods to interpret the unknown sequences. They interpret, organize and retrieve the sequences to give the final result. The combined power of NGS and bioinformatics is vital for diagnostics, medical treatment and epidemiological research.

References

Autodock.scripps.edu. (2016). AutoDock — AutoDock. [online] Available at: http://autodock.scripps.edu/

Baldi, A. (2010). Computational approaches for drug design and discovery: An overview. *Syst Rev Pharm*, 1(1): 99-105.

Berjanskii, M., Liang, Y., Zhou, J., Tang, P., Stothard, P., Zhou, Y., Cruz, J., MacDonell, C., Lin, G., Lu, P. and Wishart, D.S. 2010. PROSESS: aprotein structure evaluation suite and server ", *Nucleic Acids Res*, 38(Web Server issue): W633-40.

Berjanskii, M., Zhou, J., Liang, Y., Lin, G. and Wishart, D. 2012. Resolution-by-proxy: a simple measure for assessing and comparing the overall quality of NMR protein structures. *Journal of Biomolecular NMR*, 53(3): 167-180.

Bhattacharya, A., Tejero, R. and Montelione, G. 2006. Evaluating protein structures determined by structural genomics consortia. *Proteins: Structure, Function, and Bioinformatics*, 66(4): 778-795.

Biograf.ch. 2016. Biograf 3R - Computational Alternatives to Animal Testing - Projects - Virtualtoxlab. [online] Available at: http://www.biograf.ch/index.php?id=home

Blaszczyk, M., Jamroz, M., Kmiecik, S. and Kolinski, A. 2013. CABS-FOLD server for the de novo and consensus-based prediction of protein structure. *Nucleic Acids Res.*, 41: 406-41.

Blundell, T., Sibanda, B., Montalvao, R., Brewerton, S., Chelliah, V., Worth, C., Harmer, N., Davies, O. and Burke, D. 2006. Structural biology and bioinformatics in drug design: opportunities and challenges for target identification and lead discovery. *Philosophical Transactions of the Royal Society B: Biological Sciences*, 361(1467): 413-423.

Chang, D.T.H., Oyang, Y.J. and Lin, J.H. 2005. MEDock: a web server for efficient prediction of ligand binding sites based on a novel optimization algorithm", *Nucleic Acids Res*, 33(Web Server issue): W233–W238.

Cronin, M. and Madden, J. 2010. In silico toxicology, *Cambridge: Royal Society of Chemistry*, pp.481-495.

Eadmet.com. (2016). eADMET:: OCHEM. [online] Available at: http://www.eadmet.com/en/ochem.php

Esposito, F., Zinzula, L. and Tramontano, E. 2014. Antiviral drug discovery: HIV1 RNase H, the next hit target. *Communicating Current Research and Educational Topics and Trends in Applied Microbiology*, 799-809.

Fitzpatrick, R. 2008. CPDB: Carcinogenic Potency Database. *Medical Reference Services Quarterly*, 27(3): 303-311.

Grosdidier, A., Zoete, V. and Michielin, O. 2011. SwissDock, a protein-small molecule docking web service based on EADock DSS. *Nucleic Acids Res*, 39: 270-277.

Hopf, T.A., Schärfe, C.P., Rodrigues, J.P., Green, A.G., Kohlbacher, O., Sander, C., Bonvin, A.M. and Marks, D.S. 2014. Sequence co-evaluation gives 3D contacts and structure of protein complexes. *Elife*, 3: doi: 10.7554/eLife.03430.

https://www.schrodinger.com/suites/biologics-suite.

Hughes, J., Rees, S., Kalindjian, S. and Philpott, K. 2011. Principles of early drug discovery. *British Journal of Pharmacology*, 162(6): 1239-1249.

Kapetanovic, I.M. 2008. Computer-aided drug discovery and development (CADDD): in silico-chemico-biological approach. *Chem Biol Interact.*, 171(2): 165-176.

Kelley, L.A., Mezulis, S., Yates, C.M., Wass, M.N. and Sternberg, M.J. 2015. The Phyre2 web portal for protein modeling prediction and analysis. *Natural Protocols*, 10: 845-858.

Kim, D.E., Chivian, D. and Baker, D. 2004. Protein structure prediction and analysis using the Robetta server. *Nucleic Acids Research*, 32: 526-553.

Lewis, R. 2005. A General Method for Exploiting QSAR Models in Lead Optimization. *Journal of Medicinal Chemistry*, 48(5):1638-1648.

Maupetit, J., Derreumaux, P. and Tuffery, P. 2009. PEP-FOLD "an online resource for *de novo* pepetide structure prediction. *Nucleic Acids Res*, 37: 498-503.

Mignani, S., Huber, S., Tomas, H. and Rodrigues, J. 2016. Why and how have drug discovery strategies in pharma changed? What are the new mindsets? *Drug discovery today*, 21(2): 239-249.

Morris, G., Huey, R. and Olson, A. 2008. Using AutoDock for Ligand-Receptor Docking. *Current Protocols in Bioinformatics*. 8: doi: 10.1002/0471250953.bi0814s24.

Nielsen, M., Lundegaard, C., Lund ,O. and Petersen, T.N. 2010. CPHmodels-3.0-remote homology modeling using structure-guided sequence profiles. *Nucleic Acids Res*, 38: 576-581.

Pastor, M., Cruciani, G., McLay, I., Pickett, S. and Clementi, S. 2000. GRid-INdependent Descriptors (GRIND): A Novel Class of Alignment-Independent Three-Dimensional Molecular Descriptors. *Journal of Medicinal Chemistry*, 43(17): 3233-3243.

Pugalenthi, G., Shameer, K., Srinivasan, N. and Sowdhamini, R. 2006. HARMONY: a server for the assessment of protein structures. *Nucleic Acids Research*, 34(Web Server issue): W231-W234.

R. Bank, "RCSB Protein Data Bank - RCSB PDB", Rcsb.org, 2016. [Online]. Available: http://www.rcsb.org/pdb/home/home.do.

Rdchemicals.com. (2016). Molinspiration Cheminformatics - R&D Chemicals. [online] Available at: http://www.rdchemicals.com/chemistry-software/molinspiration.html

Richard, A. and Williams, C. 2002. Distributed structure-searchable toxicity (DSSTox) public database network: a proposal. *Mutation Research/Fundamental and Molecular Mechanisms of Mutagenesis*, 499(1): 27-52.

Šali, A. and Blundell, T. 1993. Comparative Protein Modelling by Satisfaction of Spatial Restraints. *Journal of Molecular Biology*, 234(3): 779-815.

Schwede, T. (2003). SWISS-MODEL: an automated protein homology-modelling server. *Nucleic Acids Research*, 31(13): 3381-3385.

Sutcliffe, M., Haneef, I., Carney, D. and Blundell, T. 1987. Knowledge based modelling of homologous proteins, part I: three-dimensional frameworks derived from the simultaneous superposition of multiple structures. *Protein Engineering, Design and Selection*, 1(5): 377-384.

Teodorescu, O., Galor, T., Pillardy, J. and Elber, R. 2004. Enriching the sequence subsitution matrix by structural information. *Protein: structure, Function and Bioinformatics*, 54: 41-48.

Tetko, I.V., Gasteiger, J., Todeschini, R., Mauri, A., Livingstone, D., Ertl, P., Palyulin, V.A., Radchenko, E.V., Zefirov, N.S., Makarenko, A.S., Tanchuk, V.Y. and Prokopenko, V.V. 2005. Virtual computational chemistry laboratory- design and description", *J. Comput, Aid, Mol. Des.*, 19(6): 453-463.

Vyas, V., Ukawala, R., Chintha, C. and Ghate, M. 2012. Homology modeling a fast tool for drug discovery: Current perspectives. *Indian Journal of Pharmaceutical Sciences*, 74(1): 1-17.

Wang, C., Zhang, H., Zheng, W.M., Xu, D., Zhu, J., Wang, B., Ning, K., Sun, S., Li, S.C. and Bu, D.4. 2015. FALCON@home: a high-throughput protein structure prediction server based on remote homologue recognition. *Bioinformatics*, 32(3): 462-464.

Wang, J.C., Chu, P.Y., Chem, C.M. and Lin, J.H. 2012. idTarget: a web server for identifying protein target of small chemical molecule with robust scoring functions and a divide-and- conquer docking approach. *Nucleic Acids Res*, 40: 393-399.

Wu, S. and Zhang, Y. 2008. MUSTER: improving protein sequence profile-profile alignments by using multiple sources of structure information. *Proteins*, 72(2): 547-556.

Xu, D. and Zhang, Y. 2012. Ab initio protein structure assembly using continous structure fragments and optimized knowledge-based force field. *Protein*, 80(7): 1715-1735.

Yang, J., Yan, R., Roy, A., Xu, D., Poisson, J. and Zhang, Y. 2015. The I-TASSER Suite: Protein structure and function prediction, *Nature Methods*, 12: 7-8.

Yang, Y., Faraggi, E., Zhao, H. and Zhou, Y. 2011. Improving protein fold recognition and template-based modeling by employing probablistic-based matching between predicted one dimensional structural properties of query and corresponding native properties of templates. *Omic Tools*, 27(15): 2076-2082.

Zhang, W., Yang, J. and He, B. 2016. Integration of QUARK and I-TASSER for Ab Initio Protein structure Prediction in CASP11. *Proteins*, 84: 76-86.

Chapter 29

Nutraceuticals: An Emerging Era in Natural Product Studies

Manisha Bhatia[1,*], Vipin Saini[2] & Prerna Sarup[1]

[1]*MM College of Pharmacy, Maharishi Markandeshwar Deemed University, Mullana-Ambala*

[2]*Vice-Chancellor, Maharishi Markandeshwar Deemed University, Solan, H.P.*

**Corresponding Author E-mail: manisha181979@gmail.com*

Abstract

Human lifestyle is changing drastically because of which consumers are now deeply concerned about their health care. People's focus is shifting to positive approach for prevention of diseases leading to newer concepts such as nutraceuticals, nutritional therapy, phytonutrients and phytotherapy. The most descriptive term used now days is nutraceuticals, which is a combination of two words "nutrient" meaning a nourishing food and "pharmaceutical" which means a medical drug. Nutraceuticals can be defined as a natural bio active substance whose constituents are of known therapeutic activity which would contribute to promote wellness, prevent malignant processes, have health promoting, medicinal properties thereby encompassing prevention and treatment of diseases. The nutraceutical market in India is estimated to grow toUS$2,731 million in 2016 at a CAGR of 13%. The great void remains though for a systematic, thorough review of scientific data that provides a basis for their use. So, an attempt should be made to fill-up this gap and emphasizes their need& importance for maintaining health.

Keywords: *Health care, nutraceuticals, nutrients, herbals, dietary supplements*

1. Introduction

Over the last few decades human lifestyles is changing considerably mainly because of industrialization and varying work cultures. Newer concepts are coming up in the market, such as nutritional therapy, phytonutrients and phytotherapy which play positive roles in enhancing health and improving immunity. The most descriptive term used for them is nutraceuticals, whose concept began in the early 1980s and possess high status for its numerous uses& health benefits. "Nutraceutical" was formed by the combination of two words "Nutrition" meaning a nourishing food & "Pharmaceutical" (Das *et al.*, 2012), a medical drug by Stephen DeFelice MD in 1989 (Kalra, 2003). The term implies to any substance that is a food or any of its part, providing medical or health benefits for the prevention as well as treatment of disease leading to overall health promotion (Pandey *et al.*, 2010). It is a natural bio active substance whose constituents are of known therapeutic activity or a chemically defined substance which would contribute to the therapeutic activity of the drug. The common synonyms used for them are designer foods, health foods, fortified foods, medifoods, vita foods, pharma foods, functional foods and dietary supplements (Martirosyan and Singh, 2015). These are found in a mosaic of products emerging from the food industry, the herbal market, dietary supplement market, pharmaceutical industry, and the newly merged pharmaceutical conglomerates (Doke *et al.*, 2013). Market preparation includes various forms as pills, capsules, powders and tinctures both as a single substance or combination.

2. Regulation Parameters

Regulation parameters of Nutraceuticals are totally ambiguous because they are treated differently in different jurisdictions. Food labelling regulations do not allow food labels to carry health claims in many countries. This makes it hard for companies marketing nutraceuticals to advertise the benefits of their products without a medicine license, so they may either not do any research or may research a new product and obtain a patent (Brower, 1998).

Article 1 of Directive 65/65EEC defines a medicinal product as "any substance or combination of substances presented for treating or preventing disease in human beings or animals" (Devla *et al.*, 2011). It states that "Combination of substances which may be administered for diagnosis as well as modifying physiological functions in human beings or animals is considered a medicinal product." This definition describes a product which has a medical function irrespective of whether these claims have been stated or not, which would include many supplements being taken for medical purposes.

In Canada, a functional food has been defined as being similar in appearance to conventional foods, consumed as part of a usual diet whereas a nutraceutical is a product produced from foods but sold in pills, powders and other medicinal forms which is not associated with food (Fitzpatrick, 2004). Under Canadian law, a nutraceutical can either be marketed as a food or as a drug. Nutraceutical and functional food have no legal difference, referring to a product isolated or purified from foods that is generally sold in medicinal forms, part of our everyday diet & is demonstrated to offer health benefits beyond their dietary effects (Kumar and Pal, 2015). In Britain, the Ministry of Agriculture, Fisheries and Food has developed a definition of a functional food as "a food that has a component incorporated into it to give it a specific medical or physiological benefit, other than purely nutritional benefit" (Cockbill, 1994). According to US law there is no proper definition of the term nutraceutical. Depending on their ingredients and marketed claims, a product can be put into different categories as a drug, dietary supplement, food ingredient or food. In America, "medical foods" and "dietary supplements" are regulatory terms, however "nutraceuticals", "functional foods" are determined by consultants and marketers, based on consumer trends (Aarts, 1998).

In the global market, there are significant product quality issues with nutraceuticals. In the international market they may claim to use organic or exotic ingredients, but there is severe lack of regulation which may affect the safety and usefulness of products, regulatory issues depend on the country in which they are marketed, nature of ingredients and their categorization (**Hasler, 2005**). Companies looking to create a wide profit margin may create unregulated products overseas with low-quality or ineffective ingredients.

3. History

There are only few civilizations that have used food as medicine such as The Indians, Egyptians, Chinese, and Sumerians (Singh and Sinha, 2012). Hippocrates (460-377 BC), known as the father of modern medicine stated, 'Let food be thy medicine and medicine be thy food' to establish the relationship between appropriate foods for health and their therapeutic benefits. The growth rate of nutraceutical market in India is remarkable increasing significantly at US$2,731 million in 2016 at a CAGR of 13% (Gupta *et al.*, 2013).

4. Advantages of Nutraceuticals

Nutraceutical have the potential to play a role in healthy eating and to contribute to the prevention and treatment of diseases, delay the aging process and increase life expectancy (Nasri *et al.*, 2014). The advantages of these nutraceuticals is that rather than just containing one or two vitamins or minerals, they contain all the natural compounds that occur in the food that is used to make it. Nutraceuticals offers a wide range of supplements to help to keep our body in check. They are not difficult to discover, easily available and less expensive, so there is undoubtedly a very huge and growing market. These "food like substances" are either harmless or less toxic as compared to conventional pharmaceuticals (Thakur *et al.*, 2010). They possess multiple & increased health benefits. They provide food for populations with special needs eg: nutrient-dense foods for the elderly and many more.

4.1 Pharmaceutical and Nutraceutical

Pharmaceuticals and Nutraceuticals differ from each other in terms of Ingredients, Regulation, Marketing limitations, Quality of research & Public Perception (Pise *et al.*, 2010) (Table 29.1). Pharmaceuticals are medications which are used to treat diseases; however, nutraceuticals are the substances which are considered to prevent diseases (Mishra, 2012). This distinction between them is very erroneous and superficial and only pharmaceuticals have governmental sanction. Now a day's more exhaustive regulations on nutraceuticals is being worked out so as to provide descriptions of regulatory hurdles for them, (Zhao, 2007).Pharmaceuticals usually require expensive testing whereas nutraceuticals do not need these testing documents but their regulations are rigid thereby improving their efficacy and safety(Singh and Gujar, 2013).

Table 29.1: Difference between Pharmaceutical and Nutraceutical (Chauhan *et al.*, 2013).

S. No	Pharmaceuticals	Nutraceuticals
1.	They have high chances of side effects	They come with low chances of side effect.
2.	They always need a prescription from a doctor	Nutraceuticals do not generally require a prescription.
3.	Pharmaceuticals are backed by medical research making their claims justified.	Nutraceuticals are often not backed by medical research or clinical trials which limits their claim for the manufacturer.
4.	Pharmaceuticals will simply treat the disease or health issue but not prevent it.	Nutraceuticals aid in disease prevention as well as management.
5.	License is needed from regulatory body for selling the pharmaceutical product.	Nutraceutical product does not require any license for sale.
6.	Pharmaceuticals usually require patent protection	No patent protection is required for nutraceuticals.

5. Categories of Nutraceuticals

Three broad categories which are included in this group includes substances with established nutritional functions, such as vitamins, minerals, amino acids and fatty acids known as Nutrients, herbs or botanical products as concentrates and extracts known as Herbals & steroid hormone precursors serving specific functions, such as sports nutrition, weight-loss supplements and meal replacements known as Dietary supplements (Dureja *et al.*, 2003). The most commonly known nutrients are antioxidant, water and fat-soluble vitamins (Sapkale *et al.*, 2012). Nutraceuticals holds a great promise to improve health and prevent chronic diseases with the help of herbals e.g are willow bark (*Salix nigra)*, Parsley *(Petroselinum cripsum*), Peppermint (*Mentha piperita),* Lavender (*Lavandula angustifolia) &* Cranberries *(Vaccinium erythrocarpum)* (Kumar and Kumar, 2015). Dietary supplements are products administered through mouth that contain a dietary ingredient intended to add something to the foods you eat.

5.1 Functional Ingredients of Nutraceuticals

Table 29.2: Functional Ingredients of Nutraceuticals (Kumar *et al.*, 2015).

S. No.	Functional Ingredients	Source	Medicinal Use
1.	Carotenoids	Fruits, vegetables	Neutralizes free radicals, protection of cornea against UV Light, inhibit carcinogenesis in tumour cells
2.	Plant Sterols	Corn, soy, wheat, wood oil	Maintains blood cholesterol levels
3.	Soya phytoestrogens	Soyabean	Protect against heart disease, cancers, lowers LDL and total cholesterol
4.	Lignans	Vegetables, rye & Flaxseed	Protect against development of cancer like colon and breast cancer.
5.	Lutein	Green leafy vegetables	Reduce degeneration of muscles
6.	Lycopene	Tomatoes	Minimizes the risk of prostate cancer

S. No.	Functional Ingredients	Source	Medicinal Use
7.	Beta glucan, soluble fibre	Oats, barley, psyllium	Reduce the risk of cardiovascular disease.
8.	Fatty acids: Long chain omega-3 fatty acid	Salmon and other fish oil	Improve mental functions & vision improving agent
9.	Conjugated linoleic acid	Cheese, meat	Decrease risk of certain cancers & improve body composition
10.	Tannins	Cranberries, coca, chocolate	Aids in preventing cardiovascular diseases
11.	Prebiotics & Probiotics	Present in Yogurt, other dairy and non-dairy applications, onion powder	Improve quality of intestinal microflora promoting systematic immunity & gastrointestinal health
12.	Polyols Sugar alcohols xylitol, sorbitol	Present in Fruits	They may reduce the risk of dental caries & cavities
13.	Phenolics	Fruits, Tea, citrus, vegetables, Flax, rye	Prevention of renal failure
14.	Stilbenes	Obtain from grapes, peanuts, and pines	Induces apoptosis and inhibits the growth of various human tumour cells
15.	Minerals	Food	Important constituent of balanced diet
16.	Dietary fibre & Insoluble fibre	Wheat, Bran	Reduce breast or colon cancer

5.2 Plants as Nutraceutical (Srivastava *et al.*, 2015)

Table 29.3: Plants as Nutraceutical (Srivastava *et al.*, 2015).

S. No.	Plant Species	Common Name	Therapeutic Uses	Plant part used
1.	Allium sativum	Garlic	Antiseptic, diuretic	Leaves sap
2.	Foenicum vulgare	Fennel	Stomach bloating, stimulant	Whole seed, capsule, tincture
3.	Gingko biloba	Gingko	Age-related memory loss, fatigue, tinnitus	Dried leaf, tea
4.	Citrus limon	Lemon	Kidney stone, nausea, scurvy	Pulp extract
5.	Allium cepa	Onion	Diabetes, obesity, high blood pressure	Whole onion
6.	Sambucus	Elderberry	Arthritis, diuretic effect	Berries
7.	Olea europaea	Olive	Breast cancer, diabetes, itchy skin	Dried leaves and oil of the fruit
8.	Aesculus hippocastanum	Horse Chestnut	Poor blood circulation, itchiness, water retention	Seed and leaves
9.	Paullinia cupana	Guarana	Fever, headache, diarrhea,heart problem, Chronic fatigue syndrome	Guarana seeds

S. No.	Plant Species	Common Name	Therapeutic Uses	Plant part used
10.	Bacopa monnieri	Brahmi	Memory enhancer, aphrodisiac, reduces stress	Leaves
11.	Vaccinium	Cranberry	Kidney stones, neurogenic bladder	Fruits
12.	Vitis rotundifolia	Muscadine grape	Anticancer, antioxidant, cardiovascular disorders	Grape skins, seed extract
13.	Spinacia oleracea	Spinach	Diabetes, cancer, reducing blood pressure, bone health, asthma	Leaves
14.	Andrographis paniculata	Kalmegha	Bacillary dysentery, respiratory tract infection	Shoot powder
15.	Berberis asiatica	Barberry	Ulcers, urethral discharges, ophthalmia, jaundice, fevers	Roots and berries
16.	Phyllanthus emblica	Amla	Stress, diuretic, liver function, anti-ageing, diabetes	Fruit pulp fresh or dry
17.	Puereria tuberose	Vidarikand	Eases bowel movement, constipation, skin diseases	Tubers
18.	Matricaria chamomilla	Chamomile	Sedative, indigestion, insomnia, nausea, inflammation, wound healing	Dried flowers, capsules, cream, salve, tea, tincture, bath products
19.	Plantago ovate	Psyllium	Constipation, lowering cholesterol, type 2 diabetes	Dried seed, husk, capsules
20.	Murraya koenigii	Curry leaf	Anti-diabetic, antioxidant, antimicrobial, anti-inflammatory, anti-carcinogenic & hepatoprotective	Leaves
21.	Camellia sinensis	Green tea	Cancer, heart disease, liver disease, antioxidant & fat reducing	Leaves
22.	Daucus carota subsp. Sativus	Carrot	Reduces risk of cancer, cardiovascular disease, vision	Root
23.	Mangifera indica	Mango	Prevent cancer, lowers blood cholesterol, improves eye, cleans skin &diabetes	Fruit
24.	Actinidia deliciosa	Kiwi	Stroks, heart disease, diabetes, better sleep, obesity, lowering blood pressure, constipation	Fruit
25.	Trigonella foenum-graecum	Methi	Improves digestive problems, lowers blood cholesterol, reduces inflammation & arthritis	Seeds

5.3 Role of Nutraceutical in Diseases

Nutraceutical are beneficial in heart disease (Ramaa *et al.*, 2006, Rissanen *et al.*, 2003), obesity (Hil and Peters, 2002), diabetes (Thomas *et al.*, 2006), cancer (Willis and Wians, 2003), osteoporosis, arrhythmias, Parkinson's (Losso, 2003), Alzheimer's diseases, hypertension, high cholesterol, excessive weight, macular degeneration, cataracts, angina and hyperlipidemias, menopausal symptoms, insomnia, diminished memory, concentration, constipation, headaches, varicose veins, alcoholism, depression and lethargy. They show promising results in various pathological complications, aid as a vision improving agents (Rajasekaran *et al.*, 2008), dental caries, antioxidant, arthritis (Magrone *et al.*, 2013) etc. A wide array of biological processes, such as activation of antioxidant defenses, signal transduction pathways, cell survival-associated gene expression, cell proliferation and differentiation are involved in the mechanistic actions of these natural compounds (Mandel *et al.*, 2005). Garlic, ginger, soya bean has been found to have cancer-preventing activities. *Toucrium polium*, cinnamon, bitter melon and lipoic acid act as universal antioxidant used for the treatment of diabetic neuropathy. Dietary fibers from psyllium have been used to aid weight reduction in diabetic patients and to reduce lipid levels in hyperlipidemia. Conjugated linoleic acid, capsaicin, *Momordica charantia* possess possible antiobese properties, glucosamine and chondroitin sulfate improve symptoms of osteoarthritis, *Ginkgo biloba, Zizyphus jujube, Lavandula officinalis* used for vertigo, short-term memory loss, lack of attention and vigilance, *Eleutherococcus senticosus* to enhance general health, longevity & appetite, *Ocimum sanctum* have antistressor effects, *Withania somnifera* as an aphrodisiac, liver tonic, anti-inflammatory agent and astringent, *Aloe vera* in the treatment of wounds, for genital herpes, psoriasis, have antibacterial and antifungal activities and ginseng as an anti-inflammatory molecule and tea polyphenols have antithrombotic and anti-platelet properties and effective in lowering risk of developing Coronary Heart Diseases. Nutraceuticals have proven their health benefits and disease prevention capability, playing major role in therapeutic development, but they needed to be explored further for their quality, purity, safety and efficacy.

5.4 Nutraceuticals in Different Diseases (Chintale *et al.*, 2005)

Table 29.4: Nutraceuticals in Different Diseases (Chintale *et al.*, 2005).

S. NO.	Diseases	Example
1.	Alzheimer	Vitamin E and C, alpha-lipoic acid
2.	Cardiovascular	Flavonoids (onion, black grapes)
3.	Parkinson	Vitamin E
4.	Obesity	Chitosan, Vitamin C, fenugreek
5.	Diabetes	Calcium, Vitamin D, Emblica officinalis
6.	Osteoarthritis	Glucosamina
7.	Constipation	Buck wheat

5.5 Marketed Nutraceuticals

Table 29.5: Marketed Nutraceuticals.

S. No.	Marketed nutraceutical	Category	Ingredients	Manufacturer
1.	ROX®	Energy drink	Taurine, caffeine	ROX America, SA
2.	Proteinex®	Protein supplement	Predigested proteins, vitamins, minerals and carbohydrates	Pfizer Limited

S. No.	Marketed nutraceutical	Category	Ingredients	Manufacturer
3.	Calcitriol D-3®	Calcium supplement	Calcium and minerals	Cadilla Healthcare Limited
4.	Red Bull®	Energy drink	Taurine, caffeine	Austrian Red Bull Gmbh
5.	Revital®	Daily health supplement	Vitamins and minerals	Ranbaxy Lab Ltd
6.	Vitamin D3 Drop	Nutritional supplement	Vitamins	NUTRA
7.	Melatonin Time Release	Nutritional supplement	Vitamin, mineral and hormone	Nature's Bounty, Natural
8.	L-Carnitine	Nutritional and deficiency supplement	L-Carnitine and minerals	Ever Pure, BRI Nutrition
9.	SAMe	Nutritional supplement	Minerals	Natural factors
10.	Cosamin DS	Nutritional and deficiency supplement	Minerals and vitamins	MNX wellness Innovations
11.	Omacor	Nutritional and immune supplement	Minerals and fish oil	Ferrer Group
12.	RE:Activate	Nutritional and immune supplement	Vitamins, minerals, plants	Vollara

6. Conclusion

Nutraceuticals provide all the important substances that should be present in a healthy diet for humans. They provide energy and nutrient supplements to body, which are required for maintaining good health. Nutraceutical industry is growing at a much faster rate in both food and pharmaceutical sector. These functional food plays an important role in the promotion and care of human health to prevent diseases. Their use is a key to obtain therapeutic outcomes with reduced side effects. They have gained high interests because of their possible nutritional and safety profile, other than therapeutic capability. Fulfillment of appropriate diet requirements aids in healthy, proper functioning of GI tract which results in overall attainment of proper human physiology, thereby promoting healthy living. The great void still remains though for a systematic, thorough review of scientific data that provides a basis for their use. There is a severe lack of information of phytochemicals that are used in them. So, an attempt should be made to fill-up this gap. The present investigations propose the same and emphasize the need for their safety evaluation.

References

Das, Lipi., Bhaumik, E., Raychaudhuri, U. and Chakraborty, R. 2012. Role of nutraceuticals in human health. *J. Food. Sci. Technol.*, 49(2):173–183.

Kalra, E.K. 2003. Nutraceutical- Definition and Introduction. *AAPS Pharm. Sci. Tech.*, 5(3): 27–28.

Pandey, M., Verma, R.K. and Saraf, S.A. 2010. Nutraceuticals: new era of medicine and health. *Asian J. Pharm. Clin. Res.*,3(1): 11-14.

Martirosyan, D.M. and Singh J. 2015. A new definition of functional food by FFC: what makes a new definition unique? *Funct. Foods Health Dis.*,5(6): 209-223.

Doke, R.M., Mahale, A.M., Karanjkar, A.R. and Amge, S.A. 2013. Recent trends for neutraceuticals in Indian market. *Indo Am. J. Pharm. Research.*, 3 (5): 4001-4009.

Brower, B.1998. Nutraceuticals: Poised for a healthy slice of the market. *Nat. Biotechnol.*, 16: 728-33.

Devla, M.N., Acharya, S.R., Acharya, N.S. and Kumar, V. 2011. Dietary supplements: a legal status in India & in foreign countries. *Int. J. Pharm. Pharm. Sci.*, 3(3): 7-12.

Fitzpatrick, K.C. 2004. Regulatory issues related to functional foods and natural health products in Canada: possible implications for manufacturers of conjugated linoleic acid. *Am. J. Clin. Nutr.*, 79: 1217S–1220S.

Kumar, J. and Pal, A. 2015.Research article an overview of prospective study on functional food. *Int. J. Recent Sci. Res.*, 6(7): 5497-5500.

Cockbill, C.A. 1994. Food law and functional foods. *Br. Food J.*, 96: 3-4.

Aarts, T.A. 1998. How long will the "Medical Food" window of opportunity remain open? *J. Nutr. Funct. Med. Foods.*, 1: 45-57.

Hasler, C.M.2005. Regulation of Functional Foods and Nutraceuticals. *A Global Perspective. IFT Press and Blackwell Publishing. ISBN 0-8138-1177-5.*

Singh, J. and Sinha, S. 2012. Classification, regulatory acts and applications of nutraceuticals for health. Department of Biotechnology, Panjab University, Chandigarh 160 014, India. *Int. J. Pharm. Biol. Sci.*, 2(1): 177-187.

Gupta, S.K., Yadav, S.K. and Patil, S.K. 2013. Nutraceutical–a bright scope and opportunity of Indian healthcare market. *Int. J. Res. Dev. Pharm. L. Sci.*, 2(4): 478-481.

Nasri, H., Baradaran, A., Shirzad, H. and Kopaei, M.R. 2014. New Concepts in Nutraceuticals as Alternative for Pharmaceuticals. *Int. J. Prev. Med.*, 5(12): 1487–1499.

Thakur, N., Gupta, B.P., Nagariya, A.K., Jain, N.P., Banweer, J. andJain, S. 2010. Nutraceutical: New Era's safe pharmaceuticals. *J. Pharm. Res.*, 3 (6): 1243-1247.

Pise, A.G., Pise, S., Ligade, V., Janodia, M. and Udupa, N. 2010. Nutraceuticals and pharmaceuticals: a comparative analysis. *Int. J. Curr. Res. Rev.*, 02(8): 3-6.

Mishra, B. 2012. Status of nutraceuticals in European Union with special emphasis on botanical sources. *J. Nat. Rem.*, 12(2): 103- 107.

Zhao, J. 2007. Nutraceuticals, Nutritional Therapy, Phytonutrients, and Phytotherapy for Improvement of Human Health: A Perspective on Plant Biotechnology Application. *Recent Pat. Biotechnol.*, 1(1): 75-97.

Singh, M.C. and Gujar, K.N. 2013. Nutraceuticals: uses, risks and regulatory scenario. *Int. J. Pharm. Pharm. Sci.*, 5(3), 23-26.

Chauhan, B., Kumar, G., Kalam, N. and Ansari, S.H. 2013. Current concepts and prospects of herbal nutraceutical: A review. *J. Adv. Pharm. Technol. Res.*, 4: 4–8.

Dureja, H., Kaushik, D. and Kumar. V. 2003. Developments in Nutraceuticals. *Indian J. Pharmacol.*, 35: 363-372.

Sapkale, A.P., Thorat, M.S., Vir, P.R. and Singh, M.C. 2012. Nutraceuticals-Global status and applications: A Review. *Int. J. Pharm. Chem. Sci.*, 1(3): 1166-118.

Kumar, K. and Kumar, S. 2015.Role of nutraceuticals in health and disease prevention: A review. *South Asian J. Food Technol. Environ.*, 1(2): 116-121.

Srivastava, S., Sharma, P.K. and Guru, S.K. 2015. Nutraceuticals: A Review. *J. Chr. Drug Delivery.*, 6(1): 1-9.

Ramaa, C.S., Shirode, A.R., Mundada, A.S. and Kadam, V.J.2006. Nutraceuticals - An Emerging Era in the Treatment and Prevention of Cardiovascular Diseases. *Curr. Pharma. Biotechnol.*,7 (1): 15-23.

Rissanen, T.H, Voutilainen, S., Virtanen, J.K., Venho, B., Vanharanta, M., Mursu, J. and Salonen, J.T. 2003. Low Intake of Fruits,Berries and Vegetables Is Associated with Excess Mortality in Men: the Kuopio Ischaemic Heart Disease Risk Factor (KIHD) Study. *J. Nutr.*, 133: 199-204.

Hil, J.O, Peters, J.C. 2002. Biomarkers and functional foods for obesity and diabetes. *Br. J. Nutr.*, 88: 213–18.

Thomas, B., Ghebremeskel, K., Lowy, C., Crawford, M., Bridget, O. and Shore, R.N. 2006. Nutrient intake of women with and without gestational diabetes with as pecific focus on fatty acids. *Nutr.*, 22:230-236.

Willis, M.S., Wians, F.H. 2003. The role of nutrition in preventing prostate cancer: A review of the proposed mechanism of action of various dietary substances. *Clin. Chim. Acta.*, 330:57–83.

Losso, J.N. 2003. Targeting excessive angiogenesis with functional foods and nutraceuticals. *Trends Food Sci. Technol.*, 14: 455-468.

Rajasekaran, A., Sivagnanam, G. and Xavie, R. 2008. Nutraceuticals as Therapeutic Agents-Review. *Research J. Pharm. Technol.*, 1(4): 328-340.

Magrone, T., Perez de Heredia, F., Jirillo, E., Morabito, G., Marcos, A. and Serafini, M. 2013. Functional Foods and Nutraceuticals as Therapeutic Tools for the treatment of diet related diseases. *Canadian J. Physiol. Pharmacol.*, 91(6): 387-396.

Mandel, S., Packer, L., Youdim, M.B., Weinreb, O. 2005. Proceedings from the "Third International Conference on Mechanism of Action of Nutraceuticals". *J. Nutr. Biochem.*, 16(9): 513-20.

Chintale, A.G., Kadam, V.S., Sakhare, R.S., Birajdar, G.O. and Nalwad, D.N. 2013. Role of nutraceuticals in various diseases: a comprehensive review. *Int. J. Res. Pharm. Chem.*, 3(2): 290-299.

Chapter 30

Response of Arbuscular Mycorrhizal Fungi and *Trichoderma Harzianum* on Growth and Flowering Attributes in *Calendula Officinalis* Linn.

Karishma[1], Anju Tanwar[2], Kuldeep Yadav[3] and Ashok Aggarwal[1,*]

[1]*Department of Botany, Kurukshetra University, Kurukshetra, Haryana-136119 India*

[2]*Department of Botany, Government. P.G. College, Ambala Cantt., Haryana-133001*

[3]*Department of Botany, Gandhi Memorial National College, Ambala Cantt., Haryana*

**Corresponding Author E-mail: ashokbotanykuk@gmail.com*

Abstract

An experiment was conducted under polyhouse condition to find out the response of dominating AM fungi (Glomus mosseae & Acaulospora laevis) and Trichoderma harzianum on growth and flowering in Calendula officinalis Linn. Among all the growth parameters, maximum shoot length was recorded with dual treatment of T. harzianum plus A. laevis while maximum bud formation was noticed in dual treatment of G. mosseae + A. laevis. However, all the treatment showed significant effect in all parameters in comparison to the control. The vase life under different treatments was found to be maximum up to 12 days with A. laevis treatment. Therefore, AM fungal treatment has not only improved the growth parameters but has also increased the vase life of the Calendula crop.

Keywords: *Calendula officinalis, Glomus mosseae, Acaulospora laevis, T. harzianum.*

1. Introduction

Calendula officinalis, known as Pot Marigold or Scotch Marigold, is a winter season annual grown for its loose and cut flowers owing to its orange- yellow coloured flowers, which makes very good indoor arrangements. Medicinally, it is used for the treatment of skin disorders and pain and acts as a bactericide, antiseptic, antifungal, antihaemorrhigic and anti-inflammatory in nature. Adverse effect of continuous use of fertilizers on soil and environment has paid attention towards the alternative sources. Beneficial microbes associated with plant roots are of paramount importance in agriculture and horticulture (Yadav *et al.*, 2013). Biofertilization became the most important practice required to substitute chemical fertilizers for the production of many ornamental plants (El-Khateeb *et al.*, 2010). In floriculture, arbuscular mycorrhizal fungi (AMF), shows positive effects on flower quality, flower growth and flowering time (Asrar *et al.*, 2012; Deljou *et al.*, 2014; Xie and Wu, 2017).

The increase in the inorganic nutrient uptake in AMF inoculated plants is mainly because fungal hyphae provide the large surface area for nutrient acquisition to external root surface as compared to uninfected roots. As the AMF mycelium grows through soil, it scavenges for mineral nutrients and is able to make contact with uninfected roots, sometimes of different host species (Bhat *et al.*, 2017). AMF fungi also impart many other benefits to plants including improved plant growth, biomass, and productivity particularly in extremely P-limited soil (Liu *et al.*, 2017), enhance plant resistance to drought (Kadian *et al.*, 2013) and heavy metals (Tanwar *et al.*, 2015)

Several microorganisms are known to benefit mycorrhiza formation and functioning (Kadian *et al.*, 2013). Some fungal strains have been reported to have both antagonistic and stimulatory effects on plants. The fungus *Trichoderma* is considered as wonder organism because different species act as biological control agents of plant diseases; produce cellulolytic, hemicellulolytic, chitinolytic and proteolytic enzymes; capable of biodegrading chlorophenolic compounds helpful in soil bioremediation and also produce secondary metabolites with antibiotic activity (Vinale *et al.*, 2008). Arbuscular mycorrhizal fungi and some fungal strains are found to be associated with most of the plants and their beneficial effects on plant growth, nutrient uptake and tolerance to environmental stressors have been extensively reported.

Keeping in view, the present experiment was aimed to reduce the amount of chemical fertilizers by substituting AMF alone and with *Trichoderma* in Calendula crop.

2. Material and Methods

2.1 Study Site

The present investigation was undertaken in polyhouse of Botany Department, Kurukshetra University, Kurukshetra, India.

2.2 Soil Sample and Collection

Composite soil sample from rhizospheric soil of *C. officinalis* was collected. It was done by digging out a small amount of soil close to the plant roots up to the depth of 15-30 cm.

The soil samples were wet- sieved for spores using the technique of Gerdemann and Nicolson (1963) and quantification of AM spores was done by grid line intersect method (Adholeya and Gaur,1994). Root samples were rinsed with tap water and cut into 1 cm. pieces and performed with trypan blue according to rapid clearing and staining method of Philips and Haymann (1970). The percent AM root colonization was calculated by using the following formula:

$$\text{Percent AM root colonization} = \frac{\text{Total no. of root segments colonized}}{\text{Total no. of root segments examined}} \times 100$$

2.3 Mass Production of AM Spores

Dominant AM spores i.e. *Acaulospora laevis* and *Glomus mosseae* isolated from rhizosphere of *C. officinalis*, were mass produced by Funnel technique (Menge and Timmer, 1982) using sorghum as host.

2.4 Mass Multiplication of *T. harzianum*

T. harzianum was isolated from the soil and then further mass produced in the medium of wheat bran, saw dust and distilled water prepared in the ratio of 3:1:4 (Sharma *et al.*, 2007).

2.5 Inoculation of AM Spores and *T. harzianum*

Moreover, seedlings of *C. officinalis* were grown in earthen pots (Size 25 25) under poly house conditions in sterilized soil. To each pot 10% of inoculum of AM fungi and *T. harzianum,* alone and in combination was added and following treatments were taken: Control, *T. harzianum, G. mosseae, A. laevis, G. mosseae + T. viride, A. laevis + T. viride* and *G. mosseae + A. laevis*. The effects of different treatments were recorded after 45 and 90 days of inoculation on various growth parameters i.e. change in height, leaf area, fresh & dry shoot weight, fresh & dry root weight, root length, AM spore number, AM root colonization, phosphorus content and number of bud formation. Roots and shoots were harvested and weighted separately for their fresh weight and the oven dried to 70°C for dry weight.

2.5 Leaf Area and Phosphorus Estimation

Leaf area was studied using leaf area meter (Systronics 211) and estimation of phosphorus was done by Vanadomolybdo phosphoric yellow colour method, in nitric acid system (Jackson, 1973).

Statistical analysis was performed using one-way analysis of variance (ANOVA) followed by post hoc test.

3. Results and Discussion

AM fungi and *T. harzianum* were screened for better growth performance of *C. officinalis* under poly house conditions. Different growth parameters were observed after 45 and 90 days of inoculation.

3.1 Mycorrhizal Spore Count and Percent Root Colonization

It is envisaged from the Table (1 & 2), that maximum spore count was found in the dual combination of *G. mosseae + A. laevis* (142±5.68) after 45 days and in *A. laevis +*

T. harzianum (254±6.80) after 90 days respectively and highest AM root colonization was recorded in the treatment of *A. laevis* (46.7±15.28) and *A. laevis* + *T. harzianum* (81.51± 13.33) after 45 and 90 days of inoculation. Root colonization by AM fungi was directly related to nutrients uptake by plants. The increase in AMF spore count and percent root colonization could be attributed to enhanced spread of AM fungal hyphae, which may lead to increase uptake of minerals and nutrients. Positive response in respect to spore number and root colonization has been reported by Earanna *et al.* (2001) in *Coleus aromaticus.*

3.2 Shoot Length

AM inoculation significantly increased the shoot length of *C. officinalis* after 45 and 90 days. Increase in shoot length was maximum in dual combination of *A. laevis* + *T. harzianum* (Table 1 & 2). The significant increase in shoot height could be attributed to the enhancement in the capacity of plants to absorb more nutrients and secreting plant growth hormones. Kungiu *et al.* (2008) while studied on *Senna spectabilis*, observed increase in shoot length with AM inoculation. Mehrotra and Baijal (1995) also studied the significant effect of mixed inoculation of AM fungi on growth and yield of sunflower. Similar was the assessment made by Gaur and Adholeya (2005).

3.3 Shoot Weight

Table (1 & 2) showed that biomass of all the inoculated plants of *C. officinalis* increased significantly in terms of fresh and dry shoot biomass yield after 45 and 90 days. Maximum increase in shoot biomass was found in plants inoculated with *G.mosseae* after 45 days and after 90 days of inoculation, it was found maximum in dual inoculation of *G. mosseae* + *T. harzianum*. Increase in shoot biomass may be due to enhanced inorganic nutrition, uptake and accumulation of nutrients and increase in surface area for absorption. Enhanced biomass of plant by AM fungi and *T. viride* has been reported by Kumar *et al.* (2008), while worked on *Spilanthes acmella*. The stimulated growth and increase in shoot dry weight by dual inoculation of AM fungi and Rhizobacteria has been reported by Rao and Aseri (2004) in *Emblica officinalis*.

3.4 Root Weight

The result (Table 30.1 & 30.2) show that the amount of root biomass has been found to be increased significantly irrespective of treatments over control. After 45 days, the increase in root biomass (fresh & dry) was observed maximum with *G. mosseae* + *T. harzianum* and *A. laevis* respectively and 90 days after inoculation (DAI), the increase in root biomass (fresh & dry) was found maximum with *A. laevis*. The increase in root biomass could be due to the greater surface area of mycorrhizal roots, which leads to acquisition of more nutrients. Significant increase in root biomass with AM fungi was observed by Stevens *et al.* (2002) in *Lythrum salicaria*. Similar was the assessment by Boureima *et al.* (2007).

Table 30.1: Influence of Arbuscular mycorrhizal fungi and *T. harzianum* on growth performance of *C. officinalis* after 45days.

Treatment	Change in height (cm)	Leaf area (sq cm)	Root length (cm)	Fresh shoot weight (g)	Dry shoot weight (g)	Fresh root weight (g)	Dry root weight (g)	AM spore number /10 g of soil	AM root colonization (%)	No. of buds formation
Control	*29.86±1.42	12.73±0.15	8.1±0.60	6.2±0.2	0.37±0.25	0.22±0.09	0.06±0.02	74.0±4.72	12.53±2.28	0.0±0
Glomus mosseae	34.63±1.55	17.73±0.54	16.0±1.12	15.72±3.40	1.48±0.14	0.67±0.20	0.16±0.11	102.0±3.0	44.33±7.50	1.33±1.52
Acaulospora laevis	34.56±0.41	18.8±0.45	13.3±1.16	11.5±1.11	0.92±0.09	0.58±0.07	0.52±0.44	112.0±9.0	46.7±15.28	1.41±1.73
Trichoderma harzianum	39.23±2.20	16.4±0.41	11.6±1.11	15.4±0.78	0.98±0.13	0.61±0.01	0.19±0.01	55.33±4.5	29.16±8.77	0.0±0
G. mosseae + *T. harzianum*	34.86±1.42	17.4±0.25	12.76±0.55	14.83±0.70	1.37±0.07	1.36±0.12	0.31±0.02	121.33±6.61	35.83±5.20	2.0±1.0
A. laevis + *T. harzianum*	40.7±1.53	16.53±1.42	13.86±0.66	10.33±1.05	0.86±0.20	0.70±0.02	0.12±0.01	103.33±8.61	32.5±10.89	2.0±1.0
G. mosseae + *A. laevis*	35.03±0.80	18.43±0.41	11.6±0.62	12.83±2.28	0.99±0.16	0.60±0.09	0.30±0.06	142.66±5.68	34.16±8.2	2.33±0.57

According to one-way analysis of variance (ANOVA) of data, the mean difference is significant at 0.05 levels.

* Each value is an average of three replicates

÷ Standard Deviation

Table 30.2: Influence of Arbuscular mycorrhizal fungi and *T. harzianum* on growth performance of *C. officinalis* after 90 days.

Treatment	Change in height (cm)	Leaf area (sq cm)	Root length (cm)	Fresh shoot weight (g)	Dry shoot weight (g)	Fresh root weight (g)	Dry root weight (g)	AM spore number /10 g of soil	AM root colonization (%)	No. of buds formation
Control	*47.01±1.87	12.93±0.15	13.11±0.17	8.9±1.14	0.61±0.14	0.87±0.18	0.33±0.10	77.0±4.58	26.66±2.8	1±1.0
Glomus mosseae	62.66±1.34	19.0±0.1	19.15±0.88	20.45±1.04	2.33±0.50	1.80±0.34	0.56±0.02	176.0±8.71	37.7±2.37	2.6±1.15
Acaulospora laevis	78.76±4.10	22.43±0.49	18.84±0.73	23.26±4.50	3.07±0.11	5.14±0.20	1.73±0.10	179.6±10.0	42.7±2.9	1.66±1.52
Trichoderma harzianum	61.13±2.68	19.1±0.2	13.73±0.56	18.05±1.63	2.05±0.10	1.59±0.26	0.49±0.09	133.0±11.26	31.76±8.27	1.66±1.52
G. mosseae + *T. harzianum*	68.9±3.30	18.1±0.96	20.03±0.21	27.04±1.00	3.19±0.22	2.75±0.38	0.57±0.07	170.0±8.93	43.33±5.37	3.0±2.64
A. laevis + *T. harzianum*	79.1±3.50	18.53±1.42	15.53±0.86	22.35±2.68	3.17±0.25	2.12±0.28	0.68±0.22	254.0±6.80	81.55±13.33	2.66±2.08
G. mosseae + *A. laevis*	56.86±6.8	19.13±1.05	14.70±1.34	20.98±1.48	2.67±0.50	2.08±0.61	1.03±0.06	192.6±13.18	54.99±4.24	3.66±2.08

According to one-way analysis of variance (ANOVA) of data, the mean difference is significant at 0.05 levels.

* Each value is an average of three replicates

÷ Standard Deviation

3.5 Root Length

It is evident from the Table (30.1 & 30.2) that mycorrhizal plant showed increase in root length in comparison to control. The maximum root length was found in the inoculation of *G. mosseae* (16.0±1.12) after 45 days and 90 DAI, it was found maximum in the dual inoculation of *G. mosseae* + *T. harzianum*. The increment in root length could be attributed to the mycorrhizal colonization. Increase in root length upon mycorrhizal colonization has been reported by Liu *et al.* (2000).

3.6 Leaf Area

Table (30.1 & 30.2) indicate that AM inoculation significantly increased the leaf area of all treated plants as compared to control plant. It was found that after 45 and 90 days of inoculation, the maximum leaf area was found in the plant treated with *A. laevis*. The higher increment of leaf area registered with inoculated plant could be as a result of enhanced inorganic nutrient absorption (Cooper, 1984) and greater rate of photosynthesis (Allen *et al.*, 1981). A progressive increase in leaf area to AM inoculation has also been reported in Sesame (Boureima *et al.*, 2007).

3.7 Number of Buds

It is envisaged from Table (30.1 & 30.2) the plants inoculated with *G. mosseae* + *A. laevis* showed higher number of buds after 45 and 90 days of inoculation. The increase in bud formation might be due to significant increase in P and K uptake in bioinoculated plants with AM fungi. Progressive increase in number of buds and earlier flowering due to inoculation of AM fungi has been reported by Gaur *et al.* (2005) and Usha *et al.* (2005). Chauhan and Kumar (2007) also reported the beneficial effect of nitrogen and VAM on different growth parameters of *Calendula*.

3.8 Phosphorus Content

The result (Table 30.3) showed that the amount of phosphorus in shoot and roots significantly increased in inoculated plants as compared to control. Forty-five DAI, the increase in phosphorus content in shoots and roots was found maximum in the plants treated with *A. laevis* + *T. harzianum*. After 90 days of inoculation, the maximum increase in P- content in shoot was recorded in *A.laevis* (1.80±0.015), while in roots it was observed maximum in *G.mosseae* (2.29±0.06). In symbiosis with AM fungi faster P absorption occurs in treated plant roots than non- treated roots (Jacobson *et al.*, 1992). Satter *et al.* (2006), who reported nearly 4- fold increased phosphorus uptake in seedlings inoculated with AM alone, while 3-folds increased phosphorus uptake was observed as compared to control while using AM fungi, *Rhizobium* and *Trichoderma* species. Maiti *et al.* (2007) also utilized AM association for improving phosphorus nutrition in upland rice under rainfed conditions. Charaon *et al.* (2007) also found a significant increase in phosphorus content in the secondary roots of Tobacco seedlings when inoculated with *T. harzianum*. The external AM hyphae reach beyond the depletion zone around the root hair, absorb soil P and translocate it, may be in the form of polyphosphate granules, to the arbuscules, where P is transferred to the plant cell in the exchange of carbon (Mago and Mukerji, 2003).

Table 30.3: Influence of Arbuscular mycorrhizal fungi and *T. harzianum* on phosphorus content of *C. officinalis* after 45 and 90 days.

S.No.	Treatment	% Phosphorus content			
		45 days		90 days	
		Shoot	Root	Shoot	Root
1.	Control	*0.08 ± 0.025	0.18 ± 0.152	0.32 ± 0.015	0.38 ± 0.01
2.	*Glomus mosseae*	0.51 ± 0.025	1.40 ± 0.015	0.88 ± 0.015	2.29 ± 0.06
3.	*Acaulospora laevis*	0.5 ± 0.015	1.21 ± 0.02	1.8 ± 0.015	1.73 ± 0.48
4.	*Trichoderma harzianum*	0.1 ± 0.015	0.7 ± 0.015	0.17 ± 0.0351	0.43 ± 0.03
5.	*G. mosseae + T. harzianum*	0.81 ± 0.025	1.09 ± 0.02	0.84 ± 0.01	1.36 ± 0.035
6.	*A. laevis + T. harzianum*	1.13 ± 0.21	1.54 ± 0.025	0.89 ± 0.09	1.01 ± 0.015
7.	*G. mosseae + A. laevis*	1.03 ± 0.11	1.40±0.0152	1.19 ± 0.01	1.29 ± 0.025

According to one way analysis of variance (ANOVA) of data, the mean difference is significant at 0.05 levels.

* Each value is an average of three replicates

÷ Standard Deviation

In the present study, the effect of two strains of AM fungi along with *T. harzianum* was analyzed on different growth parameters of *C. officinalis*. As mycorrhizal symbiosis is a highly evolved mutualistic relationship for plant establishment, so, this study provides a great future for utilizing the efficient strains of mycorrhizal fungi to exploit them for the beneficial effects in establishment of seedlings, increase in productivity and reduce the fertilizer application required for obtaining economic production of *C. officinalis* plant under field conditions.

References

Adholeya, A., and A. Gaur, 1994. Estimation of VAM fungal spores in soil. Mycorrhiza News, 6(1): 10-11.

Allen, M.F., W.K. Smith, T.S. Moore, and M. Chiristensen, 1981. Comparative water relation and photosynthesis of mycorrhizal and non-mycorrhizal Boutelua Gracilis H.B.K. New Phytologist, 88: 683-693.

Asrar, A.A., Abdel-Fattah, G.M. and Elhindi, K.M. 2012. Improving growth, flower yield and water relations of snapdragon (Antirhinum majus L.) plants grown under well-watered and water-stress conditions using arbuscular mycorrhizal fungi. Photosynthetica, 50: 305-316.

Bhat, R.A., Dervash, M.A., Mehmood, M.A., Skinder, B.M., Rashid, A., Bhat, J.I.A., Singh, D.V. and Lone, R. 2017. Mycorrhizae: A Sustainable Industry for Plant and Soil Environment. A. Varma et al. (eds.), Mycorrhiza - Nutrient Uptake, Biocontrol, Ecorestoration. Springer International Publishing AG.

Boureima, S., M. Diouf, T.A. Diop, M. Datta, E.M. Leye, F. Ndiaye, and D. Seck, 2007. Effects of arbuscular mycorrhizal inoculation on the growth and the development of sesame (Sesamum indicum L.) African journal of Agricultural Research, 3(3): 234-238.

Charaon, M.R., O. Rodrigvez- Galap, T. Benitez, S. Sousa, M. Rey, A. Llobell, and J. Delgado-Iarana, 2007. Transcriptase análysis of early colonization of tomato roots by T.harzianum. Microbiology, 10(1):19-37.

Chauhan, A., and V.Kumar, 2007. Effect of graded levels of nitrogen and VAM on growth and flowering in calendula (Calendula officinalis Linn.). Journal of Ornamental Horticulture, 10(1): 61-63.

Cook, E.L., and V.J. Staden, 1985. Senescense of cut carnation flowers: Ovary development and CO fixation. Plant Growth Regulation, 1: 221-232.

Cooper, K.M. 1984. Physiology of VAM associations. VA Mycorrhiza, (eds. C.L. Powell and D.J. Bagyaraj). Published by CRC Press Inc., Boca Raton, Florida, pp.155-186.

Deljou, M.J.N., Marouf A.and Hamedan, H.J. 2014. Effect of inoculation with Arbuscular Mycorrhizal Fungi (AMF) on gerbera cut flower (Gerbera jamesonii) production in soilless cultivation. Acta Hortic., 1034: 417-422.

Earanna, N., R.R. Mallikarjuniah, D.J. Bagyaraj, and C.K. Suresh, 2001. Response of Coleus aromaticus to Glomus fasciculatum and other beneficial soil microflora. Journal of Spices and Aromatic Crops, 10(2): 141-143.

Gaur, A., and A. Adholeya, 2005. Diverse response of five ornamental plant species to mixed indigenous and single isolate arbuscular- mycorrhizal inocula in marginal soil amended with organic matter. Journal of Plant Nutrition, 28: 707-723.

Gaur, R.B., R.N. Sharma, and R.R. Sharma, 2005. Shelf life of talc-based formulation of Trichoderma and soil application for biological control of dry root rot of chickpea. Journal of Mycology and Plant Pathology, 35(2): 380-384.

Gerdemann, J. W., and Y.H. Nicolson, 1963. Spores of mycorrhizae Endogone species extracted from soil by wet sieving and decanting. Transactions of the British Mycological Society, 46: 235-244.

Jackson, M.L. 1973. Soil Chemical Analysis. Published by Prinlice Hall, New Delhi, pp. 485.

Jacobsen, I., L.K. Affott, and A.D. Robson, 1992. External hyphae of VAMF associated with Trifolium subterraneum L.I. Spread of hyphae and phosphorus inflow into roots. New Phytologist, 120(3): 371-379.

Kadian, N., Yadav, K. and Aggarwal, A. 2013. Significance of bioinoculants in promoting growth, nutrient uptake and yield of Cyamopsis tetragonoloba (L.) "Taub.". European J. Soil Biol, 58: 66-72.

Kumar, A., A. Aggarwal, S. Sharma, and S. Kaushish, 2008. Interaction of Arbuscular mycorrhizal fungi and Trichoderma viride on growth of Spilanthes acmella Murr. The Journal of the Indian Botanical Society, 87(1&2): 120-124.

Kungiu, J.B., R.D. Lasco, L.U. Dela Cruz, R.E. Dela Cruz, and T. Husain, 2008. Effect of vesicular arbuscular mycorrhiza (VAM) fungi inoculation on coppicing ability and drought resistance of Senna spectabilis. Pakistan Journal of Botany, 40(5): 2217-2224.

Liu, R., M. Li, X.M.X. Liu, and X. Li, 2000. Effects of AM fungi on endogenous hormones in corn and cotton plants. Mycosystema, 19(1): 91-96.

Liu, S., Guo, H., Xu, J., Song, Z., Song, S., Tang, J. and Chen, X. 2017. Arbuscular mycorrhizal fungi differ in affecting the flowering of a host plant under two soil phosphorus conditions. *Journal of Plant Ecology*, 11:623-631.

Mago, P., and K.G. Mukerji, 2003. Mycorrhizal technology in forestry and agriculture. Compendium of Mycorrhizal Research Vol II: Role of Mycorrhiza in biotechnology, (eds. K.G. Mukerji and B.P. Chamola), A.P.H. Publishing Corporation, New Delhi, pp.129-146.

Maiti, D., M.K. Barnwal, R.K. Singh, and S.K. Rana, 2007. Approaches to utilize arbuscular mycorrhizal association for improving phosphorus nutrition in upland rice under rainfed ecosystem. The Mycorrhizae: Diversity, Ecology and Applications. (eds. M. Tiwari, and S.C. Sati). Published by Daya publishing House, Delhi, pp. 193-199.

Mehrotra, V.S., and V. Baijal, 1995. Effects of single and mixed inocula of VAM fungi on the growth and yield of sunflower (Helianthus annus L.) Philippines Journal of Science, 124: 126.

Menge, J.A., and L.W. Timmer, 1982. Procedure for inoculation of plants with VAM in the laboratory, greenhouse and field. Methods and Principles of Mycorrhizal Research.(ed. N.C. Schenck). Published by American Phytopathology Society, St.Pauls, USA, pp. 59.

Mengue, A., and E. Usta, 1994. Research on the effects of silver thiosulphate + sucrose pretreatment on the cold storage period and post storage vase life of cut flowers of carnation cv. Astor harvested at different maturities. Acta Hort, 368: 802-807.

Philips, J. M., and D.S. Hayman, 1970. Improved procedures for clearing roots and staining parasitic and VAM fungi for rapid assessment of infection. Transactions of the British Mycological Society, 55: 158-161.

Rao, A.V., and G.K. Aseri, 2004. Effect of bioinoculants on seedlings of Indian gooseberry (Emblica officinalis Gaertn.). Indian Journal of Microbiology, 44(2): 109-112.

Satter, M.A., M.M. Hanaif, T.M.M. Mahmud, and H. Azizah, 2006. Influence of arbuscular mycorrhizae and phosphate on uptake of major nutrients by Acacia mangium seedlings on degraded soil. Biology and Fertility of Soils, 42(4): 345-349.

Sharma, S., A. Aggarwal, and S. Kaushish, 2007. Effect of two arbuscular mycorrhizal fungi on the growth of Stevia rebaudiana Bertoni. The Journal of the Indian Botanical Society, 86(3&4): 100-104.

Stevens, K.J., S.W. Spender, and R.L. Peterson, 2002. Phosphorus, Arbuscular mycorrhizal fungi and performance of the wetland plant Lythrum salicaria under irrigated conditions. Mycorrhiza, 12: 277-283.

Tanwar, A., Aggarwal, A., Charaya, M.U. and Kumar, P. 2015. Cadmium Remediation by Arbuscular Mycorrhizal Fungus–Colonized Celery Plants Supplemented with Ethylenediaminetetraacetic Acid. *Bioremediation Journal*. 19: 188-200.

Usha, K., R. Mathew, and B. Singh, 2005. Effects of three species of arbuscular mycorrhiza on bud sprout and ripening in grapevine (Vitis vinifera L.) ev. Perlette. *Biological Agriculture & Horticulture*, 23: 73-83.

Vinale, F., Krishnapillai, S., Ghisalberti, E.L., G., Marra, R., Woo, S.L. and Matteo, L. 2008. Trichoderma–plant–pathogen interactions. *Soil Biol. Biochem*. 40: 1–10.

Xie, M.M. and Wu, Q.S. 2017. Mycorrhiza Modulates Morphology, Color and Duration of Flowers in Hyacinth. *Biotechnology*, 16: 116-122.

Yadav, K., Aggarwal, A. and Singh, N. 2013. Arbuscular mycorrhizal fungi (AMF) induced acclimatization, growth enhancement and colchicine content of micropropagated Gloriosa superba L. Plantlets. *Industrial Crops Products*, 45: 88- 93.